DISCRETENESS,
CONTINUITY,
& CONSCIOUSNESS

PETER LANG
New York • Washington, D.C./Baltimore • Bern
Frankfurt am Main • Berlin • Brussels • Vienna • Oxford

ALAN M. LAIBELMAN

DISCRETENESS, CONTINUITY, & CONSCIOUSNESS

AN EPISTEMOLOGICAL UNIFIED FIELD THEORY

PETER LANG
New York • Washington, D.C./Baltimore • Bern
Frankfurt am Main • Berlin • Brussels • Vienna • Oxford

Library of Congress Cataloging-in-Publication Data

Laibelman, Alan M.
Discreteness, continuity, and consciousness:
an epistemological unified field theory / Alan M. Laibelman.
p. cm.
Includes bibliographical references and indexes.
1. Knowledge, Theory of. 2. Unified field theories—Miscellanea. I. Title.
BD161.L29 121—dc22 2006101050
ISBN 978-0-8204-9528-6

Bibliographic information published by **Die Deutsche Bibliothek**.
Die Deutsche Bibliothek lists this publication in the "Deutsche Nationalbibliografie"; detailed bibliographic data is available on the Internet at http://dnb.ddb.de/.

Cover design by Lisa Barfield

Dedication

Like the previous efforts, this one is dedicated, as always, to Roberta, Linda, and Christine. I ask you to remember:

> Nothing is at last sacred but the integrity of our own mind.
> —Ralph Waldo Emerson, *Self-Reliance*

Table of Contents

Preface		ix
Acknowledgments		xv
Chapter One	Skepticism and Knowledge	1
Chapter Two	Anthropocentrism	35
Chapter Three	Discreteness and Continuity	65
Chapter Four	Metaphysics	97
Chapter Five	Physics	129
Chapter Six	Self-Consciousness	161
Chapter Seven	Mysticism	193
Chapter Eight	Parapsychology	229
Chapter Nine	Psychophysics	259
Notes		291
Citation Index		345
Subject Index		355

Preface

> The rational faculty moves freely over all things and passes an indisputable and truthful judgment upon them. Inward mysteries are apparent to it, and hidden meanings are disclosed to it.
> —Abū Hāmid Muhammad al-Ghazālī, *The Niche of Lights*

This volume is an extension of ideas expressed in prior published works. The first release, *The Other Perennial Philosophy: A Metaphysical Dialectic* (*TOPP*), was an exploration of metaphysics; the second, *Going Against the Flow: An Exercise in Ethical Syncretism* (*GATF*), devised an ethical system grounded in the speculative metaphysics of the first book. Consequently, a question immediately arises: "Is it necessary to read those materials prior to absorbing the information in the present one?" The answer is, "Yes and no."

Having read and retained the abundance of arguments presented in the earlier volumes will not be a prerequisite for perusing the content of the one in hand. Chapter four will present a severely condensed version of what has come before in much the same way it is done for serial installments of a television show, or for the science fiction/science fantasy book genre. This useful reprise will come after discussion of several key preliminary topics in the first three chapters. Upon dissemination of essential background material, the remainder of the volume, chapters five through nine, can stand on its own merits. On the other hand, much of the material in that fourth chapter will appear outlandish on first reception, and seemingly worthy of nothing other than immediate dismissal unless those previous treatises are dissected for the wealth of supporting evidence they contain, the purpose of which is to endow the incorporated metaphysical and ethical theses with substantive merit not otherwise obvious.

Collectively, the three publications present a coherent view of a grand theory unifying metaphysics, ethics, and epistemology. To see how this is so in outline, an extended metaphor first presented in *GATF* concerning the relationship between the general subdisciplines comprising the philosophical enterprise in its entirety will be repeated. If philosophy is represented by a tall building, a New York City skyscraper, then the underground foundation, exterior walls, and roof all jointly comprise the domain of metaphysics. The different rooms in the structure, each with its own purpose and design, represent the diverse ethical principles collectively establishing the overall reason for the existence of the building itself. The layout of each room with respect to the arrangement of its furnishings reveals the specific role of that particular room in context of the entire building's function in precisely the same manner moral applications of ethical principles are indicative of the general character and personality of the individual invoking those ethical and moral guidelines. The various elevators and hallways connecting room to room, floor to floor, invoke a sort of structural integrity to the building in the same manner epistemological constructs are used to convey the cohesiveness of a person's philosophy.

As befits the continuation of a series of related constructs, *Discreteness, Continuity and Consciousness: An Epistemological Unified Field Theory* (henceforth known as *DCAC*) will exhibit two methodological characteristics found in both *TOPP* and *GATF*. First, it will be abundantly referenced with citations from relevant literature sources; second, it will introduce material from a wide range of disciplines and provide needed connective tissue so that eclecticism is transformed into syncretism—a notion to be explained shortly. *It will be important for the reader to refer to the Notes section in the back of the book to get a full understanding of some of the ideas presented in the text.* The information contained therein constitutes a necessary supplement to the documentation offered within the body of each chapter. Inclusion of this material in the narrative flow would make the volume even more difficult to read than it already might be. Elimination of the extra evidentiary sources altogether would have diminished the intelligibility and coherence of some of the statements presented.

In *TOPP*, subject areas that were deemed pertinent to an expression of metaphysical ideas included philosophy, religion, mythology, mathematics, astrophysics, physical chemistry, and biochemistry. In *GATF*, philosophy, religion, and psychology were critical to an elaboration of a system of ethics. In the present instance, meaningful domains for exploration are philosophy, religion, mathematics, physics, chemistry, neuroanatomy, neurophysiology, and parapsychology. As first presented in *GATF*, eclecticism is to syncretism as a chemical mixture is to a chemical compound. In the former, the identity of the component elements or aspects of the composite are retained, whereas in the latter, the included ingredients are transformed to create an entirely new species with unique properties and characteristics.

The common practice is for specialists in singular disciplines to develop abstract theories and concrete applications relating information pertinent to their own fields of interest. This epistemological approach to the creation and development of knowledge is necessary. It has led to extraordinary advances in all fields of endeavor. However, this technique is not sufficient of itself. Bridges between these disciplines must be constructed such that the resultant whole coherently unites the disparate subjects of inquiry. If the connective links are haphazardly conceived, the outcome is eclectic—here employed as a pejorative assessment. If the product creates new areas of thought and application by virtue of its coming into existence, as well as successfully incorporating the demands of comprehensiveness and self-consistency, then this *metatheory* is a syncretistic advancement on that which has come before.

In the creation of a syncretistic scheme, the role of authoritative citation is somewhat different than its usual one. Typically, a scholarly specialist in any discipline must be extremely careful to cite precedent thoughts with exactitude not merely through usage of the precisely literal vocabulary used in the original documents—especially pertinent where translation from one language to another is concerned—but also with sincere intent to capture the context of each reference so as not to either deliberately or accidentally misconstrue the meaning of those prior statements. The charge of "taking so and so out of context" potentially levied against some commentator is a devastating critique of his or her competence. For the elaboration of some metatheory, however, extracting a novel usage for an old piece of scientific evidence or abstract intellectual construct, *especially* by taking it out of context, is permissible and possibly necessary.

In other words, the development of a comprehensive syncretistic scheme does not demand adherence to prior constraints on the proper boundaries of a particular notion. It is the very possibility of applying old data in new ways that gives credibility to the worthiness of a proffered synthesis. If complete faithfulness to the original were of paramount importance to the innovator, the result would be only a reproduction of thought, a stagnation of ideas. It simply doesn't matter whether the author of a citation would have applied his or her statements in the same manner as the inventor of bridging concepts. The virtue of creation of a syncretistic thesis is its capacity to unify disparate notions that initially do not seem to belong together, and the principal means one can accomplish this is by deliberately "taking out of context" what has come before.

A few words must be added in explanation of the form of this work as it relates to the content. The point is made in chapter three that the best three-dimensional image unifying all aspects of reality is the helix: a structure that repeats with regularity its overall design, yet at any particular location along its unceasing curve demonstrates unique features. The style of this book is arranged as a helix: ideas and themes mentioned at one point are frequently reviewed later, though from a different perspective, so as to articulate a new way to comprehend important information. Hence, the book may appear to be poorly organized, in need of an editor to cut and paste statements made about a particular notion together in one locale and to segregate apparently unrelated subjects into separate chapters. While such an approach might be clearer to the reader, it would defeat the overall purpose: to demonstrate that allegedly unrelated topics are integrally connected; that isolation into discrete units is a false conceptualization of the nature of what is; that continuity is the hallmark of ultimate essential Being.

It may seem odd to propose that an elaboration of epistemological theory necessitates the investigation of such diverse realms as religion, physics, and parapsychology. However, if epistemology is defined as the means adopted to obtain and assimilate knowledge, then the justification for the inclusion of these domains becomes comprehensible. Religion, in the form of revelation accompanied by the practice of disciplines encapsulated as meditative or contemplative in nature, evokes as its ultimate expression ecstatic mystical states universally proclaimed by its practitioners as inducing in the receiving subject vast complexes of information in extraordinarily short passages of time. Physics enters the discussion in a relevant way due to its involvement in the operational mechanisms of the brain as the premier biological center for data receipt, analysis, and storage for the purpose of future retrieval. Parapsychology, although considered to be highly implausible as a means of knowledge acquisition by many investigators, has nonetheless gained a large number of adherents to the claim that such phenomena as telepathy and precognition, unexplained by commonly accepted physical principles, are not merely legitimate avenues of information-gathering, but as representative of evolutionarily developmental abilities, are invested with significance beyond their present actuality.

Each of these three avenues is distinct. The mystical path to information access is not a cognitive approach dependent upon coupling of a ratiocinative process with a sensory data collection network, the latter being traditionally classified under a heading we call "thinking." Psychic phenomena result in comprehension of events occurring within normal spacetime, but apparently proceed by a mechanism allegedly contrary to the accepted laws of energy dissemination through space. Although there is a recognized relationship between mystical and parapsychological abilities in the sense that the latter are occasionally conceived as preliminary stages of accomplishment on the path to the former, the literature sources are in agreement that they are not interchangeable modes of apprehension. Indeed, there are warnings not to confuse the two. Given these three divergent and mutually exclusive means for the generation of knowledge, any plausible epistemological metatheory must consider all pertinent approaches.

What, then, is epistemology? It is an inquiry into knowledge: its general and conceptually abstract methodology for attainment; its scope or domain of coverage once its guiding rules for possession have been articulated; its specific content once its boundaries have been resolved; its possible or probable arenas of exclusion if any such exist. In order to proceed in this exploration, one must first prove knowledge is a potential accomplishment conceptually and not a chimera, a fantasy originating in *hubris*, an irrational pride in one's proclaimed abilities precisely because eternally incapable of affirmative and definitive substantiation. Hence, this volume begins with an investigation of the philosophic doctrine of skepticism: the thesis stating that knowledge is ultimately unattainable by the human species for a variety of reasons. From there, discussion broadens into an ongoing critique opposing anthropocentrism, a dogmatism professing reality is fundamentally guided by human conceptualization of it, which is the focus of chapter two.

The third chapter, which gave its theme to the book's title, studies the relationship between discreteness and continuity, or as usually construed, the relation between the many and the one. It is a primary consideration of this volume that unification of any type on any disciplinary subject ultimately derives from employment of idea constructs having transcendental origin and meaning. Several such notions are explored, and the ultimate unifier of the complete class of unifiers is denoted as Consciousness. After chapter four validates this latter proposition based upon an inquiry into metaphysics, the following chapter demonstrates how Consciousness acts as coordinator for fundamental concepts in physics, which is to say, chapter five explains the means by which metaphysics grounds physical theory and its associated laws as we have come to understand them.

If it is metaphysically true that Consciousness is both the same as, and distinct from, immanent existential self-consciousness, then this topic must be addressed, which it is in chapter six. Most pertinent is the mind-brain connection. Here is where metaphysics and physics collide, or to be more accurate coalesce, and just how this happens is pursued within these pages. As mentioned previously, mystical and parapsychological abilities have been conceived by many as two means of gaining knowledge about reality that are different in typology from the more universally recognized cognitive model. Consequently, these topics are the subjects of the seven and eighth chapters, respectively. Finally, the last chapter undertakes to develop a preliminary *quantitative* psychophysical vocabulary. The principal means to accomplish this task is through the tentative formulation of mathematical models relating Consciousness to manifest activity. This effort is intended to serve as a foundation in support of a more extensive exploration of these ideas by interested readers, with the goal being the future elaboration of a complete psychophysical science that exhaustively links transcendental and immanent reality. Recognition of this ambition is underscored by the book's titular reference to "an epistemological unified field theory."

Acknowledgments

I am grateful permission has been obtained to quote from:

Laibelman, Alan M., *Going Against the Flow: An Exercise in Ethical Syncretism*, New York: Peter Lang Publishing, Inc., copyright 2004. Reprinted by permission of the publisher. All rights reserved.

Laibelman, Alan M., *The Other Perennial Philosophy: A Metaphysical Dialectic*, Lanham: University Press of America, copyright 2000. Reprinted by permission of the publisher. All rights reserved.

For their efforts in bringing this book into print, my sincere gratitude goes to Phyllis Korper, acquisitions editor, and Richard Atkins, production editor, at Peter Lang Publishing, Inc. Nonetheless, I alone am responsible for all the material contained within the covers of this volume.

Chapter One: Skepticism and Knowledge

> The...structure of...reality...compels the distinction...between apprehending truly or falsely and between truth or falsity in our propositions.
>
> —Samuel Alexander, *Space, Time, and Deity*, volume II

It is a good bet the first two human beings on Earth, otherwise strangers to each other, had an argument the first day they met. Perhaps it was about the location of the best cave in which to seek shelter, or the best means to trap and kill their dinner. As long as people have lived on this planet, there have been those who believed they knew something, while others have had doubts about the correctness of presumptive knowledge. The dispute arising between those professing these fundamentally distinct perspectives has raged ever since.

With the establishment of civilized societies and communal living came the development of religion. Acceptance of the accessibility of knowledge was universal and indisputable. Chinese Confucianists and Taoists accepted that there was a natural order of event occurrences, a Command of Heaven dictating all aspects of daily life. People could absolutely know what this omnipotent order was in a simple way: by thinking, feeling, and acting in conformity with innate processes, one would live in harmony and peace, internally within oneself and externally with others.[1] Hindus and Buddhists in India proclaimed the goal of each and every individual life was to gain liberation or enlightenment, to understand what the way of the world was in reality, and not in confusion due to the inherent power of illusion and self-deception.[2] Indian philosophy in antiquity was divided into distinct schools of thought in part segregated by their beliefs concerning acceptable methods for determining Truth.[3] Judaism proclaimed its very existence derived from being able to hear, comprehend, and obey the Word of God as the legitimate source of knowledge. The tribal Jews of the Middle East were convinced that their holy men, their prophets, spoke the Word of the Law and that all males and females could come to know without doubt what was expected of them, and how to fulfill those expectations on a daily basis.[4] The Zoroastrians of Persia, and the polytheists in Greece or Egypt, all prayed and sacrificed to gods in temples dedicated to reception of authoritative and unimpeachable information, as well as benevolence, from the deities in the hope the lives of the citizens of those societies would be spared unnecessary harm.[5]

In light of what will be discussed in this chapter, Nikolai Berdyaev's comment in *The Beginning and the End*—"The world presented itself in different ways to Hellenism, Judaism, to the Persians and the Indians. The criteria adopted by general consent for the establishment of truth, of which great use has been made by socialized religion and its armory of theologies, are not criteria of truth; they are merely standards of what is useful for society"—was valid, but missed the essential point.[6] Of relevance is not whether the respective philosophies and theologies of each group was or was not absolutely True; what is of maximum significance is that each society separately abided by the tacit presupposition and overt acknowledgement of the supreme importance and reality of Truth.

So strong was presumption of the existence of Truth that contemplation of its nature with respect to the observable physical world led directly to the development of philosophy as a stand-alone intellectual tool of inquiry in the Western world—in contrast to the Orient, where to this day, philosophy and religion remain thoroughly intertwined. It should, therefore, not be surprising to discover that the earliest practitioners of philosophy conformed to a standard incontrovertible view: the existence of Truth meant that knowledge of its precise contents was to be had for the taking subject to the constraint of clearly reasoned thought, eventually denominated as *logic*.

The paragon of this approach was Socrates (469–399 B.C.E.) who, according to Friedrich Nietzsche in *The Birth of Tragedy*, was "the turning point of world history": "Once we see...how after Socrates, the mystagogue of science, one philosophical school succeeds another, wave upon wave; how the hunger for knowledge reached a never suspected universality in the widest domain of the educated world, became the real task for every person of higher gifts... How this universality first spread a common net of thought over the whole globe, actually holding out the prospect of the lawfulness of an entire solar system; once we see all this clearly...we cannot fail to see in Socrates the one turning point and vortex of...world history."[7] Nietzsche's judgment has received echoes from Eduard Zeller, in his *Outlines of the History of Greek Philosophy*, and Warren Ashby, in *A Comprehensive View of Western Ethics*.[8] Socrates insisted that knowledge was attainable, even if limited in scope—the ways of the gods could not be ascertained—while modestly, and inaccurately, claiming he didn't have any.

Given the proposition that argument has as lengthy a history as assertion, it comes as no surprise that the influence of Socrates in the Greek world was soon partially counterbalanced by the rise of philosophical doubters, those choosing not to accept any reasoned set of claims as correct, as indisputably True. Though there were "professional" doubters of knowledge acquisition before Pyrrho of Elis (365–275 B.C.E.), he was credited with formalization of the approach. Concerning the group of adherents to Pyrrho's attitude, Richard Popkin wrote in *The History of Skepticism from Erasmus to Spinoza*: "The Pyrrhonists doubt and suspend judgments of all propositions, even that all is doubt. They oppose any assertion whatsoever, and their opposition, if successful, shows the opponent's ignorance; if unsuccessful, their own."[9]

The basis for argument by these perpetual disbelievers was that the data obtained from personal observation, or the process of reasoning employed in the development of complex conclusions, were not sufficiently steeped in certainty to properly allow for the professed conclusions drawn by their opponents during public dispute. By discounting the validity of all presented evidence, these *skeptics*, as they came to be called, were proposing that the attainment of knowledge required the presentation of propositions possessing absolute Truth, and that this was not to be had under any circumstances. It is clear from this recounting how their views were contrary to the Socratic thesis: Socrates preached that Truth existed and knowledge was possible; the skeptical Pyrrhonist doctrine was that we can't *know* that Truth exists, and without that guarantee, knowledge is not ascertainable.

Framed in this way, it seems like a "Catch-22," an endless loop: to have knowledge, Truth must exist; but in order to *know* that Truth exists, we must have knowledge. Even for the modern age, the issue ultimately comes down first to the question of whether Truth as a classification can be shown to be real and not illusory, and only secondarily whether human beings possess the attributes enabling apprehension of its component parts.

Can ignorance be said to exist if the possibility of knowledge is rejected? All language demands the inclusion of opposites to be interpretively useful. When Pyrrhonists denied human beings could know Truth, were they also denying that Truth existed and was known to the Greek pantheon of deities? Were these skeptics religious blasphemers as well as philosophical doubters? If they held such opinions, that information was intentionally withheld from public disclosure. Perhaps this explains why Julia Annas hedged her bet in describing this group in *The Morality of Happiness*: "The skeptic (allegedly) begins like everyone else, searching for the truth, especially in those matters with which philosophy is concerned. But whereas most people think that after some investigation and argument they are entitled to believe in some substantial theses...the skeptic is not satisfied. He not only investigates, but identifies himself as a continual investigator and enquirer."[10]

In actuality, skepticism is the "emperor without clothes." Despite its lengthy history in philosophic discussion, when contemplated with respect to its theoretical and practical implications, it is a perspective absolutely devoid of self-consistency. To continue the linguistic argument, skepticism requires language—Greek, Latin, German, English, it does not matter—to enunciate its perspective and how that orientation differs from the viewpoint held by the unrepentant aspirant to knowledge. All languages demand the use of definitions for the words it employs; a failure to do so, or an unwillingness to do so, can lead only to gibberish, to incomprehensibility on the part of both the speaker and the listener.

The singular distinguishing feature of any language—oral or written, and involving the use of words, pictures, or any other representative form—is the necessity to employ some form of the word we denote as "is." That is, all definitions employ at a minimum the relationship "A *is* B." When applied to skepticism, the usage is encountered at the outset: the doctrine of skepticism *is* blah, blah, blah. Why this inclusion presents an insurmountable flaw for the skeptical position should be obvious: the word "is" can only imply a positive assertion of relationship between two or more concepts, or between one or more concepts and one or more physically existent objects. Thus, we have the pertinent words of Mary Beth Ingham and Mechtild Dreyer in *The Philosophical Vision of John Duns Scotus* to frame the argument: "Prior to logic, being is the foundational and necessary concept for any cognitional act... Nonbeing...cannot be conceived in a manner other than the absence of being."[11]

Note the phraseology: "The skeptic *is* not satisfied," or "The skeptic identifies himself *as* a continual investigator and enquirer." If this description were accurate, then the Pyrrhonist-type skeptic, to be self-consistent, should proclaim he or she is neither satisfied nor dissatisfied, neither a continual nor an intermittent investigator. To vocalize any other sentiment is to make a positive claim. Such a response would undercut the doctrine of skepticism, since that thesis asserts the universal inappropriateness of *all* affirmations. Also, if the Pyrrhonist-type skeptic *is* "searching for truth," this suggestion necessarily leads to the conclusion that such a skeptic acknowledges Truth as an existent with significant interpretive meaning.

It is no surprise that Annas qualified the assertion with the parenthetical "allegedly": an unqualified acceptance of the notion of absolute Truth would render the skeptical position moot at the outset. To have any pretense toward being a viable approach to the cognition of reality, this form of skepticism must not merely assert that Truth is unknowable by humans or deities, but that it does not even exist at all in the world. As mentioned, the Pyrrhonists probably didn't go this far: if they had been religious blasphemers, and their views became known, it is highly likely that they would have been banished or killed by the citizens of the Greek communities in which they promoted their ideas.

Since Popkin's book preceded that of Annas in print, it can be said the latter reiterated the proposition of the former regarding the self-identification of skeptics of the Pyrrhonist type, for there is little interpretive linguistic difference between the phrases, "Pyrrhonists *doubt* and *suspend* judgments of all propositions," or "They *oppose* any assertion whatsoever," with the previously mentioned, "The skeptic *is* not satisfied," and "identifies himself *as* a continual investigator and enquirer." Pyrrhonist skeptics who accepted the propriety of these literal characterizations would not only have been stating something positive about the relationship between themselves and some abstract intellectual discourse, namely that they did not adjudicate between the Truth content of rival propositions on any topic, but they would have also been proclaiming something positive about their relationship of themselves to themselves: they *positively knew* that this was their attitude and approach to all philosophic disputation. Hence, the position is innately inconsistent because self-contradictory.

Pyrrhonist skepticism is seen to be most convincing when allied closely with nihilism, the notion of nothingness as having ontological priority to the somethingness most frequently designated as *reality*: Truth doesn't exist; knowledge doesn't exist. This must be so. The acceptance of something as real, even if we uphold the skeptical claim that detailed comprehension of the nature of that real is intrinsically unavailable to us, is to make the definitive assertion, "reality is." That concession is sufficient to disprove the doubter's contention claiming they make no positive assertions, but hold suspended judgment over *everything*. Of course, if reality doesn't possess the attribute of existence, then neither does the skeptic nor the skeptical thesis: they cannot be held to be any more real than reality itself. The doctrine is barren at its root. Similarly, this approach to the issue also explains why solipsism, a specialized form of skeptical nihilism applied to the questioned reality of other people, is equally devoid of significance. If the solipsist were truly alone, to whom would he or she be making the argument that he or she was alone? The act contravenes the assertion.

Advocates of the skeptical position have long been aware of this logical difficulty encountered in the approach taken by the original Pyrrhonists as well as by their intellectual descendents. Possibly to alleviate this criticism of skepticism, in *Skeptical Philosophy for Everyone* Popkin articulated a key difference between Pyrrhonism, a species of "radical skepticism," and a variant he called "mitigated skepticism":

> What all forms of skepticism have in common are reservations about the attainment of knowledge. But the radical skeptic doubts that any piece of information is any better than any other. If the basis for any belief is fallible, then no belief is better founded than any other.... The mitigated skeptic is committed to the thesis that information is to be described in probabilistic terms: the more probable, the more

> reliable. But in saying this, he is also saying that no degree of probability, no matter how high, is equivalent to knowledge.[12]

Unfortunately, this tactic fails: the approach adopted by the mitigated skeptic is also self-contradictory. The claim, "some information is more reliable than other information" is tantamount to the adoption of a positive assertive stance towards information in general: at least some information tells us something intelligible. It presumptively demands a valid procedural means to evaluate the quality of specific information bits with respect to reliability. This latter condition can only be constructed out of a previously held belief in an *a priori* realm of Truth to which some information bits adhere more rigorously, more reliably, than other pieces of data. "No degree of probability...is equivalent to knowledge" demands "knowledge implies certainty" holds, which is to say, conforms to unit probability.

On what rational grounds does the mitigated skeptic propose and defend this judgment? If this equation is in fact True, and the proponent *knows* it to be True, then the holder of such a view is admitting they know something to be absolutely True. He or she has definite knowledge about this statement's veridicality allowing him or her to adopt their approach with conviction. Yet, the knowledge fact destroys the skeptical position. If this equation is in fact True or if the equation is in fact false, and the proponent *doesn't know* it to be True or *doesn't know* it is false, then the holder of such an opinionated view is admitting they have an unwarranted bias, and there is, therefore, no reason anyone else needs to conform to the dictum, "knowledge is equivalent to certainty," in declaring they are in possession of knowledge. Hence, the skeptic in this situation does not succeed in gaining adherents to that overall perspective. Such a person is merely adopting an attitude, a methodological approach, which is not convincing precisely because it is arbitrary.[13]

Is the stance taken by the radical skeptic a better, more viable position? According to Popkin, the central determinant of the argument in favor of radical skepticism is, "if the basis for any belief is fallible, then no belief is better founded than any other." The conditional introductory term *if* is crucial to the skeptical orientation. In practice, however, the radical skeptic seems to conveniently drop the conditional and tacitly *assumes* that the basis for any belief *is* fallible, and that because this is so, the skeptical approach is the wisest approach to theoretical and experiential interactions between self and surroundings.

The contention here is the skeptic must *prove* that the conditional "if" is legitimately removable. The problem thus engendered for the radical skeptic is transparent: any attempt at proof undercuts the foundational claim of no assertive statements, such as that they "doubt and suspend judgments of all propositions." In actuality, skeptics are those who have *predetermined* that the basis for all belief is fallible, with "fallible" meaning, "subject to error in formulation or to be overturned by further forthcoming evidence." However, fallibility, like ignorance, cannot be a criterion unless there is a possibility for infallibility, the opposite to fallibility, or of knowledge, which is contrary to ignorance. Similarly, "subject to error" must posit "avoidance of error" as a counterpart—otherwise the term *error* has no meaning—and the latter, whether actual or potential only, demands an absolutely invariant standard of judgment against which evidence can be held accountable.

The proposal need not be made that such a standard does in fact exist; its potential existence is sufficient to subvert the radical skeptical stance. How? The skeptic is faced with two conflicting *theoretical* conditions: a standard of Truth serving as a reference standard does exist or it does not exist. Since the consistent skeptic must accept that there is no incontrovertible evidence, then he or she must allow for either condition to *factually* hold. In other words, the possibility exists that the skeptic is wrong when the translation is made from theory to practice. Having a plausibly incorrect approach to life—and the adopted methodology not only lacks certitude of being correct, but also denies all possibility of gaining a *known* correct approach to life irrespective of the specific characteristics of that approach—results in the maintenance of an attitude conferring no demonstrable benefit.

Of greater utility is the opposite approach: positing that Truth exists and that knowledge is possible. By establishing a position on all matters, the propriety or impropriety of all decisions will ultimately be determinable by seeing what accrues as a consequence of any particular action or thought. The Taoists and Confucianists were right: attaining reproducible consistency in desirable outcomes based upon hypothesized constructs serves as both unambiguous and unimpeachable evidence for the acquisition of knowledge. The compendium of precepts attributed to Mo Tzu (~479–438 B.C.E.) made the argument with precision and brevity: "Therefore for any doctrine there must be the three standards. What are the three standards? There must be a basis or foundation. There must be an examination. There must be a practical application."[14]

A Pyrrhonist radical skeptic cannot maintain neutrality during concrete experience. Pragmatic activities *force* judgment in any situational encounter. The scenario thus developed reifies the status of *Buridan's ass*: a donkey is situated exactly midway between two piles of grain upon which to feed, and if it has no substantive basis upon which to choose to move in one direction, it will die of starvation. If it stands still, it is because it has chosen to do so. In practice, the donkey chooses arbitrarily, but nonetheless *chooses*, one of three options: stand still and starve, move towards pile A and eat, or move towards pile B and eat. The radical skeptic is subject to identical constraints, and in fact he or she makes a choice every time based upon some assertive assessment of preference, arbitrary though the grounds of preference might be, and in opposition to their claim that they suspend all judgments.

It was in this vein that Hannah Arendt offered a salient comment in *The Life of the Mind* about the role of human will in practical activity: "The will always wills to *do* something and this implicitly holds in contempt sheer thinking, whose whole activity depends on 'doing nothing.'… The willing ego, looking forward and not backward, deals with things which are in our power but whose accomplishment is by no means certain."[15] In fact, Popkin reversed his course on the neutrality of radical skepticism when its relevance to a practical philosophy of life was taken into account:

> The Pyrrhonists call this attitude *epoché*. It is the total suspension of judgment. Epoché is consistent with a kind of passive acceptance of the world; one lives in the world, acts in it, takes it as it is without reflection. Epoché is important because it is the first step toward *ataraxia*—a special kind of mental tranquility. Pyrrhonism is… a philosophy of life, a formula for living well. It reverses the Socratic point of view that the unexamined life is not worth living.[16]

The cat is now out of the bag. Far from being value neutral, Pyrrhonism, or radical skepticism preaching suspension of judgment and the impossibility of knowledge, is actually a flagrant predisposition to the belief in a specific "formula for living well." "Living well" is feasible only if living less well, less optimally, is also conceivable. Faced with the choice, there again arises the unavoidable issue of plausible, definable, and constructible techniques for elucidating the necessary features pertinent to "living well." This is not possible without a positive, assertive, preliminary conviction, based on tacit knowledge or an intuitive hunch, that one is capable of differentiating characteristics leading to *ataraxia* from those contrary to that presumptively worthy goal.

Annas admitted Pyrrhonism as a way of life is wholly dependent upon ultimately arbitrary "appearances": "The Pyrrhonists in fact frequently see their intellectual task as one of rejecting enquiry into the nature of things in favor of resting with the appearances.… The skeptic suspends judgment as to how things really are, but can live by the appearances—things inevitably impress themselves on him as being one way rather than another."[17] In this context, what is meant by "appearances"? A key word employed by Annas in the passage is "impress." In ancient Greece, impressions meant much the same as now: sense-based information about what seems to be occurring moment-by-moment in the external world.[18] Hence, a plausible interpretation of the literal text is that radical skeptics of the Pyrrhonist type accepted the data received by their sensory interactions with the environment as sufficient grounds for determining what to think and do.

Despite their awareness that sensory input could be flawed in some circumstances—which, ironically, conformed to the usual skeptical view that "the basis for any belief is fallible"—skeptics still made a positive *choice* on how to utilize, in normal living circumstances, what their sense organs told them. Clearly, this attitudinal approach overruled their advocacy of "the total suspension of judgment": they accepted without reservation, and with less than exhaustive inquiry, that the external world from which sensory data accrued was inalienably real.[19] Moreover, the doubters elected to employ not only positive physical assertions but mental ones also: the ethical thesis that the end justifies the means. They arbitrarily chose from a host of appearances without concern for reliability or validity because the outcome, *ataraxia* or untroubledness, was deemed desirable. Such is the inalienable inconsistency of the skeptical approach when attempts are made to wed its theoretical methodology with its practical instantiation.

Popkin's attempt at a distinction between classes of skeptics raises a fascinating question. In a debate between a radical skeptic and a mitigated skeptic, on what grounds would each argue the correctness of that approach *vis à vis* the alternative? The suspicion arises that any argumentative basis used by either would ultimately emanate from, at a minimum, an overt belief in the appropriateness of their position, if not covertly deriving from a strong conviction akin to knowledge. The mere possibility of such a debate further disproves the contention of either the radical skeptic or the mitigated skeptic that he or she can be consistent in his or her approach to all aspects of human experience by maintaining a value-neutral attitude.

In a sense, it was the radical skeptic's willingness to compromise the rigidity of the abstract intellectual approach with the concrete reality of lived experience that facilitated the perpetuation of skepticism as a philosophic methodology through the two thousand plus years since its inception. With time's passage the skeptical view necessarily shifted: an inability to logically deny the existence of physical and mental reality altogether—consequently and inevitably resulting in acceptance of the existence of absolute Truth—meant only particular assertions about the exact nature of that Truth and its manifestation in the world could be criticized on a claimed basis of a lack of sufficient evidential support. In other words, history has seemed to promote the contention of the mitigated skeptics over that of their more extreme brethren.

The dispute between proponents of the possibility for successful human conceptualization of Truth and moderate doubters continued throughout the first sixteen hundred years of the Common Era. During this time, religion was usually at the center of the disagreement. In one sense, the advent of Catholicism and Islam allowed believers to join with Judaism in promoting a united front by claiming the absolute nature of Truth was to be found overtly in scriptural testaments and covertly in the practices of mystical-oriented groups within each faith: Catholic orders of monks or nuns established in compliance with pre-defined Benedictine, Franciscan, or Dominican rules concerning communal practices; Islamic Sufis; Jewish Kabbalists. On the other hand, religionists of a philosophic bent who were at all versed in the other Western religions pointed to the discrepancy between various specific tenets of each tradition and vilified the other two practices for preaching heresy. Hence, history records the warfare between Catholics and Moslems known as "The Crusades"; the Islamic conquest of the Middle East, North Africa and Spain; and the general ostracism and subjugation of Jews all over the world. All these events could be viewed as theological battles for control of Truth conducted in the form of sociopolitical militaristic expeditions.

In addition, numerous short-lived splinter groups formed around these principal creeds: Zealots, Gnostics, Neoplatonists, Manichaeans, Sabellians, Pelasgians, the Islamic Brethren of Purity (*Ikhwan al-Safa*), and so on. All claimed to know Truth, yet rarely was any major or minor belief system credited by the others with validity in any constituent doctrinal part. Richard Tarnas offered a representative example of this attitude in *The Passion of the Western Mind*: "A sacred doctrine had been entrusted to Christians, and fidelity to that trust, as well as the integrity of that doctrine, needed to be maintained at all costs.... Safeguarding the faith was...the first priority in any question of philosophical or religious dialogue; hence that dialogue was often curtailed altogether lest the devil of doubt or unorthodoxy gain a foothold in the vulnerable minds of the faithful."[20]

In retrospect, here is the soil from which the philosophical doctrine that came to be known as *relativism* eventually grew: knowledge about reality was associated with truth, defined as that which one chooses to believe under a given set of conditions, and not that which is unimpeachably True.[21] If there had been such a personage as an independent outside observer, it would have seemed as if the mitigated skeptics were right: precisely because there was such a divergence of religion-inspired views on issues metaphysical, ethical and moral, knowledge was devalued to mean semi-educated guesses about the proper relationship between man and God, man and man, or man and the universe.

Science mushroomed into existence in the seventeenth century—slightly earlier for astronomy—and heralded a major revolution in thought and act extending deeply into the realms of religion and philosophy. Science, more specifically scientific method, was envisioned as a technique for answering questions about what was real and how we could know it. Experimental science—beginning with astronomy and physics before extending through chemistry into biology—preached coherence between theory and practice, precisely the arena in which radical Pyrrhonist skeptical philosophy fails as a viable mode of investigation. In conjunction with the expanding display of latent explanatory power within mathematics, scientific method allowed the conclusive demonstration of Truthful propositions presented in the form of precise formulas describing some of the relationships existent in nature.

Naïvely, one might have thought that the development of science, with confirmation of its surviving theories elicited by an unceasing accumulation of evidentiary support, would have permanently put to rest all proclamations of skeptical doubt about the existence of absolute Truth and its ultimate comprehensibility in the form of certain knowledge acquisition. Over the passage of time, however, philosophers learned the *wrong* lesson from these demonstrations: they did *not* conclude science revealed potential and actual Truth. They decided continuous alteration and refinement of interpretive meaning into ever-more elaborate theoretical frameworks, derivable from the vast accumulations of data obtained from planned research, meant truth was *eternally* contingent, dependent upon circumstances defined by the research variables, and was not an absolute reconstruction of the detailed workings of the universe. This is a highly dubious claim. It apparently never occurred to the philosophers to realize that *contingent truth could only be contingent if there were available an ultimate absolute Truth serving as an invariant standard of reference*. Simply put, truth is the discrete particular, whereas Truth is the continuous universal.

To understand how this errant conclusion came about, a cursory review of philosophical history commencing with the birth of science is necessary. Witnessing its arrival on the world stage was René Descartes (1596–1650). Until Descartes asserted his famous *cogito* in *Discourse on Method*, Western philosophers had conceived few substantive epistemological theses of any weight with the major exception of Aristotle's enunciation of the Principles of Noncontradiction and Excluded Middle.[22] They had concentrated instead almost entirely on metaphysical and ethical issues as defined by prevailing religious doctrines. Descartes changed all that forevermore:

> I had noticed for a long time that in practice it is sometimes necessary to follow opinions which we know to be very uncertain, just as though they were indubitable, as I stated before; but inasmuch as I desired to devote myself wholly to the search for truth, I thought that I should take a course completely contrary, and reject as absolutely false anything of which I could have the least doubt, in order to see whether anything would be left after this procedure which could be called wholly certain.... I decided to suppose that nothing that had ever entered my mind was more real than the illusions of my dreams. But I soon noticed that while I thus

> wished to think everything false, it was necessarily true that I who thought so was something. Since this truth, I think therefore I am, was so firm and assured that all the most extravagant suppositions of the skeptics were unable to shake it, I judged that I could safely accept it as the first principle of the philosophy I was seeking.[23]

The key point is not his conclusion—which Sören Kierkegaard (1813–1855) quite appropriately called a tautology—but the Cartesian assumption that truth must be "wholly certain."[24] The Islamic philosopher al-Ghazālī (1058–1111) had made a similar remark about the supposedly inextricable tie concerning knowledge *vis à vis* the relationship between certainty and doubt over half a millennium prior to the Cartesian proclamation.[25]

The acceptance of this idea by virtually all thinkers since the publication of *Discourse on Method* is ironic: the claim that unimpeachable knowledge cannot be had, or must be incontrovertibly proven to be possible, is itself a declaration of fact without proof. To be logically consistent, one should assert only that knowledge and doubt are equally plausible at the beginning of any inquiry in every instance. Therefore, the rational conclusion must be that the burden of proof is equally on doubters as much as on the upholders of knowledge.

Descartes never understood that in taking a position *starting* from doubt, tacitly accepting that doubt had some epistemological priority with respect to knowledge, he had already compromised his perspective on the questions he was attempting to answer. Why? To seek resolution to a query is to weigh in on the appropriateness of the quest: one who desires that which intrinsically cannot be had is not rational. In other words, the *cogito* was inevitable, just as Kierkegaard noted. If Descartes' conclusion had not been that he existed, it would have been that existential reality existed; or abstractly, that doubt and knowledge both potentially and factually existed.

Hans Küng summarized the remaining components of the Cartesian innovation in *Does God Exist?*:

> Descartes had in fact found the key ideas of his new philosophy, which would set the trend for the...subsequent period and its mathematical-technical understanding of reality: the idea of a higher plane of truth...of evidence free from doubt or error, of clear and well-differentiated concepts; the idea of a knowledge based not on uncertain sense data, distorted images, or...authorities, but on understanding, which alone produces certainty; the idea of methodical thinking, proceeding by evident steps from the known to the unknown, from the simple to the complex; the idea of an analogy between the order of mathematics and the order of nature, which obeys mathematical laws, is revealed by mathematics, and thus can be controlled by mathematics.[26]

Some of these postulates are both reasonable and correct: "a higher plane of truth"; "methodical thinking"; "analogy between the order of mathematics and the order of nature." Where he went grievously astray, as did centuries of thinkers after him, was in his quest for "evidence free from doubt," for the elaboration of conclusions "based not on uncertain sense data, distorted images, or authorities." These utterances reveal a complete ignorance of, or appreciation for, scientific method. Even though science was merely in its infancy at the time, Descartes should have been familiar with the demands of its development, especially in light of the release into print of *Novum Organum* in 1612 by Francis Bacon (1561–1626). It was this book, more than any other, which established the grounds for the possible acquisition of knowledge through the utilization of scientific procedures:

> There are and can exist but two ways of investigating and discovering truth. The one hurries on rapidly from the senses and particulars to the most general axioms, and from them, as principles and their supposed indisputable truth, derives and discovers the intermediate axioms.... The other constructs its axioms from the senses and particulars, by ascending continually and gradually, till it finally arrives at the most general axioms, which is the true but unattempted way.... Each of these two ways begins from the senses and particulars, and ends in the greatest generalities. But they are immeasurably different; for the one merely touches cursorily the limits of experiment and particulars, whilst the other runs duly and regularly through them; the one from the very outset lays down some abstract and useless generalities, the other gradually rises to those principles which are...the most common in nature.[27]

Science beneficially *uses* information obtained from sensory functioning and recognized authorities. The former has led to the creation of numerous arrays of instrumentation employed in the acquisition of experimental output ever since the development of Galileo's (1564–1642) telescope. The latter appropriates the research results of previous scientific investigators in order to construct its theorems and unifying laws so that one need not always start from the beginning, just as Galileo's results confirmed and extended those of his predecessors, Nicholas Copernicus (1474–1544) and Johannes Kepler (1572–1630).

Descartes and his intellectual kin have never understood that the glory of science does not rest primarily on its compelling conclusions but on its grounding routine: science takes human liabilities such as unreliable physical senses and limited memory capacity and converts them into assets in coming to grips with our understanding of ourselves and our environment. Sensory perception and the authoritative views of predecessors were among the readily accepted epistemological techniques known to religious theorists throughout the ages.[28]

Yet, precisely because Descartes has been so influential, the proposition "knowledge equals certainty"—the renewal of the Pyrrhonist doctrine in the modern era—again demands investigation. Is it altogether unreasonable to allege contingency as an acceptable form of knowledge? Does knowledge require permanence and inalterability as invariant characteristics? Another way to put the case is to wonder whether there is meaningful distinction between *Truth* and *truth.* If Truth is the conceptual meaning affiliated with permanence and inalterability of valid information, is truth as temporarily valid under current, but changing, conditions totally without meaning? With a classical acceptance of Truth as knowledge, must truth as knowledge be rejected automatically? Was Descartes correct?

Certainly, there is room for *a priori* consideration of the legitimacy of the argument that Truth *and* truth are of epistemological utility. Augustine (354–430) admitted, in *On Free Choice of the Will,* the invariance of Truth as a standard of measurement for contemplation and understanding of what is real: "Truth makes no progress when we see it better and suffers no setback when we see it less. It remains whole and undefiled."[29] By way of contrast, Edmund Husserl's (1859–1938) phenomenological method defined truth as a characterization of the relation between intentionality and meaning: "Truth involves identity between meaning intention and fulfillment.... All acts of knowing possess a degree of fulfillment and can be fulfilled adequately if imperfectly. In most of our knowledge, we have evidence only of the relative probability of the proposition being true."[30]

Though not a skeptic—whose grounding construct has been shown to be "all data is uncertain, therefore no knowledge is possible"—was Descartes' movement to the opposite extreme—"knowledge, to be possible, must be certain"—overdone and equally inaccurate? In this form, his thesis is the logical contrapositive, and therefore logically equivalent in terms of Truth value, to that of Pyrrho of Elis: "uncertain...no knowledge" is identical to "knowledge...certain." Consequently, if the radical skeptics were wrong, so was Descartes.

Earlier, it was stressed that for the skeptical attack to be sustainable, the conditional introductory term "if" in the proposition, "if the basis for any belief is fallible, then no belief is better founded than any other" must be proven to be "legitimately removable" and not "conveniently dropped" when applied to experiential situations. Descartes and all adherents to the *credo,* "knowledge, to be knowledge, must admit no doubt," were/are in the same spot: they must *prove* that knowledge is properly characterized as "absolute certainty beyond all possible doubt," and not merely assume it to be True.

Aside from the burden of proof placed on supporters of the Cartesian thesis, David Hume (1711–1777) suggested another difficulty encountered by proponents of "Cartesian doubt." He made the cogent observation, in *An Enquiry Concerning Human Understanding,* that the premise of certainty equating to knowledge, once accepted, could never be removed:

> There is a species of skepticism *antecedent* to all study and philosophy...as a sovereign preservative against error and precipitate judgment. It recommends a universal doubt, not only of all our former opinions and principles, but also of our very faculties; of whose veracity...we must assure ourselves, by a chain of reasoning, deduced from some original principle, which cannot possibly be fallacious or deceitful. But neither is there any such original principle, which has a prerogative above others, that are self-evident and convincing: or if there were, could we advance a step beyond it, but by the use of those very faculties, of which we are supposed to be already diffident. The Cartesian doubt, therefore, were it ever possible to be attained by any human creature would be entirely incurable.[31]

The Cartesian thesis left no avenue open to acquire information beyond doubt, since Descartes attempted to void sensory input and prior authority as acceptable methods. He insisted on "methodical thinking, proceeding by evident steps from the known to the unknown." He left unanswered how an intellectual process could be "evident" or the starting datum "known." This conundrum reveals precisely why the Cartesian view is logically identical to the Pyrrhonist doctrine: self-consistency of argument demands *all* sources of information, *all* means of collection, *all* means of interpretive analysis, and *all* means of extrapolation "from the known to the unknown" be rejected as uncertain in form and content.[32] Just as Pyrrhonists eventually accept some piece of arbitrary data as legitimate input upon which to build further thoughts or actions in the form of "appearances—things inevitably impress themselves on him as being one way rather than another," so must Descartes and his intellectual heirs.

In the latter case, Descartes was forced to accept his *cogito*: "From the very fact that I doubted the truth of other things, it followed evidently that I existed."[33] As already stated, he *had to* conclude thus; his presupposition of doubt as a necessary precursor to knowledge automatically implied that the removal of doubt was plausible, for doubt cannot exist in a vacuum. Rather, it demands existence of verifiable Truth as background reference standard. He chose to ground plausibility in the contention of his own existence: a willful though arbitrary choice, which served his purposes as adequately as any other positive affirmation would have done.

What proceeded from his revelation? Thoroughly contrary to his desire for "methodical thinking," he leapt to the conclusion of the existence of God as the source of perfection:

> I reflected that I doubted, and...in consequence, my spirit was not wholly perfect, for I saw clearly that it was a greater perfection to know than to doubt. I decided to ascertain from what source I had learned to think of something more perfect than myself, and it appeared that it must have been from some nature which was in fact more perfect.... Thus the only hypothesis left was that this idea was put in my mind

> by a nature that was really more perfect than I was, which had all the perfections that I could imagine, and which was, in a word, God.[34]

For present purposes, it doesn't matter whether this specific conclusion was correct or not. Of primary concern here is not whether God exists, or is perfect, or was the ultimate revelatory source of the Cartesian *cogito*. What is at issue is whether or not the method by which Descartes arrived at this thought a case of "methodical thinking, proceeding by evident steps from the known to the unknown." What, exactly, is *evident* about the transition from "my doubt" to "God's perfection"? His procedural steps are clear: (a) "my doubt" becomes (b) "my existence" becomes (c) "my imperfect existence because I doubt" becomes (d) "potential perfect existence" becomes (e) "actual perfect existence" becomes (f) "God as perfect existence." Some of these transitions are questionable: (b) to (c), (d) to (e), and (e) to (f) either have been or are philosophically debatable; none are unassailable. Descartes had a preexistent unexamined and unsubstantiated belief in God as innately perfect that must ultimately be the source of everything in manifestation. Paul Weiss made this abundantly obvious in *Being and Other Realities*:

> When he turned to God as the guarantee of truth of his clear and distinct ideas, he took Him to be a mediator, enabling Descartes to pass from his clear and distinct ideas to what had an existence apart from himself. He never did show that there was such a mediating God.... Since, on his view, only clear and distinct ideas were true, and then only because such ideas were guaranteed by God, he found a sanction for his clear and distinct ideas in that for which he had no sanctioned clear and distinct idea.[35]

It is worth emphasizing: *it doesn't matter whether his conclusion was correct or not.* Of relevance is his method: like that of radical skepticism, it could not be consistently and rigorously followed. In this instance, he had a starting point he assumed was legitimate, the aforementioned "preexistent unexamined and unsubstantiated belief in God as innately perfect," which fatally undercut his quest for "methodical thinking."

Hume was correct: to be a Cartesian-type doubter is to fall into an inescapable black hole. Abraham Joshua Heschel expressed the corollary to Hume's *riposte* to Descartes in *Man is not Alone*: "First we see, next we judge and form an opinion and thereafter we doubt.... But if we must know in order to question, if we must entertain a belief in order to cast doubt upon it, then doubt cannot be the beginning of knowledge."[36] Heschel's statement is correct in precisely the same manner Duns Scotus' declaration of *being* as linguistically antecedent to *nonbeing* holds True: nothing intelligible can be said about any subject without an *a priori* acceptance of the reality of being and knowledge.

Descartes and the skeptics before him were wrong: *knowledge does not demand certainty as its hallmark.* The arena of science was demonstrating that limited contingent conclusions obtained from careful observation led to ever greater inclusiveness through elaboration of theories devised to justify the data obtained. Moreover, the act of theorizing precipitated an unexpected development: it led to predictions of the results to be accrued from future carefully designed experiments. When such procedures were enacted, the information collected either supported or contradicted the expectations of the scientists. If confirmatory, the original theories gained greater credibility. If contradictory, the search was on for a new proposal to explain *all* previous explorations in a consistent manner. The cycle then repeated: all declarations led to further predictions of application demanding the carrying out of new experiments—a seemingly never-ending sequence. Data collection through the performance of controlled procedures was thus perceived as a win-win situation regardless of whether the new information served to verify or reject the currently accepted view of how things were in reality.

As stated previously, philosophers learned the wrong lesson from their distorted contemplation of scientific method. They eventually yielded on their insistence that knowledge demanded Cartesian-type certainty because they could not deny the science community's ability to display a progressive movement in understanding of man and the universe through apprehension of contingent truths. In reaction, philosophers leapt to the opposite extreme from their former attitude: they declared *all* knowledge was embodied in a set of contingent truths, with contingency implying that under competing conditions, multiple truths had to be acknowledged to hold. The doctrine of *relativism* was born. Donald Rothberg detailed the current trends in doctrinal approach in "Contemporary Epistemology and the Study of Mysticism," which is worth citing at length:

> There has been...a considerable epistemological shift.... The impetus to this shift came from work...showing that observations are theory-laden and that there are...no independent criteria either for verifying theories or for choosing between...different theories.... Naturalistic inquiry aims at empirical *explanation*, conceived of as the development of theories that identify lawful...regularities and causal connections between variables.... Recent accounts of naturalistic inquiry...frequently give up earlier claims about there being some direct relation between naturalistic theories and "reality"... Interpretive inquiry aims...at the understanding of meanings, whether subjective or intersubjective.... There are major differences between... contextualists (or cultural relativists) and universalists. The former insist...there are no fixed meanings or rules...in the world or embedded...in the subject.... Although most...universalists...agree...knowledge and experience are to a significant extent culturally mediated or constructed, they emphasize..."deep" structures that are universal, nonrelative, and not arbitrarily constructed.[37]

There is much to contest in this passage. First, "observations are theory-laden" is true some of the time, but not always. In disciplines too immature to have even a tentative theory formulated and disseminated, observations are primarily responsible for development of that initial theory as well as for all future modifications. Second, why should it be deemed detrimental to the creation of a scientific theory to have "no independent criteria"? Such a proclamation argues one of two points: (a) it assumes there is, or needs to be, a chasm inherently separating abstract theory from any practical empirical methodology employed in its formation or testing; (b) it presumes we are not, and will never be, capable of attaining independent confirmation of our belief systems from other species.

With respect to the first notion, why should an intrinsic epistemological duality be regarded as valuable? Data serving as the ultimate adjudicating authority between rival theories is a priceless asset, and should not be treated as a liability to be circumvented. With regard to the second proposition, it is simply untrue in the present tense in the evaluation of the propriety of some theories, and offers an unfounded bias for future assessments of reality. An illustration of a scientific endeavor in which we currently obtain independent confirmation is meteorology: projections of weather pattern developments often receive substantive support from observation of the behavioral actions of animals in their native habitat.

Third, the alleged concession by "naturalistic inquirers" of abandonment of the idea of "some direct relation between naturalistic theories and 'reality'," is untrue with respect to the individual scientists actually engaged in the research. *Every* physical scientist currently engaged in investigating fundamental topics in physics, chemistry, biology, astronomy, or geology believes their work reveals valuable information about the nature of reality as it, in Truth, exists.[38] Only professional bystanders, philosophers of science, treat scientific research as a hypothetical disconnection of *theoria* and *praxis* from "the real world."

As for the dispute between the contextualists and the universalists, it is fairly easy to resolve in favor of the latter. Is the contextualist argument that "there are no fixed meanings or rules…in the world" a fixed rule? If yes, then the proposition is self-contradictory. If no, then there must exist at least one situation in which its thesis doesn't apply, which is to say that there must be at least one condition under which a fixed rule is potentially existent. It is the attitude of all natural scientists that their efforts are expended towards verifying that the potentially existent is actually existent. On what grounds would any contextualist *demonstrate* the falsity of the conclusions derived by scientific method? Even the attempt at rebuttal of the claim of discovery of fixed rules in manifestation would result in the implosion of the "no fixed rules" doctrine, for the thesis is self-reflexive. With its elimination, we arrive at the universalist's conclusion that a fixed rule does—or many fixed rules do—possess the attribute of existence, which is to say that Truth exists in at least one instance, and that we know that to be the case as an incontrovertible fact. The absolute reality of Truth allows us to then postulate, as was done several pages earlier: "*Contingent truth could only be contingent if there was an ultimate, absolute Truth serving as an invariant standard of reference.* Simply put, truth is the discrete particular, whereas Truth is the continuous universal."

In alignment with Rothberg's declaration, relativism in all shades seems to be the current fashion: sociological cultural relativism proclaims that the institutions created by diverse communities cannot be judged functionally appropriate or inappropriate using any standards other than that society's own; scientific relativism asserts that the appearance of physical laws is dependent upon the observer's perspective and will consequently differ for different frames of reference; ethical relativism professes that conceptual notions of acceptable and unacceptable guidelines for personal behavior lack absolute foundation. Charles Hartshorne noted the obvious problem with this generalization in *The Divine Relativity*: "If all things are relative, if they depend upon and vary with varying relationships, then...relativity itself must be something invariant, nonrelative or 'absolute'."[39] If *all* knowledge without exception is composed of contingent truths and nothing but contingent truths, the Truth of the universal law of contingency itself becomes constant and inalterable, a self-contradiction. This is what was meant by the statement that the "no fixed rules" thesis was self-reflexive.

A case could be made that the philosophical grounding for relativism having received such widespread currency derives from John Locke's (1632–1704) emphatic declaration in *An Essay Concerning Human Understanding*: "Knowledge then seems to me to be nothing but the perception of the connection and agreement, or disagreement and repugnancy of any of our ideas."[40] According to Locke, if we perceive a connection between ideas, we agree with them and accept them as true. If the perception is missing, the agreement is missing, and then such ideas are viewed as false.

It is easy to comprehend how this claim could be employed in the service of relativism in all its variety of possible expression. It is equally obvious that John Stuart Mill (1806–1873) assented to Locke's proposal when he alleged ideas were opinion, not knowledge having an established connection to truth or Truth: "Yet it is as evident in itself, as any amount of argument can make it, that ages are no more infallible than individuals; every age having held many opinions which subsequent ages have deemed not only false but absurd; and it is as certain that many opinions, now general, will be rejected by future ages, as it is that many, once general, are rejected by the present."[41] It is also conceivable Locke's declaration strengthened the resolve of William James (1842–1910) to promote pragmatism as a useful antidote to epistemological relativism in *The Meaning of Truth*:

> No relativist who ever actually walked the earth has denied the regulative character in his own thinking of the notion of absolute truth. What is challenged by the relativists is the pretense on anyone's part to have found for certain...what the shape of that truth is.... The anti-pragmatist, in postulating absolute truth, refuses to give any account of what the words may mean. For him they form a self-explanatory term. The pragmatist...articulately defines their meaning. Truth absolute, he says, means an ideal set of formulations towards which all opinions may in the long run of experience be expected to converge.[42]

Radical Pyrrhonist skepticism, the modified form known as mitigated skepticism, and the doctrine of relativism all assert exactly the same thing: absolute Truth is a theoretical mirage equated by definition with certainty, and unattainable in practice. They differ only insofar as skeptics will allege the identification of knowledge with Truth, making knowledge unavailable also, whereas relativists will claim that knowledge is affiliated with truth and consists of a series of transitory perspectives. The last sentence in the above passage is noteworthy, but it suffers from two major deficiencies: (a) its depiction of the goal as "ideal," implying an inability or an unlikelihood of accomplishment; and, (b) its emphasis on "opinions." Circumvention of these flaws will lead to resolution of the issue concerning the relationship among Truth, truth, and knowledge.

Locke, Mill, and James all failed to take a long-term view of the nature of epistemology in arriving at their theses. As information becomes available through some combination, different in every instance, of sensory input, abstract thought, consideration of the thoughts of others, and emotional experience, interpretive propositions are created to explain the data.[43] The fact that some interpretations are later eschewed in favor of superior ones is not due to random chance. It is not due to simple agreement of person B with the statement of person A because they like the way it sounds or because it makes them feel good. Alterations in interpretation of facts become superior by virtue of being more comprehensive in the type and/or quantity of information capable of incorporation under the newly advocated umbrella. Consequently, there is a distinction to be drawn between opinion, relative truths, and absolute Truth.

Opinion is any idea at which a person arrives at as a product of his or her own thought. It need not be supported by experience or external evidence. *Relative truths* are opinions grounded in experience or external evidence of a personal nature; my experience and the evidence I employ to formulate the opinion may be unique to my situation and not directly applicable to your experience and the evidence you utilize to produce your opinion. *Absolute Truth* consists of practical and theoretical propositions that are pertinent in a universal sense not merely for human beings, but for all creatures and objects within manifest existence. Science employs relative truths as a mechanism for the elucidation of Truth. Hartshorne recognized relativism couldn't be all there is; it inherently logically demands the connectivity offered by invariant Truth. To be relative is to be relative to something that is not itself relative because comparison demands a standard.

Throughout human history, three distinct generic approaches to Truth have been pursued: (a) *science*, which ascertains relationships among diverse physical phenomena through formal experimentation; (b) *religion*, which investigates interactions between overt manifest reality and a hypothesized transcendental reality through extremes of worship resulting in revelation; (c) *philosophy*, which explores the combined, as well as specific singular, values of information products obtained from the other two modes through concept formation leading to abstract doctrines.[44] Heschel's historical summary was as accurate as it was succinct: "The Greeks learned in order to comprehend. The Hebrews learned in order to revere. The modern man learns in order to use."[45] From the Hebrews, people discovered monotheistic religion; from the Greeks, philosophy; and from modernity, science.[46]

It is—the Rothberg citation made clear—valid to conceive of scientific method as empirical, as a mode of investigation thoroughly grounded upon the notion of temporally transitory, contingent information. Under conditions 'ABC,' it is true that 'XYZ' occurs. It is axiomatic within science that if the conditions are altered to 'DEF,' then 'XYZ' may or may not still be true, and that finding out the answer is precisely why one does an experiment. Even scientific laws are acknowledged by all scientists to be contingent, awaiting constant re-verification as new pieces of data derived from experiment are incorporated into some momentarily comprehensive scheme. The recognition of outcomes as dependent upon inputs serves as an unending motivational source to continue experimentation, to renew the effort to recover or discover valid relationships between various parameters. Given this approach to the universe, truth can never be ascertained as Truth until *every* fact is revealed and placed in its proper contextual relation to all other facts. Certainly, it can be contended that the very last piece of data will incontrovertibly reveal the previous web of interconnections is flawed somehow. Does that outcome, unlikely as it would seem, overturn the entire scientific edifice, rendering as *necessary* the conclusion that none of what came before is knowledge, Truth as well as truth?

The method of religion derives not from the confluence of mathematics applied to physical forces under *quasi*-controllable conditions as in science, but rather from the merger of specialized, unique personal experiences with metaphysical unknowns under uncontrollable conditions. Major techniques employed are prayer, meditation, and contemplation. To those who claim to have successfully accomplished the interaction between the mundane and the sacred, the information gained from such experiences is inherently construed as Truth with no contingently truthful aspects.[47] To most observers not privy to inner spiritual workings, the results are not independently verifiable as they are in most scientific experiments, and pronouncements by the mystics are at best taken as subjective interpretations, as truth statements, if their utility is acknowledged at all. These contrasting evaluations would only be reconcilable if *everyone* had a mystical unfolding and shared experiences verbally or in writing. Who, then, is in a proper position to judge whether the specific information bits obtained are properly classified as Truth or truth? Does the uncertainty in authoritative voice imply that none of the data gained from mystical experience can be regarded as knowledge?[48]

As is appropriate for a bridging discipline, philosophy investigates and attempts to unify both avenues of experience, the objectively scientific and the subjectively religious, through distillation: remove the inessential, the context within which the primary explorer collects their data, and purify the remainder, the universal lessons seemingly applicable to all persons always. For subjective experiences, Edmund Husserl formalized the technique used to convert those unique event inputs into general conceptual outputs under the categorical umbrella of *phenomenology*: "Phenomenology is understood as contemplation of pure essences on the basis of exemplary individual intuitions of experiences"; it "examines the essence of perception, judging, feeling as such, not as in this or that animal organism."[49]

It is significant that a key ingredient in phenomenological method is the skeptical *epoché*, and in fact, this is the proper use of the technique: by not holding any particular rigid opinion on the reasonability or the value or the degree of acceptability of specific elements found in reports made by mystics concerning the nature of experiences, the dispassionate phenomenologist is able to analyze statements for internal and external consistency.[50] Contrary to the radical skeptical enshrinement of neutrality as appropriately applicable to all aspects of life, as key to attainment of *ataraxia*, the phenomenologist employs the tool selectively. As a consequence, the latter is able to pursue the collection and assessment of *religious* information in a *scientific* manner.

For those experiences that are primarily objective because they demand that the person serve only as an *im*personal recording device, philosophy has enshrined *pragmatism* as the means to incorporate the scientific findings into broad-based constructs. It is, therefore, *not* a coincidence that pragmatism, as a *philosophical* mode of thought, was unveiled by Charles Sanders Peirce (1839–1914) in a journal devoted to *science*: "If there were any part of a thought that made no difference in the thought's practical consequences, then that part would be no proper element of the thought's significance. To develop a thought's meaning we need...determine what conduct it is fitted to produce; that conduct is for us its sole significance."[51]

A proponent of the procedure, William James, explained the essence of the technique in the appropriately titled monograph, *Pragmatism*: "'Truth' in our ideas and beliefs means the same thing...it means in science. It means *...ideas (which themselves are but parts of our experiences) become true ...insofar as they help us to get into satisfactory relation with other parts of...experience*, to summarize...and get about among them by conceptual shortcuts instead of following the...succession of particular phenomena."[52]

Although phenomenology and pragmatism claim to explore experience and the beliefs derived from them in a scientific manner, they do so by taking diametrically opposed routes. Phenomenology focuses on minute details of experience to find its essential abstract theoretical elements; pragmatism concentrates on the big picture through its insistence on determining only the pertinent concrete practical components. It can be alleged the central concern of phenomenology is attainment of understanding through comprehension of means, whereas pragmatism devotes its principal aim at procurement of knowledge by way of attention to ends. The methods are complementary and should be operationally conjoined for full disclosure of truth as well as Truth.

Within philosophy has arisen the subdisciplines of philosophy of religion and philosophy of science: these serve to critically intellectually evaluate the underpinnings of religious theory or scientific techniques strictly on the level of application to, and coherence with, more generally acceptable currents of freestanding philosophic thought. Independent of the approach philosophy takes when contemplating information obtained from religious or scientific experience, it originates no new data of its own. Philosophers manipulate findings of science and religion through establishment of logical links from previously segregated patterns. They seek to establish integrated systems of understanding as comprehensive as possible at any moment. In recognition of the significance of the intermediary role assigned to philosophy, Johann Fichte (1762–1814) conceded in *Foundations of Transcendental Philosophy*: "Philosophy, along with everything that occurs therein, is a product of the pure capacity for thought. Philosophy itself is not a fact; instead, its task is to provide a foundation for the fact of experience."[53]

It is astonishing that philosophers of science could have misconstrued the essence of the scientific enterprise as severely as the passage from Rothberg indicated. In direct contradiction of that citation, philosophy fulfills its role as connective tissue by incorporating human action, emotion, and thought into one increasingly cumulative package whereby provisional truths evolve. In doing so, the result collectively approximates ever more closely over time what can be characterized as "eternal" Truth. Hence, the original query: Must knowledge be equated with certainty, with Truth only? Is it not reasonable to include within the definition of *knowledge* information serving as a basis for future thought or action, even if those thoughts or actions are appropriate in delimited, not universal, circumstances? If so, then the ultimate accumulation of knowledge would be the additive sum of inputs that combinatorially serve as the basis for *all* thought and action. Isn't this how humanity originally formulated its quest for knowledge to be internally self-consistent and externally maximally comprehensive?

There are two possible mechanisms for the attainment of Truth. These were codified by some of the most commanding figures in the history of Zen Buddhism: as a single burst of complete enlightenment, or as a series of tiny glimmerings coherently united over time.[54] As far as we can document, the first mechanism has never occurred: no human being has, thus far in time, received the gift of thoroughgoing knowledge of the inner workings of the universe and its contents in *all* its imaginable diversity.[55] Since no general statement can be made about other species of life and their epistemological accomplishments, our species possesses no precedent for the first mechanism and only the second pathway appears plausible for *Homo sapiens*, at least for the foreseeable future.

If we attempt to collate those "tiny glimmerings" as truth, the greater the number of future thoughts and actions derive from a given set of data that also happens to be satisfactorily explanatory for past thoughts and actions, the more likely it is contingent truth approaches to the level of Truth. Two mathematical equations can be constructed that explains these relationships using the notation of calculus to express additive sums of large numbers of information bits:

$$\int_1^n \text{information} = \text{truth} \qquad\qquad \int_1^\infty \text{information} = \text{Truth}$$

In these expressions, the first relationship sums a definable number, 'n,' of discrete pieces of information into transitory contingent truth, while the second sums an indefinable number—represented by the infinity sign, ∞—of discrete pieces of information into comprehensive absolute Truth. Since the inclusion of infinity in the expression suggests a never-ending process, an immediate response to this conceptualization will undoubtedly be that Truth can never be attained because as long, with respect to time, as existence exists, new data bits will continuously be unveiled requiring future inclusion into some comprehensive scheme. Such an argument would be both a valid and invalid construal of the proposition. Why?

At issue is the definition of "new." The virtue of any overarching theory is that it serves as a platform to comprehend all phenomena occurring under its umbrella of meaning; it represents a formula conjoining discrete elements into a continuous pattern. It can be said every additional datum corresponds to a new detail. It can also be claimed no alteration in total meaning accrues as a direct result of that occurrence. As a simple illustration, assume there is a herd of one thousand cows on which observations and measurements are made in such detail that *cowhood* becomes completely understandable: the exact nature of a cow's physical, emotional, mental, and spiritual states under all conditions in which a cow can find itself are known, as well as how it functions in every one of those circumstances. If a new calf is born to a member of that herd such that there are now one thousand and one cows, it is appropriate to claim *both* (a) a new calf means a new piece of data *and* (b) a new calf does not change the concept of "cowhood" held prior to its birth. Has truth changed as a result of the calf's birth? Yes: contingent truth, the truth represented by singular discrete elements, has certainly been altered. Has Truth changed as a result of the birth? No: absolute Truth, the Truth predicated on a community of related objects as a continuous whole, has not undergone any adjustment.

Hopefully, the preceding analysis has satisfactorily made two key points. First, skepticism—radical, mitigated, or Cartesian—ultimately cannibalizes itself when theory is mated to practice. Popkin and Stoll persuasively stated the case in *Skeptical Philosophy for Everyone*: "Nothing for the skeptic will count as evidence; so no matter what evidence is produced he will continue to doubt.... Real doubt, in principle, must come to closure. But obsessive doubt does not.... Skepticism does not have to be met by counterargument. It just has to be shown to be without substance."[56] Second, knowledge can be regarded as collections of contingent pieces of information having unifying power for both prior and future thoughts and actions, and therefore need not bear the extra requirement of absolute certitude, which is only realized when all data bits are accumulated and incorporated into an overarching scheme integrating all aspects of manifest reality. Taken together, these conclusions imply *knowledge is both theoretically possible and actually accruing, and at any moment short of the time of transition from truth to Truth, only self-consistency may be reasonably demanded of knowledge as well as partial, though not exhaustive, comprehensiveness.*

The uninterrupted advance of scientific progress from its inception to the current day is directly proportional to the ever-expanding knowledge built upon broadened vistas of contingent truth. Such success forced professional philosophical doubters of knowledge accumulation to develop increasingly sophisticated forms of argument to sustain their attack. Since they were no longer able to declare that knowledge demands unimpeachable certainty at all times under all conditions, Immanuel Kant (1724–1804) inaugurated the next level of refinement of skeptical criticism by proclaiming that only *particular* kinds of knowledge were unattainable, specifically comprehension of the "thing by itself" as defined in *Critique of Pure Reason*:

> The transcendental conception...of all phenomena in space is a critical warning that nothing...seen in space is a thing by itself, nor space a form of things supposed to belong to them by themselves, but that objects by themselves are not known to us at all, and...what we call external objects are nothing but representations of our senses, the form of which is space, and the true correlative of which, that is the thing by itself, is not known, nor can be known by these representations, nor do we care to know anything about it in our daily experience.[57]

Kant asserted we could know *about* objects without knowing the actual nature of the objects because our contact with objects is only indirect through sensory input, and it is our sensuous representations within the brain that innately employ spatial form as a parameter incorrectly attributed to the object. Today, of course, more than two hundred years after he wrote this, we know that his reasoning was in error, though suitable for his time in history. Advances in technology have allowed us to pierce the surface representation of objects, enabling us to develop secure knowledge of their inner structure and radiant energetic displays. Moreover, space is now regarded as endemic, its extent and shape a product of forces predominantly, not exclusively, that of gravity as principal determinant.

Scientific advancement aside, Kant's approach logically suffered from its dogmatic either/or nature. He reasoned that *because* space is the *a priori* form of representations of objects to us, this representation being subjective necessarily implied it could not possibly correlate with the objective nature of the thing in itself.[58] There is no innate justification for this judgment. It is plausible and self-consistent to assert both/and: space is *both* a subjective representation of things to us *and* an indicator of the nature of the objective thing in itself. Even stronger is the reasonable assertion, though not herein proven, entirely opposite to the Kantian premise: it is *because* space is an intrinsic element in the characterization of external objects that necessarily leads to our usage of it as invariably pertinent to a mental representation. This proposition would conform to the original conceptualization of space as related to the position of objects found in the framework of Aristotle's (384–322 B.C.E.) ten categories attributable to substance.[59]

For the remainder of this volume, the orientation of the *transcendental realist* in opposition to one who subscribes to *transcendental idealism*, as Kant defined the terms, is the attitude to be taken: "Transcendental Idealism teaches that all phenomena are representations...not things by themselves, and that space and time therefore are only sensuous forms of our intuition, not determinations given...by themselves or conditions of objects, as things by themselves. Opposed to this Transcendental Idealism is a Transcendental Realism, which considers space and time as something in itself (independent of our sensibility)."[60]

The incorporated, but abbreviated, review of the history of skepticism in philosophy terminates in the twentieth century, which has brought with it a new approach to this construct, a critique included in the Rothberg citation under the heading of "interpretive inquiries." It consists, at least in part, of attacking idea content based on linguistic analysis. Richard Tarnas explained the roots and implications of this intellectual phenomenon in *The Passion of the Western Mind*:

> The analysis of language...has brought forth the most...skeptical epistemological currents in the postmodern mind, and...these currents...have identified themselves most articulately and self-consciously as "postmodern."... Basic to this perspective is the thesis that all human thought is...generated and bound by idiosyncratic cultural-linguistic forms of life. Human knowledge is the historically contingent product of linguistic and social practices of...local communities of interpreters, with no assured "ever-closer" relation to an independent a-historical reality.[61]

In order to rebuff the conjectures contained in this passage, a brief recitation of the philosophic meaning of "idea" must first be presented. What are ideas and where do they come from? In addressing this question, we need not consider definitional distinctions between *impressions* or *perceptions* on one hand, and *conceptions* or *ideas* on the other.[62] Plato (427–347 B.C.E.) believed they were self-existent archetypes inhabiting a transcendental realm waiting to descend to Earth and enter someone's mind-brain so as to be evoked in propositions of supposition or fact.[63] Aristotle disputed their self-existent nature; in response to his arguments, Neoplatonists placed ideas within a transcendental intelligent Mind from which they could infiltrate the human mind-brain.[64] John Locke and David Hume further transported their point of origin by relocating them entirely within an individual's mind or self-consciousness, thereby converting the construct into a wholly immanent epistemological term and denying of it any transcendental metaphysical implications.[65]

The "postmodern" critique is dependent upon the presupposition that Locke and Hume were correct in relegating ideas—the entities conveyed from person to person by language—to the individual without a connection to a transcendental domain. Thus, there are two things wrong with this attack on the possibility of acquisition of knowledge and either truth or, ultimately, Truth. First, if the passage is an accurate reflection of postmodern linguistic analysis, and if Tarnas chose his words with care, thinkers of this persuasion appear to be critical of the alleged lack of any certain connection between idea formulations and an "independent a-historical reality." In other words, they are seeking the very concept, a transcendental source for ideas, their implicit presumption denied. Yet, complaint of its absence was articulated in a manner impossible of fulfillment. How can one establish a linkage to something independent? Any type of connection would deny for the object of the search its independence. Only a return to the doctrine of Plato or his successors would allow for a postulation that there exists, at least *in potentio*, a plausible "a-historical reality." However, such a transcendent realm could in no way logically be deemed entirely independent of manifest material existence. If it were independent, we could never come to postulate, much less know, its existence.

Second, the nature of the proposition is itself the product of a linguistic-cultural relativism. According to the dictates of internal consistency, it need have no contextual connection to Truth outside its own expression. If this is the case, then why should anyone believe it to be useful beyond its own statement of itself? In other words, why should someone other than the originator of the proposal use it as the basis for unifying the perspectives obtainable from other pieces of information? Through its own relativism, it cannot serve as a unifying platform for any overarching epistemological paradigm. It would be as if we arbitrarily chose one iceberg from among a set of free-floating icebergs to serve as the connection point for the rest: being free-floating, it could not anchor the others.

This is the intrinsic problem to which all forms of skepticism are subject: by inciting a thoughtful person to doubt every piece of data's bearing on truth, much less Truth, they consistently induce that person to doubt the skeptical position as well. Skepticism, irrespective of the grounds for its argumentation, is unable to sustain itself—as Tarnas himself commented: "On its own terms, the assertion of the historical relativity and cultural-linguistic bondage of all truth and knowledge must itself be regarded as reflecting but one more local and temporal perspective having no necessarily universal, extra-historical value.... This is the unstable paradox that permeates the postmodern mind."[66]

It should be clear that skepticism in any form is ultimately ineffective; to consistently attempt to doubt the potential and actual accretion of knowledge *about* is impossible in practice.[67] The failure is ultimately logical because consistency *in* anything—speech, thought, or action—demands acceptance of the precept *of* consistency: constant application of a given principle in a proper manner. This, in turn, necessitates a positive affirmation of the reality of a guiding rule, the correctness of that reference, and the existence of a proper manner of adaptation to the approved standard. These can all be man-made: arbitrary and relativistic. However, there then need be yet another, an *absolute*, framework by which one can judge existence, adherence, and propriety. A surprisingly simple epistemological formula results: *knowledge unifies ideas about some object with words and numbers, truth tentatively combines knowledge with reality, and Truth is reality.*[68]

The greater stringency of comprehensiveness in comparison with internal and external consistency as an attribute of knowledge was underscored in a paper published by mathematician Kurt Gödel in 1931 proving that all self-consistent mathematical systems were necessarily incomplete, meaning there were propositions in that system known to be valid, but which could not be so proven using only principles innate to the system in question.[69] Although intended to be directly applicable only to mathematics, the conclusion can be expanded to encapsulate all other disciplines: construction of any abstract intellectual system unifying thought and action demands the employment of assertive statements either useful or necessary for the elicitation of other pertinent elements of that system, but which are themselves unsupported by rigorously provable logical chains of reasoning internal to the system in question. Obvious examples can be obtained from philosophy in general and metaphysics in particular, although neither science nor religion is immune to this constraint.[70]

British philosopher-archaeologist Robin George Collingwood's *An Essay in Metaphysics* called those foundational tenets that frame any further study on a theme *absolute presuppositions*.[71] Collingwood meant to establish by the term "absolute" that such constructs were necessary in all cases, and could not be circumvented based upon the subject of investigation or the nature of the approach taken by an investigator. Paul Trainor's essay on Collingwood in *Ultimate Reality and Meaning* made it clear that the usage of "absolute" as a characterization was unfortunate.

While it was true that some beginning was indeed unavoidable, any such point of departure was inevitably relativistic: "Absolute presuppositions vary from civilization to civilization, and in...time, within a civilization as well.... There is no...reason for choosing one set of absolute presuppositions...they are not self-evident nor verifiable nor transcendentally justifiable. They are simply chosen or believed."[72] Alfred North Whitehead (1861–1947) was critical of any tendency to overlook the suppositional basis of philosophy's starting constructs: "Philosophy has been afflicted by the dogmatic fallacy, which is the belief that the principles of its working hypotheses are clear, obvious, and irreformable."[73]

Having presuppositions does not invalidate data facts, collections of facts combined into tentative hypotheses, assemblages of successfully tested hypotheses united into working theories, or the inclusion of distinctly segregated theories into a system of self-consistent thought. All levels of information convey contingent truths, and at different stages of construction, all depict knowledge of reality. Hans-Georg Gadamer even claimed that unexamined presuppositions—"prejudices" according to his vocabulary—were epistemologically beneficial because innately motivational: "It is not so much our judgments as our prejudices that constitute our being... Prejudices are not necessarily unjustified and erroneous, so that they inevitably distort truth.... The historicity of our existence entails that prejudices...constitute the inner directedness of our whole ability to experience."[74]

Prejudice can indeed be where we begin our path to understanding; it can also be the point at which we end that quest. They are acceptable provided we have made transparent what they are so that at some point in the future we can reevaluate and discard them as being inconsistent with subsequent learning, if necessary. Whether Gadamer's remark is validated or not in each instance, what the Gödel paper proved for mathematics, what is applicable to all fields of study, is that at some point the original presuppositions must be examined and verified by employing the perspective of a meta-system, a higher-order orientation that allows for the enunciation and implementation of an even more comprehensive scheme than the original system.

In *Charmides*, Plato ascribed to Socrates the famous aphorism: "If a man has any feeling of what is due to himself, he cannot let the thought which comes into his mind pass away unheeded and unexamined."[75] Recall the words of Richard Popkin quoted earlier: "Pyrrhonism is thus ultimately a philosophy of life, a formula for living well. It reverses the Socratic point of view that the unexamined life is not worth living." No other combination of statements could more clearly express the difference between those who envision the existence of knowledge and Truth as a reality, and those of the skeptical persuasion who don't. The Socratic injunction applies equally to presuppositions, for there is one questionable aspect of Trainor's statement cited above: the assertion that they are not "verifiable nor transcendentally justifiable."

As a generalization, it is unacceptable unless he could justify the opinion in his own terms or by quoting Collingwood. He and/or Collingwood—it is unclear from the essay whether Trainor was reporting Collingwood's view or supporting it—never considered the possibility of a metatheory. Following the previous passage, Trainor continued: "Since absolute presuppositions are the *a priori* conditions for any science, they cannot be validated by the science itself in any straightforward way, since they are the preconditions for the canons of validation within a science. Thus, Collingwood writes, 'any question involving the presupposition that an absolute presupposition is a proposition, such as the question "Is it true?"...is a nonsense question'."[76]

Collingwood, and possibly Trainor, abided by Gödel's theorem through acknowledgement that a presupposition "cannot be validated by the science itself," but extended the theorem's implications into a region beyond what Gödel was comfortable in deducing through his mathematics: the exclusion of any possibility for the existence of a higher order system with the capacity to incorporate the presuppositions into a verifiable framework. Such an extension into uncharted waters on the part of Collingwood and/or Trainor must be justified. A failure to do so results in the formation of yet another unexamined presupposition.

For the conversion of contingent truth to absolute Truth to be realized, a hierarchy of meta-systems invoking ever-increasing degrees of inclusion of extant data facts must be not merely possible in theory, but also attainable in practice. In an attempt to support an assertion that such an unending series would never attain resolution, one might be tempted to draw an analogy with the notion of infinity: a series of ordered terms without end. The previous example with the cows demonstrated resolution is hypothetically possible in at least one significant respect even within a limited series of data points.

More importantly, the invention of calculus was designed specifically to solve that problem with regard to an unlimited series of data points: deriving mathematical relationships to precisely calculate the outer boundaries of the infinitesimally small as well as the infinitely large. Calculus can accomplish this task because although the individual terms in the series can be listed without ever coming to a final term, it is possible to determine with accuracy and precision what value the series *approaches* in a convergent manner. The mathematics of infinity is in fact what led to the previous formulations for truth and Truth. Similarly, a metatheory of sufficient power, in conjunction with all previous systems and meta-systems on any and every conceivable topic, will be comprehensive to a degree necessary to determine the limit of the series; in other words, the inescapable approach of contingent truth to absolute Truth.[77]

It might even be claimed that the forerunner of the calculus, invented in its current mathematical form in the seventeenth century, was conceived as early in history as the time of Aristotle. The manner in which he tackled the problem of creation and first cause in his *Metaphysics* can be depicted as a type of philosophical calculus: "In spite of all, there is a definite beginning, and the reasons for things are not infinite either in series or in kind. For there can be no infinite regress in the production of things from their materials."[78] In other words, Aristotle "calculated" as follows:

$$\int_{\text{first material thing}}^{\text{all material things}} \text{creation} = \text{First Cause} = \text{Prime Mover} = \text{God}$$

Neither Aristotle nor anyone else could list all the material things that have been or will be created in the physical universe, since the sequence has no discernible last term. Nonetheless, Aristotle reasoned that the series still converged to a limit that could be determined: a Creator, a Prime Mover, denoted as God. The only quarrel one can have is with the term "God": one could just as easily insert "Brahman" or "Tao" or "Zeus" or "Satan" or "Big Bang" *et cetera* and the convergence limit would still apply.

Contrary to Collingwood's judgment, a meta-system of sufficient power *would* allow for verification of his absolute presuppositions. Further, such a meta-system *would* allow for determination of the transition point from truth to Truth for the same reason: the point of convergence would be clarified such that the unique features of an all-encompassing schematic could be distinguished. In *Jnāna Yoga*, Vivekananda presented a very human, very simple, yet profound version of the process envisioned here:

> A man gets a new and better idea and then he looks back on those he has given up, and forthwith decides that they were mischievous and unnecessary. He never thinks that, however crude they may appear from his present point of view, they were very useful to him, that they were necessary if he was to reach his present state, and that every one of us has to grow in a similar fashion, living first on crude ideas, deriving benefit from them, and then arriving at a higher standard.[79]

To seek for and find this final "super metatheory" that will reveal Truth, one must make use of an epistemological principle of startling simplicity, yet of enormous importance, offered by David Hume in *An Enquiry Concerning Human Understanding*: "What never was seen, or heard of, may yet be conceived; nor is anything beyond the power of thought, except what implies an absolute contradiction."[80] The freedom to think demands of *Homo sapiens* a tremendous responsibility: not to treat this power in a cavalier manner. Some of the grossest abuses of this freedom and consequent responsibility derive from individuals who spend a lifetime trying to "educate" others to adopt the position that fruitful logical thought is impossible: radical skeptics who deny that knowledge is possible; post-modern deconstructionists who claim that the meaning of anything is whatever one arbitrarily chooses to make it; scientists committed to a philosophy of materialism professing that physical existence is the *only* possible existence; and, purveyors of religious doctrines proclaiming that all phenomena are illusory projections emanating from a transcendental reality that is the *only* reality.

Hume's principle claims imagination is not contrary to logic; rather, they are united in their capability in comprehension of the nature of the universe, everything in it, and everything lying beyond it. It can be imagined, for example—without sacrificing logic as a proper tool in shaping imaginative thought—that intelligent creatures exist whose means of interaction with their surroundings are not primarily sensuous in nature; that there is direct mental contact between being and object on the level of energy transfer, such as that indicated in conjunction with the paranormal talents claimed by some individuals. Similarly, one can imagine, especially in light of claims of actual accomplishment, that immanent and transcendent can exchange information through mystical experiences entirely non-sensuous in quality and character.

Chapter Two: Anthropocentrism

> Man's sense is falsely asserted to be the standard of things; on the contrary, all the perceptions both of the senses and the mind bear reference to man and not to the universe, and the human mind resembles those uneven mirrors which impart their own properties to different objects, from which rays are emitted and distort and disfigure them.
>
> —Francis Bacon, *Novum Organum*

The opening chapter offered a lengthy argument against the proposition that radical skepticism or its various modifications are viable epistemological methodologies with reference to their attitudes concerning the existence and availability of Truth and knowledge. Every theory of knowledge *must* begin with the premise that acquisition of knowledge is attainable to such a degree as to eventually arrive at levels from which deductions can logically be made concerning the convergence of the transitory contingent truth of things to the ultimate limit, absolute Truth. Chapter one also introduced Collingwood's allegation that the pursuit of knowledge necessarily adopts some arbitrary starting point as a presuppositional basis for achievement of this goal, and he reinforced this notion with the claim that all such points of origin are variable from civilization to civilization.

Collingwood's charge was incompletely stated. Immanuel Kant had long before convincingly demonstrated in *Critique of Pure Reason* that humanity was conditioned to frame all thoughts according to a small group of innate *a priori* templates, especially those pertaining to time or space. Collingwood's proposition should be modified to reflect the judgment that his idea refers to the content of theories while Kant's concepts involve their form. The latter's thesis concerning an innate spatial conformity inhabiting all human thought was offered in the prior chapter. With respect to time, he wrote: "If we drop our manner of looking at ourselves internally...and thus take objects as they may be by themselves, then time is nothing. Time has objective validity with reference to phenomena only, because these are themselves things which we accept as objects of our senses; but time is no longer objective, if we remove the sensuous character of our intuitions."[1]

It is universally acknowledged by both the community of scientists and philosophers familiar with the evidence that we no longer accept the template of space as derived entirely from our own intellects—thanks to the principles of physics established by Isaac Newton as well as the introduction of Albert

Einstein's General Theory of Relativity claiming space is determined by the force of gravity and the mass of objects present in a given space. No similar assessment has attained widespread approval with regard to time. Hence, Kant's insistence that it has no truly objective reality could still be correct.

Even if he is found to be in error on this point—a proposal to counteract his claim was expressed in *The Other Perennial Philosophy: A Metaphysical Dialectic* and will be reprised here in a later chapter—Kant's innovations are useful in an unanticipated manner: they represent a magnificent illustration of an *undesirable* characteristic for presuppositions.[2] Kantian doctrine—its metaphysics, ethics, epistemology—has an important unexamined conviction at its core, one that has ill-served the human pursuit of knowledge for two and a half millennia. The chapter title gives a name to this all-pervasive, yet insidious, dogma. Many pages will henceforth be devoted to its destruction as a functional idea upon which to build any proper epistemology.

It should occur as an unending source of astonishment to anyone who notices: the vast majority of all theoretical and methodological propositions historically presented in philosophy and science are ultimately grounded in one statement uttered by Protagoras of Abdera (~481–411 B.C.E.). It was he, a Sophist, who first announced: "Of all things the measure is man; of the things that are, that they are, and of the things that are not, that they are not."[3] As the opening epigraph attests, Francis Bacon had such a response. Since that first extraordinary self-promoting, self-celebratory declaration, human beings have held it as an invariant fundamental presupposition underlying many of their perceptions and conceptions about the nature of reality.

Aristotle proclaimed in the *Nicomachean Ethics* that the central notion in ethics, with all its consequent moral applications, was in conformance to the dictum of Protagoras: "Everybody who understands his business avoids alike excess and deficiency; he seeks and chooses the mean, not the absolute arithmetic mean, but the mean considered relatively to ourselves."[4] Although Chinese philosophers probably had not heard of Protagoras, they unwittingly ascribed to his thesis. Livia Kohn's *Early Chinese Mysticism* paraphrased an early Common Era commentary on the Taoist text, *Chuang Tzu*: "Originally people did not discriminate anything at all; then they began to recognize other beings. Gradually they learned to distinguish between this and that, and eventually evaluated everything according to its being good or bad, right or wrong, useful or useless. People's definitions...came to determine the place of everything in the world."[5]

The unwarranted extravagance of the Protagorean assertion, as well as of its future extrapolations, was not immediately apparent. In consequence, ever since its utterance it has behooved philosophers and scientists to discover the intellectual basis for rejection of the doctrine, and then to the practitioners of various religions to demonstrate its misguided orientation. This foundational construct even served as the underlying support for the Ptolemaic contention that the Earth was the center of the universe with the Sun orbiting around our home planet. One might think once Copernicus debunked that errant thought, human beings would reorient this view of the significance of their place in the cosmos. This has not been the case; indeed, the reverse has transpired: as a species, we are, perhaps, more self-centered than ever before in history.

Nikolai Berdyaev admitted in *The Beginning and the End*: "Medieval scholastic philosophy was fundamentally Greek; it did not pass beyond the bounds of ancient thought; it was a philosophy of the object, that is to say it was cosmocentric. Modern philosophy...has become a philosophy of the subject; it is anthropocentric and its center of gravity is transferred to man."[6] In saying that "scholastic philosophy was...cosmocentric," what he was referring to was its main thematic focus, which was essentially metaphysical thought directed toward a principal concern with the nature and attributes of deity.

These concerns were pondered almost entirely from the perspective of how we should act towards the divine, since it was generally acknowledged that God was in essence unknowable. Consequently, given the directionality of inquiry, it was the relationship of the transcendent to humanity that was deemed the primary issue. The remaining core of Greek thought inherited by Western philosophers in the Middle Ages was confined to the domains of ethics and morals—What is good and proper for us?—and also were entirely anthropocentric even as these themes faced outward toward the universe in their effects. With modernity has come the elaboration of epistemology, the study of human attainment of knowledge. Therefore, all philosophy is, and has been, essentially oriented inwards, with humanity as its principal concern and focus.

Returning to Kant, his important contribution concerning the inherency of concepts of space and time for human comprehension of the universe was unfortunately coupled to the assertion of Protagoras. Hence, Kant extended his doctrine to conclude space and time in themselves do not exist and lack intrinsic meaning for objects generally: "The transcendental conception...of all phenomena in space, is a critical warning that nothing which is seen in space is a thing by itself, nor space a form of things supposed to belong to them by themselves"; also, "Time has objective validity with reference to phenomena only, because these are themselves things which we accept as objects of our senses; but time is no longer objective, if we remove the sensuous character of our intuitions."

The overt unwarranted extension due to the hidden presumption is clear: "all phenomena" seen *by us* in space and time means space and time are not "a form of things supposed to belong to them by themselves." Because space and time are intimately conjoined to representations of objects for us, he concluded the fact of this representation being subjective necessarily implied it could not correlate with the objective nature of the thing in itself. Why? It is plausible and self-consistent to conclude space and time are involved *both* in the subjective representation of things to us *and* exist as indicators of the nature of the objective thing in itself, a point argued in chapter one. There was no justification for Kant's implication; it was an additional metaphysical presupposition influenced by Protagoras, although with complete lack of awareness on the part of Kant.[7]

Employing contemporary language, the declaration of Protagoras—and its utilization in the thought of Aristotle, Kuo Hsiang, Ptolemy and Kant—is most appropriately categorized under the heading of *anthropocentrism.* To be anthropocentric in one's perspective is to believe human logic, values, and levels of achievement appropriately serve as the gold standard for rational thought processes, justificatory motivations, and accomplishment for other intelligent creatures. Hans Küng was forthright in *Does God Exist?*: "Man's values and ideals—truth, justice, love, morality, religion—do not exist anywhere... They are entirely the product of man's invention and definition; man projected them originally into things and thus created values and ideals in accordance with his life's needs."[8] In *The Phenomenon of Man* Pierre Teilhard de Chardin asserted a view shared by many throughout history:

> Can we seriously doubt that intelligence is the evolutionary lot proper to man and to man only? If not, can we...hesitate to admit that man's possession of it constitutes a radical advance on all forms of life that have gone before him? Admittedly the animal knows. *But it cannot know that it knows*.... If it could, it would long ago have multiplied its inventions and developed a system of internal constructions that could not have escaped our observation.[9]

There are occasional whispers of disapproval and disagreement with this dominant intellectual presupposition afflicting *Homo sapiens*, such as that emanating from Francis Bacon. P. D. Ouspensky was another contrarian to this philosophic approach. He proclaimed in *Tertium Organum*: "Our chief mistake as regards the inner world of an animal lies in our ascribing to it our own logic. We think that there is only one logic... Yet, in actual fact, it is merely the laws of the relation of our inner life to the outside world or the laws which our mind finds in the outside world. A different mind will find different laws."[10]

One can disagree with Ouspensky's conclusion that there are multiple forms of logic while accepting his inference that humanity does not hold a monopoly on the powers of intellect: animals think, may employ different operating constructs than do human beings, and may even be capable of the awareness that might accompany thoughts, emotions, and behavior as well. Since these conditions, if verifiable, would significantly alter all proposed theories relating the nature of the ontological and teleological reasons for the existence of the universe, as well as the functional mechanisms by which that physical manifestation continues to be, it is clear anthropocentrism carries with it an implicit belief that what human beings accept as truth is not merely truth, but is also, *by definition*, absolute invariant Truth.

De Chardin's argument also highlights a corollary to anthropocentrism: the subsidiary doctrine of *anthropomorphism*. Whereas the former proclaims humanity as the philosophical, if no longer the astrophysical, center of the universe, the latter states that other living creatures are biologically closer to us, and therefore evolutionarily advanced, to the degree they imitate our structural and/or functional patterns. Interestingly, there are those who find philosophically repugnant the notion that any species can even be mentioned in the same breath as *Homo sapiens* regardless of the nature of the physical-emotional-intellectual construct under discussion. In other words, although anthropomorphism demands anthropocentrism as its ground—there would never arise the incentive or need to propose to categorize species according to their degree of similarity or difference from humans were it not for the presupposition that "man is the measure of all things"—some biologists and psychologists are so extreme in their man-centered orientation they perceive these constructs as oppositional.

In his essay, "The Myth of Anthropomorphism," John Andrew Fisher chose to *defend* anthropomorphism while being *critical of* anthropocentrism: "Many assume anthropomorphism is a category mistake of ascribing human attributes to something not human.... We share many features, physical, biological, and social, with other animals... It remains an empirical question which, if any, mental characteristics humans have uniquely."[11] He opposed a generic form of anthropomorphism to a radical form of anthropocentrism:

> I will call...*hard anthropocentrist*...those critics of anthropomorphism who regard ...application of...M-predicates...to *any* animals categorical anthropomorphism.... Hard Centrists are committed to a sharp division between humans and...animals.... Most of us think it *can* be categorical anthropomorphism to apply...M-predicates to some species. Also, there are some M-predicates that most of us would hesitate to apply to any nonhuman animals.[12]

Fisher defined "M-predicates" as attributions of mental states common to humans, such as fear, concern, anxiety, love, joy, desire, belief, and so on. Hence, his statement declared his belief that some animals are like us in their mental and/or emotional responses to stimuli, and some are not. The notion of a "category mistake" means some investigators choose to reject even the potential for similarity between other species and us.

Fisher did not dispute the viability of the premise of anthropocentrism enabling anthropomorphism to be its logically subordinate argument: it is proper, if not inherently obligatory, for humanity to uphold ourselves as the reference standard for judgment of qualities characteristic of other species. The only issue at hand from Fisher's perspective was whether nonhuman animals are like us in various ways; he never questioned the assumption that it is appropriate to evaluate those traits in a relativistic manner. It is certainly reasonable to determine properties possessed by living things on an absolute scale of presence or absence independent of the invocation of comparison with human beings.

Our language was constructed for and by humans. The consequence of human definitions and applied interpretations necessarily follows from this fact. Therefore, M-predicates may reasonably be applied to other species as language descriptors when their utility is readily ascertainable. However, *we must explicitly acknowledge that these concepts are borrowed terms and may inaccurately convey the essential characters of nonhumans*. In illustration, it is not logically necessary that an animal's sense of fear be induced by exactly the same conditions as, or be experienced in the same manner as, or result in similar behavior to, those implications ascribed to human fear in order to employ the term "fear" for phenomenological interpretation of the animal's condition. The alternative would be an endless multiplication of specialized linguistic attributions of animal, or even plant, behavior—a highly unwieldy procedure.

By making this pronouncement of "borrowed terms," with repetition as often as needed for the point to be rigorously absorbed into each individual's thought pattern, at least one important consequence would immediate ensue: instead of being continuously caught up in a series of fruitless debates as to whether other creatures are "like humans"—currently irresolvable precisely because we do not share a common language, and cannot communicate with these nonhumans in a mutually normal conversational mode—we would allow ourselves to observe other species and recognize it as highly probable these others have their own version of various emotional and mental states.

It is plausible that nothing about the human makeup is unique except perhaps its external appearance in form. In fact, with respect to the internal structure of the brain, variation between humans and animals is actually fairly minimal, though not without highly significant, externally observable effects engendered entirely by specific differences encountered. In *Animals in Translation*, Temple Grandin mentioned the most relevant anatomical distinction:

> When you compare human and animal brains, the only difference that's obvious to the naked eye is the increased size of the *neocortex* in people.... The neocortex is the top layer of the brain, and includes the frontal lobes as well as all of the other structures where higher cognitive functions are located. The neocortex is wrapped around all the *subcortical* or lower brain structures, which are the seat of emotions and life support functions in people and animals.[13]

Note Grandin's use of "higher." Was it employed neutrally in reference to the layered construction of the brain for all species possessing a central nervous system? Was it intended to carry an evaluatory tone to be interpreted indicating "improved," "more sophisticated," or "preferred"? *If structure is different, expectation of similar function is untenable*, and out goes the entire foundation for anatomical anthropomorphism with respect to that physical feature. On the other hand, *if there exists similar function despite substantive differences in structure, physical construction cannot be the primary source of the similarity*, suggesting anthropomorphic comparison only on the basis of mentality is feasible. One could claim either sufficient structural similarity exists between human and other species to allow for comparison, or the traits of intelligence and emotionality exhibited by diverse creatures are, literally, *meta*physically grounded. If the former is argued, it is hard to find rational substantiation for the human claim of superiority in any overall sense; if the latter is the position taken, philosophical materialism is sacrificed.

Most four-wheeled, internal combustion engine vehicles are constructed with four-, six-, or eight cylinders. In all likelihood, an eight-cylinder vehicle will be able to attain a greater maximum speed than one with four. Does that make it a superior automobile overall? Is a sports car better than a tank, or *vice versa*? Different vehicle designs often signify divergent purposes; the same is the case for brains and other physical features.

Some animals have physical attributes readily identifiable as analogous to those possessed by humans and utilized in the same general manner—eyes and vision being an obvious example. Yet, no one would claim humans have superior eyesight than all other animals. This one illustration is sufficient to demonstrate inappropriateness in employing human capacities as a standard of measurement. Analogously, animals could possess *some* similar mental or emotional traits as humans—Fisher's gross scale "M-predicates," perhaps—be deficient in others, and not be *inferior* evolutionarily, for "inferiority" is a grammatical comparative demanding an acceptable standard for judgment. If employed as the relevant guideline, the Protagorean postulate is not thereby confirmed, but merely assumed in all attempts to comprehend the animal kingdom.

At the heart of all debates on whether animals are equal, superior, or inferior to humans—like us or not like us in characteristic mental, emotional, and physical states—is the issue of self-consciousness. Since there is current controversy on what *human* self-consciousness entails, discussion of *animal* self-consciousness is bound to be highly unproductive at this point in time. Development of a suitable definition for the nature of human consciousness would immediately, substantially impact individual opinion on the feasibility of animal self-consciousness. Hence, though hardly more than an assertion, it is nonetheless meaningful that researchers such as Donald Griffin are strong advocates of animal self-consciousness. In *Animal Minds*, he wrote:

> The most basic and essential aspect of consciousness is thinking about objects and events. The content of conscious experience may ordinarily be limited to what the animal perceives at the moment about its immediate situation; but sometimes its awareness probably includes memories of past perceptions, or anticipations of future events.... A conscious organism must...experience some feeling about whatever engages its attention.... One reason to suspect that nonhuman animals do experience conscious thoughts is...the basic structures and functioning of neurons and synapses are quite similar...in all animals with organized central nervous systems.[14]

It can be inferred the anthropomorphic dispute about self-consciousness is frequently a case of optimists *versus* pessimists: those in favor of animal self-consciousness seek support from similarity in human and animal brain structure and function; those opposed point primarily to the distinctions that can be drawn between human and animal brain structure and function. Of significance is that neither side rejects the anthropomorphic assumption as a viable evidentiary basis for their respective contentions. Since both sides are correct from their particular perspective because they differ only in specific aspects of behavior they choose to illustrate their viewpoint, it is reasonable to believe the issue will not be finally resolved on this level.

Indeed, this conceptual block is one reason to posit self-consciousness, whether animal or human, as not principally an issue of physical anatomy. Self-consciousness may ultimately be decidable only upon consideration of, literally, metaphysical justificatory arguments. These contentions need not imply an appeal to some transcendental agency; they can refer to "mental or emotional states" that are in some manner phenomenologically distinct from, yet functionally related to, the physical workings of any organism's brain.[15] Barring, at least temporarily, the speculative involvement of a spiritualistic construct, the issue becomes one of determining whether human and animal personality and/or behavior are sufficiently consonant *strictly by employment of external observation* to make plausible the proposition that if humans can claim self-consciousness, then animals can also.

We thus return immediately to the assertion by de Chardin: "Admittedly the animal knows. But it cannot know that it knows.... If it could, it would long ago have multiplied its inventions and developed a system of internal constructions that could not have escaped our observation." Schematically, his reasoning, shared by a number of ethologists, philosophers, and cognitive scientists, can be delineated as the systematic premises: (a) possession of a bigger, more intricately interconnected brain than other animals means *ipso facto* superiority; (b) the existence of greater neuroanatomical capacity and complexity allows for emergent development of self-consciousness; (c) proof of evolutionary superiority to all other species on Earth is indicated by the greater degree of sophistication of the civilization we have created. These propositions allow us to claim the quality known as self-consciousness, if it exists at all, is a property exhibited exclusively by human beings, or if shared by some other species, reaches its apex of unfoldment in human beings.

Weighing in on the other side of the argument is Fisher, who generously stated: "Most philosophers have tended to think there is a sharp division between humans and other animals...and always this division is a hierarchy, with humans at the top and animals underneath separated by a gulf. Thus philosophers have not been concerned with getting an animal's mentality right...nor have they been concerned to ban anthropomorphic interpretations to make room for another sort of mentality, alien to our understanding."[16] Donald Griffin, in *Animal Minds*, was more condemnatory of the human predisposition to put ourselves at the top of the evolutionary pyramid: "The acceptance of biological evolution and the genetic relationship of our species to others was a shattering blow to the human ego from which we may not have fully recovered, for it is not easy to give up a deep seated faith that our kind is unique and qualitatively superior."[17]

While it is true human beings have developed a more complex society than any other species on the planet in terms of our ability to manipulate the physical environment to conform to our desires for comfort and security, this fact need have no bearing on the question of evolutionary superiority. It is well-established there are many species that are, in one respect or another, physically superior: hawks have better eyesight, bats better hearing, dogs a better sense of smell, deer run faster, elephants are stronger, camels have greater endurance, and so on. De Chardin's proposition is fallacious in that it conflates physical facts with facts about values: it assumes any species, or every species, would mirror our chosen preferences and goals, and therefore choose to follow the course we have set for ourselves in transforming the surroundings. It further assumes this directionality constitutes evolutionary advancement by definition.

There is no evidence to support these assertions. What is *necessary* about the direction or growth rate of human civilization—whether "growth" refers to population, knowledge, wisdom, wealth, technological skill or any other meaningful measure—especially in view of the fact these rates have not been constant over time? No independent justificational legitimacy exists for the claim that other species must adopt our set of values. Thinkers on the issue of nonhuman cognitive development often accept de Chardin's presuppositions because they conform so neatly to that ancient dictum of Protagoras: "Man is the measure of all things." The words written by Roger Crisp in "Evolution and Psychological Unity" echo those voiced by Fisher and are directly on-point:

> The "ladder" picture of evolution lies behind a common view of the classification of animals.... There is a...scale of evolutionary advancement, on which human beings are at the top, unicellular creatures at the bottom, and other animals in between.... We should then accept that a single linear scale will not suffice. Rather, we must compare living animals according to...recency of origin, phylogenetic relatedness, and adaptation to the environment.... Another thing we may do is place human beings at the top of the scale, while admitting that there is no clear biological reason for doing this (the scale is not one of pure behavioral or neurological sophistication). Our justification must be that the scale is to be constructed from *our* point of view.[18]

Note Crisp's claim that it is proper to "compare living animals according to...adaptation to the environment," and not the reverse; that is, use as the standard the degree of adaptation of the environment to suit the animals, human or otherwise. This injunction effectively nullified the anthropocentric bias: it argued our superiority in re-creating the environment to placate our specific interests was not a sufficient justification for a claim of evolutionary superiority. Also of significance is the explicit admission that our traditional consideration of evolutionary comparison is not based upon "pure behavioral or neurological sophistication."

The statement's importance is it undercuts what is probably *the* central thesis upon which the discipline of ethology, the study of the behavior of nonhuman animals, is built: *Morgan's Canon*. This cornerstone, enunciated in the late nineteenth century, exclaimed: "In no case may we interpret an action as the outcome of the exercise of a higher psychological faculty, if it can be interpreted as the outcome of the exercise of one which stands lower in the psychical scale."[19]

How is this principle applied? What is its relevance to the discussion? Hugh Wilder's "Interpretive Cognitive Ethology" explained: "How are we to interpret Morgan's 'higher' and 'lower' designations?... Lloyd-Morgan recognized three psychical faculties: from lower to higher, they were instinct, intelligence, and reason. Lloyd-Morgan hypothesized that humans were the exclusive possessors of reason."[20] Notice the descriptor, "hypothesized." This prejudicial view—in the literal meaning of "prejudice," which is to judge prior to the acquisition of factual evidence in support of the conclusion—is still highly influential over a century later.

Like Fisher, even when he criticized the Canon, Wilder was influenced by the controlling epistemological premise derived from Protagoras: "While ethologists of course need to respect differences between species, they also need to acknowledge inter-species structural...similarities and evolutionary continuities.... Where...similarities and continuities are pronounced, it may make sense for the ethologist to advance attributions of cognitive capacities to nonhumans."[21] As has been stated, making comparisons with humans is not of itself an inappropriate technique in acquiring information. However, to *presuppose* without any supporting justification that humans are rightfully to be considered the reference standard for comparison is an entirely unreliable methodology.

The generality of Morgan's Canon stands in opposition to the normative doctrine of continuity of species from an evolutionary perspective. If Lloyd-Morgan is found to be correct in his view that "humans were the exclusive possessors of reason," there exists an unbridgeable gap in the evolutionary development of this planet. The fossil record—upon which any elaboration of a thesis in support of, or in opposition to, evolutionary continuity must be ultimately grounded—is incomplete. Earth is a vast planet, much of which is still unexplored in detail, and a claim that we have already devoted sufficient archaeological effort to warrant closing the book on speculations concerning evolutionary theory suggests scholastic incompetence. Wilder firmly rejected the possibility of discontinuity, though also without proof of hypothesis.

This conflict is symptomatic of an incessant quest for universality across existent species in the face of claims of discreteness. Roger Crisp appealed to a *Principle of Psychological Continuity* to combat the inference drawn from Morgan's Canon and to facilitate resolution of the dilemma: "For any given psychological state, act, or function, P, if a given animal belongs to some species other than the lowest one and that animal is capable of P, then there is an animal of some lower species such that the lower animal is capable of some psychological state, act, or function, P', and P' is a model of P."[22]

After citing the Principle, Crisp acknowledged it suffered from the twin defects of anthropocentrism/anthropomorphism: "A number of questions are begged by the Principle of Psychological Continuity. The two most obvious concern the notions of one animal's being higher than another, and of the mental state of one animal's being a model of the mental state of another."[23] Nonetheless, he made an effort to support the construct: "Any reason we have to doubt the existence of the minds of animals...gives us a reason to doubt the existence of...minds of other humans. We are faced with a choice between attributing mental states to animals and solipsism or skepticism concerning other minds generally."[24]

Since Morgan's Canon and the Principle of Psychological Continuity are mutually antithetical on the attribution of mental states denotative of a power of rational thought applicable to both human and diverse animal species, the issue of self-consciousness cannot be resolved by theoretical justification. Final adjudication is possible on the level of concrete evidence alone. It is worthwhile to note that rationales for the theses of anthropomorphism and anthropocentrism have shifted.

Evolutionary continuity is, at least in part, a consequence of physical structure and function. Do features present in all known species demonstrate a precise gradation in form such that one knowledgeable in the field can unambiguously determine the reality of a biological continuum? An entirely separate level of investigation concerns continuity with respect to mental and emotional displays corresponding to overt behavior. Illustrations have been advanced showing superior physical attributes possessed by some species over our own with respect to optimal sensory enhancement or ability to adapt to diverse environmental conditions. Hence, it must be concluded *Homo sapiens* do not appear at the apex of the evolutionary ladder from a strictly anatomical perspective.

Since species appear to have divergent, and to some degree mutually compensating, virtues in their ability to interact with the environment in a life-preserving manner, establishing a linear scale for evolutionary advance is impossible. It was the human rush to justify its self-proclaimed status as the apex of life on the planet that led to a shift in argument from the purely physical grounds of anatomy and physiology to the non-physical grounds of mental, emotional, and behavioral states of being. This is the real importance to be ascribed to the introduction of Morgan's Canon as well as its antithesis, the Principle of Psychological Continuity.

What observational behaviors could be appropriately characteristic of self-consciousness, even for human beings? Several *should be* acceptable: (a) activities suggestive of comprehension of self as an entity distinct from others; (b) employment of a variety of novel situational responses to threats to the physical survival of self, or to kin to whom one is strongly mentally and/or emotionally attached; (c) employment of a variety of novel situational responses to undesirable events affecting oneself even when literal survival is not at stake.

What is common to these proposals is an awareness of self in relation to environment, and more particularly, as Crisp remarked, an adaptation of self to one's surroundings rather than the reverse. In other words, it is the non-anthropocentric view wherein the other, in this case the physical environment inclusive of living entities, provides the dominant focal point. To utilize the Protagorean premise is to beg the question at the outset, for it asserts the self-consciousness of *Homo sapiens* as not merely self-evident, which is hardly a proof, but as the point of reference against which all other species are to be judged.

With regard to the first characteristic, Gallup reported in *Science* that several chimpanzees were exposed to large self-reflecting mirrors, and over the course of ten days learned to recognize themselves in the mirror rather than think it another animal, which was apparently their first response.[25] They demonstrated self-recognition through use of the mirror to groom their bodies in areas not otherwise visible in the absence of the mirror, and by removing food particles from their teeth while watching their self-image to see the results of their efforts.

Does activity compatible with comprehension—possession of a mental concept of "self"—necessarily imply comprehension in fact? Any rational response reveals the fundamental methodological flaw in relying solely on externally observable behavior. From any perspective other than that of the subject's, the question is incapable of definitive answer. Here is an example of the principle of *epistemic asymmetry*: the inherent difference in definitive knowledge about mental states available from first-person and third-person viewpoints.

Güven Güzeldere admitted this epistemological gap to be problematic with respect to a discussion of *human* self-consciousness in his introduction to *The Nature of Consciousness: Philosophical Debates*: "There seems to be no...way to peek into the inner lives of others... These are the limitations of the third-person perspective: *from the outside*, firsthand explorations of the consciousness of others...seem to be out of the reach of ordinary scientific methods... In this asymmetry between the first-person and the third-person perspectives lies the epistemic duality in the study of consciousness."[26]

The conundrum posed here is pertinent to all similar investigations. It is notoriously difficult to prove a negative hypothesis: in this case, such-and-such species does not exhibit self-consciousness. It appears more rational to assume the absence of self-consciousness and seek positive proof of its existence: this approach would be to argue for the epistemological primacy of Morgan's Canon over the Principle of Psychological Continuity as an appropriate rule for judgment.

It is clear one cannot definitively conclude from the chimpanzees' use of mirrors that they had developed a real conception of "self." If we observed a human being enacting the same behavior, we would ordinarily acknowledge he or she had such an understanding. If one is to use anthropomorphism as an epistemological tool in resolving an issue, it is preferable that it be employed on a consistent basis. To not do so in the case of a nonhuman is to reject the notion simply because it failed to conform to pre-established expectations.

With respect to the difficult issue of self-consciousness, one would then be hard-pressed to subsequently define what evidence *would* constitute positive proof of its exhibition by any or all other species.

Indeed, *human* self-consciousness has been debated on similar grounds. Many have argued it is an attribute that cannot be assumed and cannot be definitively proven. Since the idea of human self-consciousness is central to any possible articulation of a human epistemology, an entire chapter will later be devoted to this topic. All needing to be said here is these advocates are in error; proof is available in abundance. For the sake of discussion, self-consciousness will be posited as intrinsic to our species independent of the factor of evolution.

Given the inherent barrier to being able to get "inside another's head" at the present time of human personal or technological sophistication, the *a priori* employment of Morgan's Canon would seem to guarantee an inability to answer the question of animal and plant self-consciousness. By virtue of epistemic asymmetry, it is impossible to accurately, or rationally, judge any set of behaviors as offering positive evidence, much less definitive proof, that such and such species was in possession of such and such psychic ability or mental state.

Hence, the Principle of Psychological Continuity would seem to be a methodologically superior guiding framework for adjudication of all issues related to self-consciousness. It allows at minimum an acceptable mechanism for interpreting evidence in the affirmative. Absent *certain* knowledge about the developmental capabilities of any being, one should treat as tentatively more plausible the conclusion derived from the dictum, emanating entirely from folk-wisdom, "what is good for the goose is good for the gander." That is, the same rules apply uniformly to all species when judging solely from behavior—or, in scientific terms, that the same effect follows from the same cause.

The intensive effort put forth to survive, to continue one's life for as long as possible, is innate to all species according to biologists and philosophers. Many go to the extreme of alleging that the behaviors cultivated by living organisms to ensure survival are genetic predispositions with no allowance made for individualized control. This hypothesis underlies the popular notion of the "four F's" as fundamental to all species: fleeing, fighting, feeding, and fornicating. Human beings have the capacity to either temporarily suspend or permanently eliminate all four elements: acts of extreme bravery, extreme passivity, self-willed starvation, and sexual abstinence are legion throughout human history. Moreover, human beings have been widely known to evoke novel circumstantially extreme behaviors that are anti-survival oriented, yet adopted in defense of kin. Consequently, there is a *prima facie* reason to consider whether nonhuman species could enact such behaviors by choice rather than being constrained by innate biological command.

Many children in the United States learn the words to "Old MacDonald's Farm" before they enter kindergarten. Each child learns that a diverse array of animals make sounds untranslatable to us, but which may be uniquely meaningful to that species. It is reasonable to consider that animal language constitutes a hallmark not merely of intelligence, but also of consciousness pertaining to self and others. Michael Philips and Steven Austad presented a list of criteria for animal linguistics transparently conforming directly to the typical characteristics of human speech:

> Communication occurs when: (a) an animal transfers information to an appropriate audience by use of signals; or (b) an animal gains information that another animal intends to convey by means of signs. By signals we mean...behaviors...that have evolved to convey information... An appropriate audience consists of the animal(s) to which the signal evolved to convey information.... By "signs" we mean any behaviors...animals have invented in order to convey that information.... When we say that signs and signals convey information, we use "information" in a non-quantitative sense, although nothing in our definition would prohibit a quantitative usage. Thus, animals use signals to convey information about their identity...their condition...their behavioral probabilities...features of their environment...and other matters. However, they communicate successfully only if an appropriate audience gains information by receiving and decoding the signal.[27]

In *Animals in Translation*, Temple Grandin summarized the details of an extensive study performed by researchers at Northern Arizona University involving prairie dogs inhabiting the American Southwest that had evolved a very effective warning system when predators approached. In other words, the prairie dogs displayed all the traits suggested by Philips and Austad:

> Prairie dog colonies have a communication system that includes nouns, verbs, and adjectives. They can tell one another what kind of predator is approaching...and they can tell...how fast it's moving. They can say whether a man is carrying a gun or not. They can also identify individual coyotes and tell one another which one is coming.... If the prairie dogs are signaling the approach of a person, they can tell one another something about what color clothing the person is wearing, as well as something about his size and shape.... Dr. Slobodchikoff was able to interpret the calls by videotaping everything, analyzing the sound spectrum, and...watching the video to see what the prairie dog making a distress call was reacting to... He also watched to see how the other prairie dogs responded.... He found...the prairie dogs reacted differently to different warnings.... Dr. Slobodchikoff also found evidence that prairie dogs aren't born knowing the calls... They have to learn them.[28]

The last sentence is critically important: prairie dog calls are apparently so diverse in composition, stressing type of predator, distance from potential prey, speed of approach, and distinguishing predator characteristics, it would be impossible to genetically constitute all the variables into a preformed set of automatic symbols. Of equal significance, the signals *are* communicative, which means not only do these animals have a notion of "self" as prairie dog *versus* "non-self" as predator, they also have a construct of "self" as prairie dog 'A' in contrast with "other" as prairie dogs 'B,' 'C,' and 'D.' They also appear to possess a *volitional* willingness—proven by their need to learn the calls as well as the situations suggestive of their use—to help others of their kind to survive by avoiding encounters with the enemy. Grandin supplied a legitimate justification concerning the origins of this sophisticated behavior conforming perfectly to the proposition that language "signals...conveyed a fitness advantage."[29]

If applied to human beings, we would not hesitate to regard this behavior as consistent with self-awareness, and interpret similar actions as illustrative of intentionality characterizing self-consciousness. Adoption of the Principle of Psychological Continuity demands the same conclusion be appropriately applied to nonhumans. It is understood anthropomorphism has unavoidably entered the evaluation. The problem of self-consciousness is so complex it is impossible at our present level of comprehension of its nature and attributes to evaluate its presence or absence in other species independent of arguments we would use in self-examination.

While it can be determined in a completely objective manner whether an animal is hunting for food, and how we *unambiguously know* it is doing so, it is impossible to state what the experience *might* seem like to that creature unless we used ourselves as a measuring stick contemplating a parallel event. For the temporal present and foreseeable near future, epistemic asymmetry necessarily distorts and colors all judgments on the mental and emotional capabilities of other creatures.

As mentioned earlier, our use of language demands that its terminology makes sense to us. This evaluation of the behavior of prairie dogs is not unremittingly anthropocentric even though anthropomorphism is a corollary to anthropocentrism. There is no supposition this species *really does* employ "nouns, verbs, and adjectives." It is highly plausible their interpretation of "self," "non-self," "other," and "enemy" does not cover the same terrain as ours does. It would be ridiculous to believe their signals *literally* correspond to a phrase such as, "a man in a white hat is carrying a gun, heading this way from the North, and will be here in seven minutes." To claim prairie dogs have language they use to cope with survival issues—*anthropomorphically* similar to the human use of language under similar specific conditions—is *not* to claim their symbol set *anthropocentrically* corresponds to that of *Homo sapiens* in form or substance, and according to which our manner of employing language is the proper standard by which to gauge whether other species possess linguistic communication skills.

Differentiation between an initial assessment of an anthropomorphic-like use of a symbol system as a kind of language for communication by other creatures and a final anthropocentric-based evaluation of the qualitative level of skill or sophistication in the use of that symbol system arises even when comparing two human languages deriving from cultures possessing variance in socio-philosophic perspectives. For example, English and Chinese do not resemble each other in any way structurally. They use distinctively different characters as symbols, and employ grammar and syntax in divergent ways. In addition, the "way of thinking" expressed by words and concepts typically employed in the two languages differ substantially. English is a language centered on individuals and reflective of Judeo-Christian ethical values and standards; Chinese aims to express notions pertinent to communal patterns of thought common to Taoist-Confucianist principles. Nonetheless, an English-speaking person would have no reluctance in attributing both intentionality and self-conscious awareness to a speaker of Chinese *solely* by observation of external overt behavior abetted with no extraneous presuppositions. An American who doesn't speak "prairie dog" with any greater degree of facility than he or she speaks Chinese should rationally follow the same procedural rules of implication when assessing behaviors whose motivations he or she cannot readily comprehend.

Even if self-survival is not at stake, animals apparently engage in self-aware behavior to aid kin that are dependent on those others. Carolyn Ristau presented evidence of such activity in her research paper entitled "Aspects of the Cognitive Ethology of an Injury-Feigning Bird, the Piping Plover":

> Because the nest is easily accessible to predators, protection of the eggs depends on camouflage, preventing potential predator's knowledge of the nest's location, and keeping them out of the nest's vicinity.... On some approaches of an intruder, the birds may do a gradation of broken-wing displays, which may perhaps begin with a fanning tail and gradually increase the awkwardness of walk until it has one and then both wings widely arched, fluttering and dragging. It may then vocalize loud raucous squawks as well.... The bird presents a convincing case for being injured, and the observer often trudges hundreds of meters after the bird only to see it suddenly fly away with agility. At that point one is far from the nest or young.[30]

Ristau offered detailed experimental data using human intruders, which convincingly demonstrated the parent bird's actions consisted of a series of deliberately premeditated responses to a novel threat to offspring. In other words, the piping plover was intentionally engaged in sophisticated strategic thinking.[31]

The final modality in which the plausibility of animal self-consciousness could become externally observably evident is through considered response to novel naturally occurring environmental alterations, which, though not life endangering to self or kin, induces the animal to choose to undertake specific adjustments as a proportional response to the situation. Note the emphasis placed on "naturally occurring environmental alterations." Numerous reports in ethological literature concerning animal responses to laboratory-controlled scenarios are not viewed as depicting meaningful conditions for evaluating animal self-consciousness however much these studies may be revealing of animal *intelligence*. That animals can be trained to learn new tricks does not bear on the issue as to whether they are aware of themselves as learning new tricks. Teaching dolphins to use sign language or respond to human linguistic commands—in general to undertake unnatural activities the animal would never perform in their usual habitat—reveals performance without intent. Contrast this undertaking with the communication skills of vervet monkeys developed in the wild.[32]

Intentionality has been considered one of the principal hallmarks of self-consciousness ever since the notion was originally framed by Franz Brentano and then modified by Edmund Husserl: "Husserl saw himself as advancing Brentano's program of 'descriptive psychology,' taking Brentano's account of intentionality as the key concept for understanding…classifying conscious acts and experiential mental processes."[33] Robert Forman had an informative response to the question, "What does intentionality mean?" in *Mysticism, Mind, Consciousness*: "An experience is intentional if and only if someone encounters sensory input and *takes it as referring* to an object. The key and novel feature of intentionality no longer is some aspect of the object itself or the object's inexistence within the subject, but is part of the structure of the knowing mind."[34]

If I perceive a direct relationship between my thoughts at any particular moment in time and an external physical object, I am viewing that object with intentionality. Intentionality demands that the species to whom it occurs be simultaneously self-aware, aware of an exterior thing, and aware of an immediate connection between them. The dolphin example just mentioned fails the strictures of intentionality because such activities do not require self-awareness; the acts in question only demand awareness of surroundings to which the dolphins are programmed to react according to classical models of behavior reinforcement.

This is not to allege all researches involving human-animal interactions are suspect. It is to say steps taken by the investigator to influence animal responses—other than in the setup of the initial problem—might render the derived data inconclusive if the aim of the research is an investigation of the presence of animal self-consciousness. To illustrate the distinction, Michael Rosenzweig's "Do Animals Choose Habitats?" discussed the behavior of a hummingbird displaying intentionality under human-controlled conditions:

> Bene performed a series of experiments on a single black-chinned hummingbird. The bird learned to come to a feeder in a particular location and then the feeder was moved a short distance. The bird returned to the original spot and hovered as if to begin feeding.... He moved the feeder to a third location. The bird went to spot two, hovered awhile, then went to spot one and repeated the hovering. The bird retraced its flight despite the fact that the backward direction was novel. Bene repeated this procedure with a fourth spot after the third had been learned and the bird retraced from three to two to one. Locations were not in a straight line, nor...equidistant. Bene continued...until nine had been established. Beyond four, a new phenomenon appeared. The bird began to take shortcuts.... Demonstrating the existence of mental habitat maps proves that animals do store information relating to habitat location and contents. Knowing this to be so, and knowing also that the information is used to determine pathways of movement, is compelling evidence in favor of interpreting those movements as a result of choice.[35]

The conclusion offered is logical. The bird had apparently related its own location to the three-dimensional spatial orientation of the feeders and was able to plot the shortest route between itself and the object of interest. Willful choice in action would be accepted as an indicator of self-consciousness in human beings, and by extension should be accorded to this bird. Another convincing illustration was obtained from Pilleri's study of beavers. Donald Griffin used this case to support his position on animal self-consciousness in *Animal Minds*:

> Pilleri reports an ingenious response to a novel situation by two beavers confined in an enclosure where the water level of a 3.5 x 2.5 meter pool 1.6 meters deep was regulated by an outlet pipe equipped with a cap perforated by three holes eight millimeters in diameter through which the water escaped. After about two weeks the two beavers began to plug these three holes with peeled twigs which had been gnawed off obliquely at both ends by the beavers and whittled down in such a way that they exactly fitted the holes.... The performance was repeated several times, always at night, after we had removed the sticks in the morning and restored the water in the pool to its normal level. Every night the beavers made new calibrated sticks and blocked up the holes.... Finally they changed their technique. In addition to the sticks they used grass and whole piles of leaves mixed with mud.[36]

The beavers' ability to adapt to unanticipated circumstances not of their own creation in order to solve a problem is plausibly interpreted to indicate their intentional awareness of themselves, their activities in repairing a dam, and the consequences ensuing from each action. There is no justification in attempting to evade this conclusion by claiming the behavior was genetically implanted in them and not under conscious control. There were too many variables introduced to believe unconscious pre-programmed response is the likely ultimate cause for activity. Such a suggestion would demand a highly sophisticated program able to induce a range of behavior in an almost infinite number of diverse conditions, a phenomenon highly improbable statistically. The same reasoning would lead one to conclude human activity is genetically innate, a product of complex neurological behavioral-linguistic programming rules. It is unlikely any scientist would construe this suggestion as a plausible and reasonable argument.

Evidence for animal self-consciousness is not built solely on inferences of mental states obtained from ethological observations. Research studies reveal animals express emotional states suggestive of self-awareness. Temple Grandin—who has established a national reputation for developing the meat industry's standards for the slaughter of cattle, pigs, and chickens—used her extensive experience to unequivocally state: "Animals feel pain. So do birds, and we now have pretty good evidence that fish feel pain, too.... Dogs, cats, rats, and horses all limp after they've hurt their legs, and they'll avoid putting weight on the injured limb."[37] The intentional avoidance of putting weight on an injured limb is potent evidence for self-awareness: it is they who feel pain and want to avoid it. Grandin did not limit documentation to that remark. She cited two research studies in support of her contention:

> Up until recently nobody knew whether fish felt pain or not, but two researchers in Scotland have shown...they almost certainly do.... First they anesthetized some fish and applied painful stimuli like heat and mechanical pressure to their bodies while running a brain scan. They found neurons...that fired in a pattern very close to pain firing in a human brain.... It shows...fish have...the *sensory* component of pain, though it doesn't tell us...the fish were...feeling it consciously.... In the second part of the study...they injected either bee venom or vinegar into the fish's lips... The fish acted the same way mammals act when they're in pain. It took the fish an hour and a half longer to begin eating again than it did fish who'd had painless saline water injections.[38]

> Francis C. Colpaert...injected rats with bacteria that produce a temporary bout of arthritis we know is painful in humans, then gave them a choice between a bad-tasting liquid analgesic and a sweet, sugary-tasting liquid rats normally like. The rats chose the bad-tasting painkiller over the sugar solution, a pretty good sign they were choosing it for its painkilling properties.... Once their arthritis cleared up they switched to the sugar drink.[39]

Yet another possible indication of similarity in mental and/or emotional states between people and animals is that some animals also appear to dream: "When human subjects are awakened...they are more likely to report they were dreaming during rem sleep. Comparable recordings from sleeping birds and mammals show similar patterns of rem sleep, indicating that they also dream."[40] Obviously, the existence of similar brain wave patterns during rem [rapid eye movement] sleep does not, of itself, establish animal awareness of dreams during, or subsequent to termination of, the sleep cycle. However, until there is available a means of direct communication with each species, such a thesis cannot be simply ignored as *prima facie* impossible.

Using the notion of anthropomorphism in conjunction with the Principle of Psychological Continuity as the foundation for comprehension of animal characteristics adheres closely to the epistemological guideline created by William of Ockham, and frequently employed in scientific investigations. This maxim claimed that all things being equal, the simplest explanation should always be tentatively preferred.[41]

It is more comprehensive to allege varieties of self-conscious creature exist and demonstrate their awareness of self, as evidenced by their internally and externally referenced behavior in response to environmental conditions, through a common mechanism than to assert the opposite: all species are constructed differently such that there exist non-overlapping processes under which they function so as to respond to the range of experiences to which they are subject. John Dupre presented an objectively balanced conclusion on this theme in his essay, "The Mental Lives of Nonhuman Animals," in which he wrote:

> To be successful, as opposed to merely logically intelligible, in attributing beliefs to other animals, we need to know a great deal about the animals we are talking about. If we do not know...their perceptual capacities...what features of the environment they are capable of discriminating; the goals...such animals often pursue; the level of intelligence they are capable of bringing to bear on...pursuit of these goals; their tendencies to stereotyped or habitual responses to certain kinds of situations...we are likely...to be wrong in our suggestions as to what their cognitive states are.... But I see no reason why, as we come to know more about other creatures, we should not come to be very successful in giving intentional stance explanations of their behavior.[42]

Various species of animal—chimpanzees, piping plovers, hummingbirds, beavers, prairie dogs, rats, fish, and cats—have thus been found to engage in behavior rationally indicative of tool use, deception, internal memory maps, self-protection, other-protection, skill in communication, emotional display, and fantasy. The premise of Morgan's Canon—implicit stratification of the psychological reality exhibited by diverse creatures such that that displayed by human beings is unique and qualitatively superior to all other species—cannot be upheld in the face of the ongoing accumulation of evidence. With its rebuff goes the weakening of the anthropocentric premise uttered first by Protagoras over 2500 years ago.

To hope it can be partially rescued, that *Homo sapiens* could be claimed to be unquestionably elevated over plant life at least, can also be met with suspicion and rebuttal. There is support in the research literature for alleging the existence of mental states in plants. These reports are, quantitatively, considerably less numerous and, qualitatively, considerably less convincing. It should be anticipated the case for the existence of self-consciousness in plants would be much harder to argue than for animals for two reasons. First, in terms of evolutionary development of features pertaining to structure and function, we are more distantly related to plants than animals. Therefore, it is likely to be far more difficult to accurately interpret real correlations among observable behaviors, known chemistry and physiology, and ideas regarding any potential psychology. Second, most plants are immobile objects—with the common exception of known phototropic responses to heat and light—which would effectively limit the range of behaviors they could undergo as a consequence of diverse environmental stimuli.

Nonetheless, some provocative information on plant self-consciousness has been accumulated. In *The Secret Life of Plants*, Peter Thompkins and Christopher Bird discussed Cleve Backster's experiments using electrodes connected to a polygraph, or an electroencephalograph, detector attached to plant leaves:

> The most effective way to trigger in a human being a reaction strong enough to make the galvanometer jump is to threaten his or her well being. Backster decided to do...that to the plant... He would burn the actual leaf to which the electrodes were attached. The instant he got the picture of flame in his mind...there was a dramatic change in the tracing pattern on the graph in the form of a prolonged upward sweep of the recording pen.... More than twenty-five different varieties of plants and fruits were tested... The phenomenon appeared to persist even if the...leaf was detached from the plant, or if...trimmed to the size of the electrodes.... The plants reacted not only to threats from human beings, but to unformulated threats, such as the sudden appearance of a dog in the room.[43]

This report, though suggestive, is inconclusive since there is an alternate explanation: the galvanometer needle was responding to the heightened state of anxiety or expectation in Backster rather than in the plant. Of course, this hypothesis has its own deficiency: one must provide a mechanism to explain how the individual's mental state was capable of being recorded when the electrodes employed to measure the instrumental changes were not directly attached to the person.

In a review article entitled "Physical Aspects of Psi Phenomena," J. H. Rush reported one such successful prototype experiment performed by R. M. Brier: "Brier attached a polygraph recorder to a leaf on a philodendron plant and studied the effects on the polygraph trace (recording the current through the leaf and...any changes in...electrical resistance) when a subject tried to influence it. Consistently positive results were obtained through five series of tests."[44] As precedence, a principal candidate as an alternate interpretation to vegetative cognition of a direct threat would be human psychokinetic ability, a proposal most scientists would be equally antipathetic to as a plausible rationale. Thompkins and Bird reported a series of follow-up experiments to Backster's performed by another scientist:

> Eldon Byrd, an operations analyst with the Advanced Planning and Analysis Staff of the Naval Ordnance Laboratory in Silver Spring, Maryland, has been duplicating Backster's experiments with some success.... Byrd believes...a voltage change in the cells is...being measured, and...it is the mechanism of consciousness which causes the change in potential. Byrd's research supports Backster's observations that plants exhibit a quality of awareness and an empathy to other organisms that are stimulated in their presence.[45]

This passage only claimed that Backster's experiments are replicable and does not resolve the difficulty in procedural interpretation despite invocation of a plant-based physiological mechanism for the effect. There still needs to be found irrefutable evidence that the initiating response is by the plant and not by the human being. Even if the human element can be factored out, the result obtained must still be shown to lead to a tentative conclusion of plant self-consciousness as opposed to a simple chemical response to changes in the plant's immediate environment.

Thompkins and Bird provided a critical piece of information that might eventually lead to such a proof: "Continuing to probe further with a sensitive high-impedance device of his own design, L. George Lawrence sought to discover whether individual cells in a quarter-inch slice of onion, attached to a Wheatstone bridge and an electrometer, would react to various stimuli. He found that they seemed to respond to irritations such as a puff of smoke, or... to his mental image of their destruction, in about 100 milliseconds."[46] Why is this relevant? According to Max Velmans' *Understanding Consciousness*, the normal human response to *external* stimuli has been measured and found to require a greater length of time, typically two hundred milliseconds, than the one hundred milliseconds determined by Lawrence for onions:

> Experiments on the timing of conscious awareness by...neurophysiologist Benjamin Libet suggest...consciousness of input does not arise until at least 200 milliseconds after stimuli arrive at the cortical surface. Libet...found that direct microelectrode stimulation of the somatosensory cortex required a pulse train of at least 200 milliseconds duration before any conscious awareness of the stimulus was reported (pulse trains 10% shorter than this were not subjectively experienced).[47]

This data is not entirely convincing by itself because the issue is whether a person could psychokinetically influence a galvanometer needle starting from zero time that was not recorded for another one hundred milliseconds. This is a possibility not eliminated by the investigator, but is, upon reflection, infeasible for the opposite reason: data on psychokinetic ability suggests it is essentially an immediate activity; there is little delay time between its initial attempt and its supposed effect.

In the 1970s, Dr. William Tiller, a material sciences engineer at Stanford University, constructed a device to measure current changes due to human psychokinetic effort. The apparatus is described in the Notes. According to the fundamental laws of electrical circuits, voltage is a product of current flow times resistance within the circuit, *Ohm's Law*, E = IR. With a constant applied voltage, an increase in current output must mean electrical resistance decreases through a circuit wire. Tiller reasoned that decreased resistance to motion—movement of electrons produced by degradation of a gas passing through the circuit—would be a valid indicator of psychokinetic response.

Tiller's *Science and Human Transformation* recorded data compiled at two voltages, but with a constant current. For several individuals tested, there were 11,335–53,972 distinct current surges above the established baseline over five minute test periods, which equates to a range of 37.783–179.906 changes per second. A more meaningful way to report these results is to calculate the reciprocals of the changes per second: 1/37.783–1/179.906 = 5.56–26.5 milliseconds per current adjustment.[48] The conclusion is clear: *deliberate* human psychokinetic influences take significantly less than the response time of 100 milliseconds observed by Lawrence for onions.

Both types of direct human involvement have been eliminated as factors. Intentional psychokinetic influence without an external stimulus is faster than the plant's recorded response time; psychokinetic influence subsequent to awareness of an external stimulus is slower. The attempt can be made to diminish the significance of the conclusion by rejecting the interpretation: the plant response to a physical stimulus being of a different time order than human response does not necessarily translate into a concomitant mental state for a plant in the way it does for a person. In other words, it is possible to posit the result is a function of automatic chemical processes initiated in reaction to the environment. This routinely occurs in photosynthesis, where a plant chemically converts carbon dioxide in the surrounding air to oxygen.

A possible argument against this latter proposal could be sustained on the basis of a pertinent detail of the original experiment: Lawrence tested "individual cells in a quarter-inch slice of onion." Do individual cells, in contrast to collections of cells, possess the chemical machinery necessary to produce the established response as reflexive acts absent willful intent? The answer is uncertain and begs for further investigation. Raising the issue indicates a simple declaration of challenge to the thesis of plant intentionality and self-consciousness doesn't *ipso facto* constitute proof, either.

In the first years of the twentieth century, Jagadis Chandra Bose showed that some functional organs in animals had their analogs in plants; they could be said, in a very loose way, to have eyes, stomachs, muscles and a nervous system. Thompkins and Bird explained: "Jagadis Chandra Bose was able to show how the skins of lizards, tortoises, and frogs as well as those of grapes, tomatoes, and other fruits and vegetables behaved similarly. He found...the ...digestive organs in insectivorous plants...were analogous to animal stomachs. He discovered close parallels between the response to light in leaves and in the retinas of animal eyes.... He proved that plants become as fatigued by continuous stimulation as animal muscles... Bose demonstrated the characteristics of a nerve system in mimosa."[49]

Bose authored several books on his findings, all of which are currently out of print.[50] If plants can be shown to have structural features that function in the same manner as some animal organs, why could they not also have the mental states accompanying and interacting with the physical performance of those organs? On whom is the burden of proof? It is reasonable to propose believers in plant mentality must offer definitive support for their claim, and those rejecting this supposition must provide rational alternate propositions to account for the results elicited in investigation other than an appeal to the anthropocentric premise as a justification. To assert plants are too far "down" the evolutionary scale to feasibly possess self-consciousness is unacceptable: use of the Protagorean presumption begs the question.

Perhaps counter-intuitively, for some areas of intensive research, physics has employed anthropocentrism as a functional premise almost as readily and extensively as biology has. Since physics, the foundational science for all others, traditionally lays claim to being wholly objective—in the literal sense of concerning itself entirely with the forces impinging upon or emanating from an impersonal object—it is ironic to discover one of the most esoteric realms of contemporary study is grounded *entirely* on human subjectivity. In particular, high energy subatomic physics holds as a core proposition: human levels of self-consciousness are so elevated, our importance in the universe so substantial, the outcome of quantum mechanical events affecting common aspects of physical reality are undetermined until a human observer collapses the mathematical wavefunction describing those events.

To provide requisite background for non-physicists, Evan Harris Walker explained the origin of quantum mechanics in *The Physics of Consciousness*: "Planck assumed...radiation emitted by the walls and objects in the box had a discrete amount of energy...proportional to the frequency of radiation....

Five years later, Einstein...proved...light is...also *absorbed* in chunks of energy.... The lower the frequency of the light, the less energy the electrons had when kicked off the surface of the metal."[51]

In a later chapter it will be shown that the Planck-Einstein *assumption*, the foundation for the discipline of quantum mechanics, is unnecessary as an explanation for the results of their experiments. Postulation of energy quanta to solve specific problems would have consequences beyond that anticipated by its initial proponents. The most significant initial implication, with respect to the present theme, was emphasized by Ted Bastin in *The Encyclopedia of Ignorance*: "The early forms of quantum theory never attempted to *explain* discreteness in the sense that they could be said to have incorporated both the discrete and the continuous within one theoretical structure. They...imposed discreteness as a mathematical constraint upon the ranges of values that were allowed to be interpreted as the results of measurements of certain physical quantities."[52]

In other words, to solve separate researches in the energetic properties of inert objects, Max Planck and Albert Einstein overturned over two centuries of theoretical physics dating from Isaac Newton. Rather than develop sound mathematical laws based *solely* on objective facts, they formulated theories and mathematical relationships according to the means by which humans could best comprehend the outcome of physical events. *They made human comprehension a primary criterion for reality, and saw fit to impose that guiding precept on the structures of knowledge.* The deviation from Newton was extreme. Newton's *Scholium* had announced the principles of his system based entirely on what Immanuel Kant would call "things in themselves," the absolute schemata of time and space:

> Absolute, true, and mathematical time, of itself, and from its own nature, flows equably without regard to anything external, and by another name is called duration; relative, apparent, and common time, is some sensible and external measure of duration by the means of motion... Absolute space, in its own nature, and without regard to anything external, remains always similar and immoveable. Relative space is some moveable dimension or measure of the absolute spaces, which our senses determine by its position to bodies.[53]

Planck and Einstein had shown wanton disregard for absolute space and time in their formulation of the properties of electromagnetic radiation. As Einstein himself revealed in his Special Theory of Relativity—published in the very same year, 1905, as his theory of the photoelectric effect—light is characterized by two parameters: *wavelength*, defined as a discrete distance within absolute space, and *frequency*, a discrete portion of time relative to an absolute spatial distance. In successive research papers, Einstein employed contrary notions of physical energy when constituted in the form of light: as discrete particle and as continuous wave. Using the Planck-Einstein concept as a springboard, Neils Bohr transferred the revolution in theory from the domain of physics to that of chemistry. The story is told in Amit Goswami's *The Self-Aware Universe*:

> Suppose, said Bohr...the orbits...electrons describe are discrete, like the quanta of energy suggested by Planck.... Only when an electron jumps from a higher-energy orbit to one of lower energy does it emit light as a quantum... The electron...can never occupy a position between orbits...when it jumps, it must somehow transfer directly to another orbit.... It seems to disappear at one rung and to reappear at the other—quite discontinuously.... We cannot tell when a particular electron is going to jump nor where it is going to jump if it has more than one lower rung from which to choose. We can only give probabilities.[54]

It is clear continuity has been abandoned; all physical reality—composed of atoms in which electrons make discontinuous movements in space—is construed as innately discrete. The loss of perspective heretofore conceived as *the* essential construct underlying the power of science to reveal the nature of reality has caused a transition in knowledge accumulation from conceptual certainty to a relative likelihood enumerated in mathematical probabilities. It was left to Werner Heisenberg to complete the transformation in orientation on the nature of the physical domain by positing that the central implication of quantum mechanics was reality's dependence on observation, notably by human beings, for its structure:

> What we know are...the...outcomes of...experiments that probe...the microworld of the atom, and so our representations...should be a mechanics of observations.... There is an ultimate limit to the accuracy of the data we can hope to get...in order to determine the path of an object or to see the inner workings of the tiny components of our world.... Heisenberg pointed out that if there is a limit to the precision with which we can know an object's path, then to that extent, *path has no meaning*.[55]

It is important to understand the scope of this reorientation in worldview, for it was grounded in an entirely novel metaphysical and epistemological perspective. Newton had developed his physics of particles in motion and the energetic consequences of the interactions between them as the outcome of predictable causal relationships objective and *independent* of the percipient.[56] His proposals adhered to the metaphysical orientation of his historical time.[57] Planck, Einstein, Bohr and Heisenberg jointly reformulated these interactions between particles as subjective and *dependent* on the percipient. Whereas Newton contemplated continuous, uniform movement allowing for confident prediction of unfolding patterns of events, Heisenberg contemplated discrete, irregular movements allowing for estimates of consequences of interactions.

Heisenberg took the extraordinary step of proclaiming that absent *our* participation in those interactions by an act of observation and measurement, the actuality of those interactions was without significance. Heisenberg thus became the quintessential pragmatist, as that construct was defined in chapter one. In essence, he had mated science with the epistemological framework conceived by Kant nearly 150 years earlier in his *Critique of Pure Reason*:

> Hitherto it has been supposed that all our knowledge must conform to the objects: but, under that supposition, all attempts to establish anything about them *a priori*, by means of concepts, and thus to enlarge our knowledge, have come to nothing. The experiment therefore ought to be made, whether we should not succeed better with the problems of metaphysics, by assuming that the objects must conform to our mode of cognition, for this would agree with the demanded possibility of an *a priori* knowledge of them, which is to settle something about objects, before they are given us.[58]

Kantian epistemology was anthropocentric—"objects must conform to our mode of cognition"—and the new physical science of Planck, Einstein, Bohr, and Heisenberg readily complied. Why? Bastin had made it clear what the justification was: "They simply imposed discreteness as a mathematical constraint upon the ranges of values that were allowed to be interpreted as the results of measurements of certain physical quantities." Henceforward, *we* would be deities with the capability of determining reality. A Protagorean profession of faith had insinuated itself in the human mind and achieved external expression with no self-limiting qualifying rational statement—"this is how reality *appears* to us"—but in full glory through the extraordinarily grandiose declaration, "this is what reality *is*." When Einstein finally realized the folly implied by the philosophical grounding for a quantum mechanical theory he helped produce, he recanted: "Einstein argued that if the trajectory of a correlated quantum object is, in principle, predictable, but quantum mechanics is unable to predict it, then there must be something wrong with quantum mechanics. Einstein's...conclusion from this dilemma was that quantum mechanics is an incomplete theory."[59]

An anthropocentric epistemological approach applied to atomic physics resulted in two pronouncements. The first, electromagnetic radiation in the form of light has properties characteristic both of a localized point-source particle and of an unlocalizable wavefront. Further, the theory proclaims, which appearance light energy takes on in any given situation is dependent upon the measurement apparatus we decide to employ to collect relevant data; that is, the physics of energy radiation is allegedly influenced by one's selection of a method to investigate various phenomena. The second relevant contention, directly related to the first, claims that when light energy is to be described as a wave, the properties to be ascertained are mathematically interpretable as a range of probabilities emanating from the general solution of an equation, which doesn't yield numerical results until we determine them.[60] The first thesis asserts physics is human-dependent; the second posits the mathematics associated with physics is human-dependent.

The classic illustration used to express this latter phenomenon to laymen is a fantasy about "Schrödinger's cat," named after Erwin Schrödinger, who developed the mathematical form of the wave equation:

> Suppose we put a cat in a cage with a radioactive atom and a Geiger counter. The radioactive atom will decay in accordance with probabilistic rules. If the atom decays, the Geiger counter will tick, the ticking will trigger a hammer, the hammer will break a bottle of poison, and the poison will kill the cat. Let us suppose...there is a fifty percent chance of this occurring within an hour.... How would quantum mechanics describe the state of the cat? Of course, if we look, we will find the cat to be either alive or dead. What if we do not look? The probability that the cat is dead is 50 percent. The probability that the cat is alive is 50 percent.... It describes the state of the cat at the end of the hour as half alive and half dead. Inside the box is... "a coherent superposition of a half-alive and a half-dead cat," to use the proper jargon.[61]

What is wrong with this image? Does human influence really determine the possibility of survival of the cat? The entire scenario is a hoax operating under one or two false premises. Standard quantum mechanical interpretation asserts the act of observation collapses the wavefunction and determines whether the cat is truly alive or dead. This assumes the principle criterion is the consciousness of the human observer. Hence, the first unjustified false premise is that the physical outcome is not self-determined by the physical construction of the system. Even if one concedes this presupposition for the sake of argument—resolution requires metaphysical intervention in the form of Consciousness—the suggestion it is the singular *human* act of inspection that is the final determinant is a second fallacious dogmatic belief. Given the information presented in the first portion of this chapter ascribing a high plausibility to the notion of animal self-consciousness, it is much more likely the *cat's* self-awareness acts as the principal element selecting the outcome of the situation. The extreme anthropocentric nature of the usual interpretive approach has immediate bearing on the sources of mischaracterization for the scenario.

Our current conviction of our own standing in the universe even extends to that which is unknown and purely speculative. Invoking loud echoes of de Chardin's dubious thesis—"The animal knows. *But it cannot know that it knows*.... If it could, it would long ago have multiplied its inventions and developed a system of internal constructions that could not have escaped our observation"—there are frequent pronouncements by reputable scientists that if other species of equal or greater intelligence to ourselves exists, we would have been in communication with them. We apparently find it inconceivable some alien culture would not have intentionally sought us out, or responded to our broadcasts of radio signals beamed omni-directionally into space. The possibility of another civilization having no interest in interacting with us is somehow judged to be implausible. Such is the extent of our delusion, "man is the measure of all things."

We *must* take seriously—not as compliment, but as warning—the words of Richard Tarnas in *The Passion of the Western Mind*: "Descartes enthroned human reason as the supreme authority in matters of knowledge, capable of distinguishing certain metaphysical truth and of achieving certain scientific understanding of the material world. Infallibility, once ascribed only to Holy Scripture or the supreme pontiff, was…transferred to human reason."[62] Our ability to think, to reason, is a powerful tool, and the innate intelligence that is its source is a marvel, but neither the tool nor the source should blind us to recognition that, at this level of development as a species, we have not yet earned the pinnacle we lay claim to in our theories about the nature of reality. Plato should be heeded when he declared in *The Laws*: "You do not seem to be aware that this and every other creation is for the sake of the whole, and in order that the life of the whole may be blessed; and that you are created for the sake of the whole, and not the whole for the sake of you."[63]

Chapter Three: Discreteness and Continuity

> What will be emphasized...is that fragmentation is continually being brought about by the almost universal habit of taking the content of our thought for "a description of the world as it is."... Since our thought is pervaded with differences and distinctions, it follows that...the world is...seen and experienced as actually broken up into fragments.
>
> —David Bohm, *Wholeness and the Implicate Order*

At first glance, the universe appears continuous: animate and inanimate objects seem to be integrated solids; time seems to pass in a ceaseless flow from past to present to future; distant locations in space seem contiguous to adjacent ones such that "here" is always connected to "there." Yet, on a more analytical level, everything appears parsed into small stand-alone fragments: one living cell is separated from another; molecules are mostly empty space whose constituent atoms are not in direct contact; time can be divided into years, months, days, and subdivided further into hours, minutes, and seconds; space disintegrates into miles, yards, feet, and inches. Bohm's epigraph, in the title of its source and in content, expresses the dilemma with clarity. Is the real nature of the universe continuous, as befits the manner in which it is qualitatively experienced? Is manifestation actually discrete, consistent with its quantitative conceptualization? Is it somehow both? If the last, how is the rampant discreteness bridged so as to form the continuous?

Chapter one suggested knowledge is constructed out of the assembly of singular information bits obtained from the manifold of actual thoughts and experiences. It was also proposed that such aggregations result in theories; only the formation of metatheories can bridge the diversity of individualized disciplinary systems; only the sum of an infinite number of discrete truths constituting knowledge of particular things can result in a continuous Truth of the whole. Chapter two posited self-consciousness as the glue uniting the mental, emotional, and behavioral states coinciding with, and inducing the motivation for, in-the-world activity of all biological lifeforms: certainly with respect to humans and animals, and probably for plants as well.

While it has not yet been demonstrated that self-consciousness is other than brain structure and function, or other than a derivative product of brain structure and function, evidentiary support will be forthcoming in another chapter to substantiate this point. When the themes of these first two chapters are combined, they can be broadened into a general thesis grounding this volume: *only concepts with transcendental import, taken literally to mean nonphysical, can ultimately provide the needed unifying connective links to transform all discrete manifest multiplicities into continuums*. The intent of the present chapter is to offer the concepts of (a) *number* as construed within arithmetic and geometry; (b) *energy* in the form of electromagnetic and gravitational forces operational in (c) *space* and (d) *time*; (e) *causality*; and, (f) *self-consciousness* as the transcendental ideas most needed to serve as mediating constructs enabling the development of continuity in the face of a normally overwhelming tendency to judge discreteness as the predominant characteristic of reality.[1]

The five senses employed to gather data about our surroundings can be conveniently distributed into two classes: those operating at long range—sight, hearing, and smell—and those requiring intimate contact—taste and touch. It is a commonplace truism that we are physicochemically structurally built to perceive only discreteness in properties of objects we encounter at a distance: our visual, auditory, and olfactory senses operate within precisely quantified energetic ranges within certain predetermined boundary limits. We know there is more than meets our eye, ear, and nose because other animal species have capacities in these domains differing from our own.[2]

Even direct exposures to objects through taste and touch revolve around *a priori* proscribed boundary conditions. While we have no difficulty in distinguishing sweet from sour, an accurate description of the distinction in taste between two bland foods is far more challenging. In *The Roots of Consciousness,* Jeffrey Mishlove explained how our nervous system could be depicted as a form of digital computer in possession of only two states in response to its surroundings: "Information about any stimulus is carried by the frequency of nerve firing and by the number of particular fibers which carry impulses, and not by the strength of any single impulse. This...is the same on/off principle by which information is coded in a digital computer."[3]

The "frequency of nerve firing" means there is a threshold energy below which, as well as an upper limit above which, the nerves leading from the surface of our bodies to our brain will not respond to their environment. "The number of particular fibers which carry impulses" is significant because even if the exterior signal is within the proper energetic range for reception by our nervous system, the stimulus must be carried through a diverse array of neural systems in order to be recognized and translated into a form useful to the individual in the sense of being a prelude to an action or reaction.

With this brief dissertation on the human structural and functional means to interact with our local environment, some concepts have unavoidably been introduced: space, time, energy, number, and causality. As mentioned in chapter one, it was Immanuel Kant who first alleged, quite correctly, that we are *a priori* committed to thinking using at least some of these constructs as a framework: thought is impossible without their employment. Immediately, the issue of discreteness *versus* continuity arises, as Franklin Merrill-Wolff explained in *The Philosophy of Consciousness Without an Object*: "A certain amount of time is required before an image can be seen, and thus the sensible motion of external objects is really no more than a series of images with gaps between.... The same is true of the other sense impressions, as there is always a time factor involved in any sensible recognition."[4]

What has just been considered is the utilization of our sensory capacities on a macroscopic level for readily discernible objects. Explored from the microscopic level, we arrive at precisely the same concerns. Objects, say the physicists and chemists, are comprised of atoms: individual units possessing a nucleus containing most of the particle's mass in the form of positively charged protons and uncharged neutrons, surrounded by a sea of negatively charged and much lighter electrons. Two elementary facts about the atom are of immediate pertinence: the nucleus has a diameter of $\sim 10^{-13}$ centimeters, while the entire atom has a diameter of $\sim 10^{-8}$ centimeters. In other words, over 99% of the mass of an atom is confined to a region comprising 0.001% of the total volume occupied by the particle; the atomic particle is 99.999% empty space.[5] Despite this, we seem to be surrounded with a variety of solid objects. Again, ideas about space, energy, number, and causality unavoidably loom in our construction of reality.

The obvious questions to be considered from this understanding of the physics and chemistry underlying our interaction with the material universe of objects are three in number. First, given that discrete atoms, prior to their combination to form objects, are mostly empty space, what accounts for the appearance of three-dimensional solids or liquids suggestive of continuity during the formation of complex molecules? Second, if the sensations by which we detect externals inherently generate discrete sources of data about our surroundings, how are we able to formulate continuity in the face of apparent discreteness? Third, why would, or should, we want to perform this conversion? The universe screams individuality; we seek collectivity. Or, as Alfred North Whitehead wisely observed in *Process and Reality*: "Continuity concerns what is potential; whereas actuality is incurably atomic."[6] What implications does this apparent truth evoke?

The natural sciences trumpet as one of their core virtues an ability to correlate quantitative information with qualitative speculations. Therefore, to begin to answer these questions, one would do well to begin with number. Starting here also allows for reinterpretation of the issue of continuity *versus* discreteness into the familiar connotation of one *versus* many, and permits elaboration of the relationship between a whole and its parts.[7] All numerical relationships are based on the distinctiveness of integers: zero, one, two, *et cetera* are inherently segregated *quantities*, and not merely linguistically independent. Despite this inherent gap between integers, George Santayana accurately identified the characteristic of continuity as the needed foundation for the fundamental process of counting numerals: "The first condition of counting...is...the sensorium should retain something of the first impression while it receives the second, or...the second sensation should be felt together with a survival of the first from which it is distinguished in point of existence and with which it is identified in point of character."[8]

The reality conjoining integers is not a feature *of* the integers: extrapolate past '1' by adding a decimal point and an infinite stream of consecutive '9s,' and one will still never reach the number '2.' Nonetheless, any five-year old will not only be able to rapidly engage in counting the first several numbers with precision, but will also proudly tell you that he or she knows, "1+1 = 2." Think about that: 'one' and 'two' are inherently wholly disjunctive, yet can be bridged through an abstract operation called *addition*. It is impossible to exaggerate the importance of addition to all quantitative operations. It is the bedrock of arithmetic: (a) when processes involving negative numbers were introduced, subtraction of one positive number from another was redefined as the addition of a negative number to a positive number; (b) multiplication has always been defined as repetitive additions of any given number; (c) division is the inverse of multiplication, or repetitive subtractions, both of which mean division is ultimately grounded in addition.

What this suggests is that *addition is a transcendental procedure within manifestation making continuity possible*. Its power derives from its ability to bridge finitude and attain the infinite. How does it do so? Quantity as mere quantity cannot be united by quantity, for all such are intrinsically separate. Addition must be a qualitative characteristic of transcendent reality. This is a necessary conclusion. A qualitative attribute of *physically manifest* origins fails to achieve the desired goal, as Arthur Lovejoy remarked in *The Great Chain of Being*: "Every purely *qualitative* difference between two things—whatever be true of differences of position, quantity...degree—is necessarily a discontinuity. The only way to save the principle of continuity, therefore, is by supposing all things to have some degree or measure of any quality which is possessed by anything."[9]

The supposition of each inhering in each is needed to make sense of our universe. It is an innate aspect of the concept of causality, discussed later in the chapter. Continuity emanates from a nonmaterial realm; in the domain of mathematical transformations, it enters the physical world as the process of addition.[10] Language, in turn, conveys the transcendental meaning of addition through use of an extremely powerful word: the conjunction 'and.' The sole function of this term is to connect two separate notions into a single coherent construct engorged with a meaning that neither notion has on its own, which is another way of stating continuity is created out of apparent discreteness within the manifest universe.

Chapter one demonstrated how integral calculus, addition of an infinite set of terms, allows for development of Truth from truth. Calculus is the pinnacle of arithmetic. It is built on the foundational methods of integration and differentiation, which is merely another way of saying that in the limit of infinite accumulation or infinite disbursement, we can add or subtract single units to attain a hypothetical absolute whole: everything or nothing. Further, chapter one paraphrased the words of John Duns Scotus asserting being as a necessary cognitional determinant prior to the expression of any proposition pertaining to nonbeing.

How do being and nonbeing relate to everything and nothing? Masao Abe conceived the two pairs as ontologically identical according to Buddhist philosophy. In "Nonbeing and *Mu*—The Metaphysical Nature of Negativity in the East and the West," he explained it in these terms:

> In the West, since being is considered ontologically prior to nonbeing, the ultimate, beyond the opposition of being and nonbeing is Being… By contrast, in Buddhism, *Śūnyatā*, as the ultimate, is realized by directly transcending the very duality of *U* (being) and *Mu* (nonbeing), which stand on an equal footing and are completely reciprocal… The ultimate…is not "Being"…but formless "Emptiness."… In order to attain true Emptiness, Emptiness must "empty" itself; Emptiness must become non-Emptiness. Thus, true Emptiness is wondrous Being, absolute *U*, the fullness and suchness of everything.[11]

Infinite accumulation and infinite disbursement lead to the same end, and we can transition from manifest discreteness to transcendental continuity in a numerical way only by employing as a pedagogical tool a postulate whereby an intrinsically indefinable *quality*, namely infinity, is mystically utilized to produce a *quantitative* certainty.

Another representation for the way being and nonbeing, everything and nothing, merge into an equality of identity is through consideration of two fundamental postulates of arithmetic. The *addition property of zero* states that zero, the mathematical equivalent to nothing, when added to a number leaves the number unchanged. Exactly the same condition is attributed to the *multiplication property of one*: one, alias everything, when multiplied by any number leaves the number unaltered. Multiplication is repeated addition, yet zero assumes the role in one instance that one has in the other. On some deeply profound level, zero and one develop interchangeable character.

Geometry, another fundamental discipline in mathematics, quantitates the relationship not of pure number, but of hypothetical or real objects in space. It is based on the notion of lines, and yet it is never explained in any mathematics classroom exactly how a series of dimensionless points, having by definition zero extension in space, can be joined together to produce a one-dimensional line, much less a two-dimensional plane object or a three-dimensional object of everyday experience.

To start, what does it mean to posit a point with zero extension in space? *Zero extension means the point is not a material object; it is an abstract, transcendental postulate of reality without intrinsic meaning for objects in physical manifestation.* It is the something that is also a no-thing, a geometric confirmation of Abe's comment. The concept of zero in mathematics is as replete with mystical power as the notion of infinity. As a transcendental, it does not in any way alter the properties of a manifest quantity. Hence, zero added to any number is the unchanged number. If you add successive zeros repetitively, it doesn't suddenly transition into a physical reality. Hence, zero multiplied by any real number remains zero.

The question remains: How does a zero-dimensional point with no in-the-world physical reality transition into a one-dimensional line potent with the capacity to become visible to manifest lifeforms? It can only be by virtue of the transcendental power of zero to unify something discrete, an *infinite* series of zero-dimensional points, into a continuum behind-the-scenes with the result that a physical one-dimensional line is endowed with immanent being. This suggestion interweaves a transcendental meaning of zero with a transcendental meaning of infinity: the no-thing and the every-thing jointly produce a concept with physical properties—what might be called "applied Buddhist philosophy." Integration of a differential, infinite summation of the infinitely divisible, is the Kantian "thing in itself" in physical spacetime:

$$\int_1^\infty dx = x$$

In fact, we never detect *lines* in the universe, only *line segments*—some discrete fraction of a wholesale continuity—because lines extend beyond the physical into the transcendental domain, while line segments, the deliberate truncation of lines, are wholly material.

After introducing line segments into the material world, explaining their elaboration of one-dimensional extension into our common conceptualization of three-dimensional space becomes a trivial exercise, although one with metaphysical significance. For example, /\/\/\/\/\ is a continuous figure. It is composed of ten discrete line segments, each of which is straight. Now, if each segment is straight, the whole should be straight, since a line segment by definition possesses only the characteristic known as *extension* and is lacking the qualities *height* and *depth*. Upon visual inspection, however, one does not regard this line as straight. Why?

The answer is the standard for judgment is of a higher order. The two-dimensional plane serves as the reference: inclusion of a second coordinate axis for measurement allows for the incorporation of height. Similarly, two-dimensional line segments extend into three-dimensional space by including depth as a quality capable of quantitative determination through the adoption of yet another, even more elevated, perspective.

Making such procedures explicit does not imply mathematical operations should be rejected even though the conceptualizations allowing us to use such techniques border on the magical. P. D. Ouspensky asserted: "The laws of mathematics and the laws of logic are the laws of the reflection of the phenomenal world in our perception and thinking."[12] Inherencies grounding physical and mathematical reality, as we understand it and live in it, offer support to a philosophical principle of first-level importance: *the frame of reference for judgment of the existence of a property demands consideration of a perspective superior to the one inhabited by the objective property being contemplated.* We have already observed this thesis at work in the discussion of Gödel's mathematics as indicative of the need for some metatheory as the means to procure higher-order Truth from truth.

Geometry is needed for an understanding of space as a whole, as well as for comprehension of some of the properties of particular objects located in space. It is to be noted *physics* is the union of philoso*phy* with mathemat*ics*.[13] Albert Einstein conflated the ideas of geometry, object mass, and space in his revolutionary General Theory of Relativity. He insisted: "The more massive a body, the more spacetime is curved and 'gravity is nothing but this curvature of spacetime.'... No longer is gravity an instantaneous action at a distance. Gravity is now the way in which mass curves spacetime, an action that travels at the speed of light."[14]

Einstein's affirmation, "no longer is gravity an instantaneous action at a distance" is a reference to the original formulation of the concept of gravity developed by Isaac Newton. Recall Newton's claim of an absolute space and time: real in themselves, needing no external justification for their universal application. Given spacetime as an absolute frame of reference, Newton's postulation of the force of gravity involved the development of the famous inverse-square law for determination of its quantitative effect: "One body exerts a gravitational pull on another with a strength determined...by the mass of the objects involved and the magnitude of their separation.... This means...if their mass or their separation should change, the objects will... *immediately* feel a change in...mutual gravitational attraction."[15]

The distinction between the theories of Newton and Einstein is clear: the former invoked spatial distance, object mass, and the force of gravity relating them as three separate components that jointly comprise physical reality; the latter conceived gravity as merely a derivative outcome, nothing other than a definitional consequence of space and mass. Mass is capable of precise quantitation, is necessarily discrete, and is an intensive property of localized matter encapsulated in a restrictive region of bounded space. For Newton, gravity was the unceasing attractive force bridging gaps in matter contained in unbounded space: "Gravity is the undividedness seen in the divided."[16] To Einstein, gravitational force was a local, discrete phenomenon: a product of individualized units of matter possessing mass inducing a particular formal geometric shape on localized space. Who was correct?

Einstein's 1915 general relativity theory was a continuation of the line of thought begun ten years earlier, special relativity theory, which construed all the essential components of manifest reality as subjectively dependent upon an observer: "The space and time in which objects exist...must be made conditional upon the vantage point of the observer who looks upon this reality."[17] Einstein's Special Theory of Relativity gave scientific support to the epistemological underpinnings of Kant's Transcendental Idealism: space and time were absolute *only* in the sense that they served as inalienable *a priori* templates for all human thought.

Kant's principle was challenged on grounds of impropriety: the idea of anthropocentrism should not be employed as a valid cognitive framework for manifest reality. The same reasoning suggests this aspect of special relativity theory cannot be ultimately correct. Does the argument extend to general relativity? No. Although the Special Theory of Relativity alleged matter to be observer dependent with respect to taking cognizance of its form, mass refers to matter's substance. The distinction is significant: the quantification of mass is an intrinsic property of material independent of any observer; its formal shape is an extrinsic property amenable to one's subjective perceptual capacity. The philosophic dispute between Newton's and Einstein's methods for doing physics, as it applies to the notion of gravitational force, must be resolved on another basis.

Resolution does arise as a consequence of special relativity theory, for it was there in which Einstein erred in his reasoning by claiming measurement was a fundamental element in understanding nature and natural phenomena. It is true the enormous power of physics derives primarily from an insistence on quantifying its constructs, and therefore, demands measurements be made on all its discernible parameters. This fact stipulates measurements and the calculations deriving from manipulation of primary data are pertinent to *our* comprehension of the universe; it need not necessarily suggest measurement is fundamental to the universe *itself*.

In *The Physics of Consciousness*, Evan Harris Walker articulated the great scientist's seemingly logical deduction: "Any limitation that affects our ability to measure a physical quantity...that persists no matter how we try to devise equipment for...measurement, not only represents a limit on us but ...shows a limitation that must exist when any two physical objects interact. This limitation on me as the observer is also a stricture on nature."[18] The last sentence is the crux of the issue. It is an example of the plague permeating the quest for knowledge and understanding for two-and-a-half millennia: Protagoras' anthropocentric presupposition of man as principal determinant for manifest reality, a judgment already discredited. It must be rejected to develop coherency between thought and experience as it refers to spacetime.

If we are innately able to measure only discreteness with our perceptual abilities, we will comprehend only discreteness, and we will presume such a characteristic naturally applies to everything. Measurement by definition imposes discreteness on things measured. It creates boundary limits on its operational realm by segregating an object to be measured from those not measured. Absent groups of individuals aiming to quantify using numbers, which become embodiments of discreteness when they become identified as inherent properties *of* substances, physical reality is always experienced as qualitatively continuous.

As a direct consequence of a commitment to measurement, there are generated an abundance of theories in the realms of chemistry and physics positing discontinuities as existent everywhere. In black hole physics at the event horizon, the physical laws inside the black hole are postulated to be discontinuous with the laws operable in the surrounding space outside the black hole. In quantum mechanical tunneling by electrons, said particles are claimed to be able to mysteriously occupy discrete, randomized locations in space without transitioning in some perceptible manner from one locale to another, as demonstrated by photon absorption and emission spectra obtained from electron transitions between orbital levels.

These theories, while not claimed to be wrong from *our* perspective, are restrictions not necessarily shared by manifest reality despite our capability to measure with refinement. In other words, the physics in a black hole may be entirely consistent with that of the surrounding space, but when phrased in the conceptual terms we ordinarily use to calculate physical parameters, a comparison of results in the regions leads to a distortion of what is actually the case. Similarly, the movement of electrons in the region surrounding the nucleus may possess the property of a continuous flight path, but our ability to perceive and measure these motions demands distinction in spatial coordinates that are ultimately illusory.

Newton may be capable of vindication: manifest reality might be most properly construed as continuous spatially and temporally, with spacetime a thing in itself. Barring re-education with the assistance of other intelligences not constrained to suffer the same perceptual limits we possess, apprehension of continuity demands we adopt a meta-perspective; to be more precise, a metaphysical perspective. *The continuity of space and time has its origin in a dimension we cannot literally envision. The forces of electromagnetism and gravity, the former as light energy, act as unique emissaries from that transcendental domain.* Is there evidence to support this assertion?

The mathematical relationship between mass and energy obtained from special relativity theory states: if a material object were to attain a velocity close to the speed of light—a constant value, 'c,' Einstein characterized as a universal maximum speed limit—the object's mass, 'm,' could be converted to pure energy, 'E,' according to the relationship, $E = mc^2$. *Light energy is without mass; it is not a substance; it is literally immaterial.* All physicists accept *energy as a continuous phenomenon: energy is omni-directionally distributed throughout the universe.* Physicists also recognize the existence of another pertinent relationship: $E = Fd$, which means energy is a product of a force, 'F,' operating over a spatial distance, 'd.' *The concept of force acts as perpetuator of that most salient property of energy, namely immateriality, and must also permeate space continuously.*

Gravitational and electromagnetic forces are the only universally acting forces known, both of which are nonmaterial in substance. They do not turn "on" and "off" in a temporal sense; nor are they active only in certain locales in a spatial sense. This continuity of energy is recognized in the formulation of one of the grounding postulates of physics, the *Conservation of Matter-Energy Principle*, which states energy is neither created nor destroyed when considered in its entirety comprising system and surroundings, but merely changes form under any given set of conditions. It is important to recognize "surroundings" does not refer solely to physical spacetime; matter-energy is conserved when the nonphysical dimension is included in the determination.

The implication is transparent. Einstein's model doesn't hold up: gravity cannot be merely the resultant outcome of operation of two *discrete* physical parameters, mass and space. A series of discrete numbers does not result in the continuity exhibited by the number line; accumulated discreteness *never* produces a whole. The conclusion *is* consistent with a Newtonian physics in which gravitational force acting over defined distances provides continuity unifying discrete object masses localized in form in a universal space. It is clear why gravity was originally conceived as "an instantaneous action at a distance." Gravity, a force derivative to universal energy, is co-extensive with three-dimensional space. While the specific *magnitude* of its effect on material objects may vary in direct proportion as such objects are perceptible to external observers, it exerts a continuous *qualitative* influence on all objects independent of any observer.

Discrete distances do not enter discussion of gravity's qualitative impact because gravity is not an object like any other object; it is always *there* in its effects exactly where any object is located within existential space. It only *appears* to act at a distance from the perspective of someone outside the interaction between force and object. Moreover, it displays an *instantaneous* effect on objects because time is also co-extensive with space: Einstein *was* correct in uniting time with space to produce spacetime. Just as the force of gravity acts everywhere in space, it always influences material objects *now*, an "every-when" with respect to time.

After Einstein had proposed that time could be mathematically portrayed as a fourth spatial dimension, resulting in fusion of the two parameters into a single spacetime comprising four dimensions, Samuel Alexander expanded this notion in his doctrine that the underlying continuity relating successive temporal instants was location—both events occur *here*—and likewise, the underlying continuity that relates contiguous spatial points was transition—both events happen *now*.[19] The influence on visible objects by gravitational force is one means by which this union of space and time is detectable.

Upon consideration of spacetime as a continuous thing in itself the way Newton did, the way science's obsession with quantitation can distort the conception of physical reality becomes comprehensible: to conclude what is real is only what one can measure with precision. *Measurement necessitates discreteness of substantial being*. When directed towards continuous, non-material notions—energy, force, space, and time—the process of measuring simply takes a limited three-dimensional slice of these metaphysical qualities and treats these portions as if they were the whole in the same manner as the geometry of immanent line segments incorporates a microcosm of the properties of transcendentally originated macrocosmic lines.

Since misunderstanding is all-too-possible upon contemplation of every difficult topic, a need to try again is transparent. It is *not* here being proposed that the concept of number is invalid, or that mathematical operations are metaphysically devoid of meaning. On the contrary, number and operations involving number are of paramount import in comprehension of existence. As will be demonstrated throughout this volume, there is useful information to be gleaned from the logical manipulation of mathematical symbols and formulae as well as from the numerical results of such efforts. What *is* being claimed is within physical reality, quantification into discrete units is one of two general aspects necessitating consideration leading to comprehension; the other is qualitative continuity. The former must not be perceived as the dominant, much less the sole, orientation of significance.

In the transcendental domain, the principal importance of number derives from employment of a specific metaphysical theory regarding the ontological origins of substantial being. The historical dispute pervading all religious or philosophic commentary surrounding the issue of unity *versus* plurality—one God under the titular heading of whatever denotative label seemed to appeal to the individual theorist, or many Gods characterized according to whatever diverse criteria for differentiation of attributes was deemed plausible—can be interpreted as being predicated on the human inability to conceptualize the potential source or sources of discrete units of multiplicity in any manner other than through the unifying influence of categories.

How could *one* Being produce a *multitude* of differentiable categories capable of organizing the inherent diversity of manifestation? Alternatively, how could there exist *more than one* Being if the idea of absolute—meaning a complete perfection guaranteeing the absence of errors, gaps, overlaps, or irregularities of any kind—is to be valid? Every systematic metaphysical philosophy/theology must introduce number as an inalienable transcendental quality with immanent quantitative repercussions. The topic will be explored in the next chapter.

An early precedent for the logical conflict between what we can measure and what we experience began 2500 years ago with Zeno of Elea (~515–445 B.C.E.) and his paradoxes of motion. Using clever illustrations of arrows in flight, or races between speedy and slow creatures to reach a given terminus, Zeno asserted every instant of time and portion of space are discrete and possess unbridgeable gaps; he "proved" motion was theoretically impossible even though motion was also obviously a component of normal experience.[20] His confounding puzzles were eventually circumvented under the postulate that space and time were in fact continuous despite our ease in degrading those concepts into discrete, though nonetheless completely arbitrary, units susceptible to measurement.

It was later suggested that resolution of his paradoxes stipulated that the spacetime continuum is only conceptually infinitely divisible, while its finite dividedness into discrete portions is a limited, though practical technique for measurement.[21] With these notions in place, it could then be demonstrated—again by use of calculus through its property of summation over unbounded time and space—that Zeno's perplexities concerning motion were false not only because they were contrary to lived experience, but because they ran counter to Newton's metaphysical theory claiming the natural properties of time and space, real in themselves, flowed continuously, thus precluding any arbitrary limitations. It was the functional appeal to *infinite summation*—two terms shown to be transcendental in origin and meaning—which overcame the logical problems Zeno constructed.[22]

Though spacetime can be analytically subdivided, its inherent continuity makes possible two common constructs human beings employ to interpret phenomenal events both logically and epistemologically: the *Principle of Noncontradiction* and the *Principle of Causality*. The first, referenced in the opening chapter, alleges 'A' and 'Not-A' cannot hold at the same time when viewed from the same perspective.[23] The corollary, less frequently explicitly employed in our mental processes, is that 'A' and 'Not-A' can both be true at the same time when regarded from differing perspectives. As applied to time and space, it is highly plausible they are continuous when considered from a meta-position, a domain outside our usual experience, yet discrete when considered by us from our traditional orientation. The next chapter will show how Time and Space, the transcendental referents for physical spacetime, are substantively meaningful.

The causality principle in human logic proclaims any given effect must have one or more discernible causes, where "discernible" is understood to mean the connection between cause and effect is, or can be made, transparent to comprehension. There is a direct relationship between causality and the notions of continuity and discreteness: "In causation, we may emphasize the emergence of the effect as something new and different, or...its necessary connection and continuity."[24] William James expressed the same relationship in the form of a question in *Some Problems in Philosophy*: "A cause and its effect are two numerically discrete concepts, and yet in some inscrutable way the former must 'produce' the latter. How can it intelligibly do so, save by already hiding the latter in itself?"[25]

James' remark should recall Arthur Lovejoy's statement expressing the same idea. Lovejoy professed this notion of "each inhering in each" as the functional significance at the root of continuity. Taken together, the resultant thought is causality must be yet another physical manifestation of continuity. Indeed, this concept was known and understood by Aristotle, who declared in his *Metaphysics*: "What is passing away must still have something of what has passed away; and of what is coming to be, something must already be. And, in general, when anything is perishing, something is present; and when anything is coming to be, there must be something from which it comes to be and something by which it is generated."[26]

In *A Treatise of Human Nature*, David Hume related causality to space and time. He tried to show the properties innate to causality: *contiguity* in the sense of two events occurring proximally in space and time; *priority* in that one of the two always seems to precede the other in time; and, *constant conjunction* in that both appear always in the same spacetime relationship sequence.[27] In formulaic terms, the sequence is: 'A+B'; 'A' before 'B'; 'A' always before 'B'; therefore, 'A' causes 'B.' Hume's binding of causality to space and time implies causal linkages are ultimately plausible if and only if space and time are in fact continuous, in alignment with common experience, despite their persistent measurability as partitionable elements in existence.

As in the previous cases, causality demands a principal orientation to some metaphysical construct to link discrete effects with a continuous chain of relational components: causal continuity demands continuity of substance. If the most essential aspect of an object does not persist over variable periods of measured discrete units of time, causality is impossible. Objects would disappear simultaneous with the appearance of new objects, and without identification of the objects as possessing some element of continuity of substance, there would be no reason to posit any relationship between the objects, much less postulate a causal influence acting among these objects.

The perpetual existence of subjects and objects—perceivers as well as things perceived being equally necessary to comprehend causality—leads to consideration of apparent continuity of being applicable to sentient creatures. Dean Radin outlined the parameters of this puzzle with respect to human beings in *The Conscious Universe*:

> The average neuron consists of about eighty percent water and about 100,000 molecules. The brain contains about ten billion cells, hence about 10^{15} molecules. Each nerve cell in the brain receives an average of 10,000 connections from other brain cells, and the molecules within each cell are renewed about 10,000 times in a lifetime. We lose about one thousand cells a day, so the total brain cell population is decimated by about ten million cells, losing in the process some one hundred billion cross-linkages.... "And yet," as Paul A. Weiss writes, "despite that ceaseless change of detail in that vast population of elements, our basic patterns of behavior, our memories, our sense of integral existence as an individual, have retained their unitary continuity of pattern." *All* of the material used to express that pattern has disappeared, and yet the pattern still exists. What holds the pattern, if not matter?[28]

If we are constantly losing brain cells in huge numbers on a daily basis, and it is the efficient operation of these cells and their connections to other portions of the brain that underlie our experience of self and surroundings, then how can our sense of self-identity causally persist? Phenomenally, cells are discrete units in the sense that there exist boundary constraints specifying distinction between one cell and another, time is measurably discrete—even if it is acknowledged to be abstractly, even necessarily, continuous—and the mathematics, as always, involves discrete numbers. In the face of so much isolation, how can we, literally, continue from moment to moment?

In ancient times, continuity of person was attributed to an individual's soul; in more modern times, it is called "self-consciousness." On this topic, William James wrote: "The proposition that consciousness feels continuous means two things: that even when there is a time-gap, the consciousness after it feels as if it belonged together with the consciousness before it, as another part of the same self; that the changes from one moment to another in the quality of consciousness are never...abrupt."[29] Fiona Steinkamp noted the epistemic confusion resulting from the concept of self-consciousness in an essay entitled, "Parapsychological Phenomena and the Sense of Self":

> My internal coherence comes from my internal perspective on myself. It might appear...that the essential self is the transcendental self.... If the transcendental self is the essential self and if it remains constant whereas the self it observes does not, there must be a sharp distinction between the transcendental self and the self that it observes.... In order for us to have a sense of internal coherence, the transcendental self needs to be able to identify itself as the self it observes.... The transcendental self therefore appears to be an absurd concept. It is both constant and changing.[30]

Note her use of the term "transcendental self": it conveys the notion that the continuity of person must emanate from a nonphysical orientation, since in the existential world matter comprising the human body always undergoes a process of conversion, just as the citation from Dean Radin explained is the case for the brain. Hence her characterization of the concept as "absurd," because the self, when contemplated as a physical object, is alterable; when viewed metaphysically, it must be constant. A way to make both perspectives intelligible must be found.

Difficulty in discovering the path of resolution has inevitably meant few terms exist more controversial than that of self-consciousness. This is highly ironic given that in antiquity the extraordinarily vague term "soul" seemed universally understood. Nonetheless, there are intense debates about the very existence of self-consciousness as a thing in itself, as well as sophisticated academic controversies over its nature and properties, if it does exist.

There are very few other notions ever conceptualized that is in its class as a contested thesis. *God*, *Truth*, and *Good* have been proposed as absolutes incapable of satisfactory definition or admitting proof of existence. Due to their seeming incapacity to be defined by other meaningful terms, they have occasionally been equated to each other, as in the formulas, "God is Good" or "God is Truth."[31] Aside from a frequently postulated identity, they have been used analogically with respect to their philosophical influence: the absolute "God" in metaphysical speculation has been taken to serve the same role as the absolute "Truth" for epistemological thought, which is, in turn, utilized in the same manner as the absolute "Good" of ethical inquiry. What will be shown to be most remarkable about self-consciousness is that *it is inalienably linked to all three of these constructs.*

If self-consciousness is the linking element in the transformation of an individual from one discrete moment to the next; *if* the only means to conjoin two discrete aspects of a phenomenon into a continuum is through adoption of a "higher" orientation, a meta-perspective; *then* self-consciousness must have some transcendental component. What is proposed is the translation of self-consciousness from a phenomenal property of a biological self into a noumenal property of a *supra*-natural self, just as Immanuel Kant declared: "No knowledge can take place in us, no conjunction or unity of one kind of knowledge with another, without that unity of consciousness which precedes all data of intuition, and without reference to which no representation of objects is possible. This pure, original, and unchangeable consciousness I shall call 'transcendental apperception'."[32]

Within phenomenal reality, causality is not discernible; everything is a series of isolated events. This fact justifies Hume's skepticism concerning the existence of causal processes, as he admitted in *An Enquiry Concerning Human Understanding*: "All reasonings concerning matters of fact seem to be founded on the relation of *cause* and *effect*.... I shall venture to affirm, as a general proposition, which admits of no exception, that the knowledge of this relation is not, in any instance, attained by reasonings *a priori*; but arises entirely from experience, when we find, that particular objects are constantly conjoined with each other."[33] While this passage merely highlights Hume's conclusion that causality is, at best, entirely empirical if it exists at all, his overall skepticism was voiced by Richard Popkin in *Skeptical Philosophy for Everyone*:

> Hume's theory of knowledge severely criticized the kinds of knowledge...we think we gain through causal reasoning. If a cause is something that has the power to produce an effect, and our knowledge of causes and effects comes from experience, do we ever actually have any experience of the causal power? Or is it that we just observe successive events?... Why cannot the causal agencies be contained within the material world itself? Hume also contended that the maxim "every event has a cause" is not self-evident or demonstrable. Our causal knowledge is gained only through experiences, and...just consists in noticing recurring sequences of events, so that when we observe one, we expect the other.[34]

"Why cannot the causal agencies be contained within the material world itself?" Quite simply, since Hume couldn't literally *see* continuity of process, he considered it appropriate to be doubtful of its reality. Another way to state Hume's case is to say *the linkage of events within manifest physical existence appear to be not causal but associational*; they appear together as discrete eventualities in discrete spaces at discrete times.[35] The implication for this observational realization returns to the theme of this chapter: *causality can only enter the picture when an appeal is made to a higher dimension than the phenomenal realm we habitually perceive*.

What is the causal linkage? Sentience. The self-consciousness of living creatures connects events, events lifeforms are able to perceive through their sensory apparatus. The transformative conversion from *actual* association to *factual* causality occurs by induction within the individual self-consciousness through reflection on multiple repetitions of some condition. Returning to the sequential form of Humean causality: self-consciousness—whether human, animal, or vegetal—is responsible for cognizing 'A' alone; cognizing 'B' alone; adding them together, a transcendental function, within an isolated spacetime nexus to obtain 'A+B'; cognizing the formal sequence as always 'A' before 'B'; and finally concluding 'A' causes 'B.'

Therefore, causality as an operable principle could be construed as an epistemological *a priori* due to constant activity of the "pure, original, and unchangeable consciousness." Contrary to Humean thought, the situation is as Kant depicted in *Critique of Pure Reason*: "If we consider that which we call nature is nothing but…a number of representations in our soul, we shall no longer be surprised that we only see her through the fundamental faculty of all our knowledge…the transcendental apperception, and in that unity without which it could not be called the object of all possible experience, that is, nature."[36]

To comprehend unity in natural events is to conceive causal relations between those events even when we are able to only perceive associational relationships. If change is omnipresent within manifestation—including, as Dean Radin remarked, all organic brain function—only that which does not change, our transcendent self-consciousness, could detect unity "behind the scenes." The error Kant made in his union of transcendental apperception to every causal process was in believing causality was subjectively an element of reality and could not be an objective one as well. He was committed to such an absolute either/or conclusion by his central epistemological premise: the *Copernican Revolution* to which reference has previously been made.[37] In the words of Stephen Palmquist, it was "the assumption that *a posteriori* objectivity is based on *a priori* subjectivity, rather than *vice versa*."[38]

In fact, there must be an origin to any *a priori*, whether it be designated as a transcendental unity of apperception or categories—such as time, space, and causality—since they are given to us by some source not of our own creation. Kant admitted there were several such transcendental constructs; he had merely to classify them as properly being transcendental analytic or transcendental synthetic in type. Whichever route he chose, he would have had to acknowledge we have definite knowledge of something transcendental beyond phenomenal appearance, and this is precisely the point at which he is self-contradictory: "As…such an intuition…is entirely beyond our faculty of knowledge, the use of the categories also can never reach beyond the limits of the objects of experience."[39] On the contrary, the mere existence of *a priori* categories, whatever they might be, *forces us* to admit the knowledge of things "beyond the limits of the objects of experience." If not, we could never come to know of the existence of the categories *as* categories that necessarily frame all our conceptual thinking.

Unfortunately, Kant was not conversant with the mystical literature, and so felt he had to content himself with positing concepts like *deity*, *freedom*, and *immortality* as mere contingent hypotheses rather than as unassailable transcendental realities. Nikolai Berdyaev was correct when he affirmed in *The Beginning and the End*: "It was a fundamental mistake in Kant that he recognized sensuous experience, in which appearances are the data, but he did not recognize spiritual experience, of which the data are noumenal."[40] Where Descartes separated man's mind and what it told him, *res cogitans*, from man's body and what it told him, *res extensa*, Kant severed man—mind and body together—from the rest of reality.

With admission of the reality of the transcendental and our ability to gain knowledge about its functional activity within the physical world—which is to say, with elimination of the anthropocentric bias basic to human belief that we are the principal determinants for the nature of reality—it is much more reasonable to posit a transcendent *a priori* objectivity available only through the *transcendental* notion of causal connectivity. The role of causality is to make possible *a posteriori* subjectivity of experience attained by individual *transcendental* self-consciousness.

The persistence of self-consciousness in the face of the constant turnover of brain matter would *necessarily* lead to the conclusion that the two are not equivalent: the brain is neither the origin nor the seat of self-consciousness. This represents the first self-consistent chain of thought in support of the contention introduced in the chapter's first page: "Self-consciousness is other than brain structure and function, or other than a product of brain structure and function."

Aside from continuity in both the perceiver and the perceived needed to establish causal processes, Hume's version invoked a temporal element, "A always *before* B," exhibiting singular directionality proceeding from past to future. The universe *does* appear committed to a unidirectional sequence, and appearance is the key descriptor. Is physical reality aligned necessarily in one preferred temporal order? One might be predisposed to propose: If Time has a transcendental aspect that unifies the manifold of discrete instants, then having the property of continuity might result in a lack of directionality. If that were the case, how could a directionality for time be introduced into the manifest universe? As an alternative, if Time is unidirectional, then it must be pondered: How could this claim be justified metaphysically?

The passage of immanent time is understood to serve as a referential guideline to mark either event occurrences—the perception of transformation of material substance from one state to another—or energy transitions—the conception of alteration of intangible process from one state to another. For sociological purposes, we determine a day to be equivalent to the time it takes, twenty-four hours, for the Earth to complete one rotation on its axis such that any particular location on the planet arrives back at its original position relative to the sun.[41] Here, the principal concern is with the Earth and the sun as fixed substances capable of alteration in space with respect to each other. For scientific purposes, a second is the basic unit of time, and is defined as the reciprocal of the frequency of light energy involved in the relocation of an electron from one atomic orbital to another occurring around the nucleus of a cesium atom.[42] Here, the principal concern is with a specific nonmaterial exchange of energy content.

Rotation on our planet's axis is a continuous motion and always occurs in the same direction. It might be considered that this directionality of spin confers directionality on temporal progress; however, other planets in our solar system—Venus, Uranus, and Pluto—exhibit retrograde spin: they each rotate on their planetary axis in a direction opposite to Earth's. Hence, spin directionality, while intrinsic to utility of the concept of *day*, is not intrinsic to the issue of time advancement with a specified directionality from past to future.

Electron transitions in cesium atoms also occur equivalently in both directions. Transition from a lower orbital level closer to the nucleus of the atom to a higher orbital level requires input of light energy due to increasing separation of electric charges—positive for protons in the nucleus, negative for electrons outside it—exactly equal in magnitude to the energy released by the atom if the electron reverses course and transitions from higher orbital to lower, more favored level. Thus, electron energy transitions, while useful for the determination of a *second*, also fail to provide any justification for our innate conceptualization of time as having an overall directional movement.

The conclusion must be we cannot solve the problem from the standpoint of the physical means by which time is measured. This implies resolution either must come from the metaphysical level, or there is another avenue by which to assess origin and meaning for time's apparent directionality within the manifest existential level of reality.

If physics, the foundational science of inanimate matter, doesn't hold the answer, perhaps one should consider biology, the counterpart of physics with respect to animate beings, as a possible source of revelation. Peter Coveney and Roger Highfield's *The Arrow of Time* revealed the futility in seeking the key to unlocking time directionality from biological processes: "Aspects of time in biology are analogous to both cyclical and linear cultural experience. Cyclical time appears in cell division and the orchestra of different rhythms in our bodies ranging from high-frequency nerve impulses to...cycles of cell turnover…. Irreversible time is manifested by aging in the passage from birth to death."[43]

Cyclical temporal experience leads to no directionality for time, while a linear progression implies directionality. This confusion is beneficial. Living creatures inherently reflecting both linear and circular two-dimensional time characteristics means that with respect to three spatial dimensions, time must be conceived as a combination of these two tendencies, as *helical*. Helical time denotes that nothing, including planetary or personal histories, is *exactly* repeated. The development of a new cycle, in contrast to precise repetition of an old one, suggests the introduction of *new information* that makes things progress in an evolutionary sense.

Helices, also known as three-dimensional spirals, are ubiquitous in the natural world. Priya Hemenway wrote in *Divine Proportion*: "Spirals are everywhere—some obvious, some less so. From embryos to galaxies, the spiral offers what is perhaps one of nature's most dynamically proportionate messages in that it arises out of the reconciliation of opposites, harmonious and unaligned. It is the middle way…never leaning too far in one direction or the other but always finding perfect balance."[44] Aside from embryos and galaxies, Hemenway mentioned as additional examples: coils of rope, watch springs, weather patterns, shellfish, flower petals, the human heart, and the DNA molecule at the root of all organic life.

There is rhythmic regularity to the numerical relationships appearing in helices, which Hemenway discussed in great detail, and is hinted at in the title of her book, a title with metaphysical implications. To quote from the book's dust jacket: "With its natural balance and elegant beauty, the Divine Proportion is a perpetual reminder…our hope for regeneration and continuity lies in realizing the meaningful and harmonious relationship of all the parts to the whole."[45]

The helix fits well with prior postulation of the development of Truth as a summation of separate contingent truths aggregated in increasingly larger bundles over time as a function of the establishment of new knowledge. What conclusion is drawn from this confluence of ideas? *The directional nature of time must exist as a means to measure the rate of incorporation of new information within existential reality*. New data could arise in two ways: as a seemingly natural progression emanating from existent information—as would arise from technological advancement—or as a creative leap beyond all expectation—such as would arise from artistic inspiration. This outcome was foreseen in the previous discussion on the attainment of Truth *via* the competing modes of gradual awakening and sudden enlightenment described by Zen Buddhist philosophic principles. Without directionality, *newness* of anything could never arise. Complete time reversibility means everything done could be undone; novelty could never then accrue as an attribute of phenomenal manifestation.

Further substantiation obtains from chemistry, the scientific intermediary between physics and biology. *Thermodynamics* is the branch of chemistry that allows for the calculation of the change in energy content in reactant and product molecules as a result of any physical or chemical transformation. In *The Arrow of Time*, Coveney and Highfield explained the significance of the two laws governing the field, in particular the second law's direct relation to time:

> The equivalence of heat and work as forms of energy is the basis of the First Law of Thermodynamics...energy will always be conserved in a physical process, even though it may be converted from one form to another.... In principle, any form of work can be converted entirely into heat. But dissipation means that the reverse is not true; heat energy always gets frittered away as it is converted back into work.... The Second Law was sharpened up in 1865 when Clausius distinguished reversible and irreversible processes by introducing the concept of *entropy*, a quantity that relentlessly grows with dissipation and attains its maximum value when all the potential for further work is spent. According to...the Second Law, the entropy change in a reversible process is zero...entropy always increases during irreversible processes.... It...furnishes an explicit arrow of time: increasing entropy coincides with time's forward movement.[46]

On one hand, the claim has been put forth that creativity results in the development of new information; it is the invocation of novelty that justifies a specific directionality for time independent of any particular measure of its passage. On the other hand, the above citation utilizes the concept of entropy, which increases when physical or chemical processes are not reversible—that is, time symmetric—as a scientific indicator of temporal directionality. Not only do irreversible processes necessarily introduce novelty, meaning these different perspectives reinforce one another, but communication theory appropriated the concept of entropy as a quantitative measure for information transfer, as John Pierce related in *An Introduction to Information Theory*:

> In communication theory we consider a message source...which may produce...one of many possible messages. The amount of information conveyed by the message increases as the amount of uncertainty as to what message actually will be produced becomes greater.... The entropy of communication theory is a measure of this uncertainty... The more we know about what message the source will produce, the less uncertainty, the less the entropy, and the less the information.[47]

To summarize: (a) increasing entropy correlates with increasing amounts of potentially accessible information; (b) entropy characterizes irreversible chemical and physical processes; (c) both acquisition of data and conversion of material result in the introduction of creative novelty in a technological or artistic sense; (d) innovation implies time directionality toward a progressive future; (e) the causal mechanism suggests preferred temporal orientations in every sequential development.

It is proposed that *causality is a transcendentally originated mechanism established for the continuous development of an ever-expanding community of contingent truths obtained through experience, the purpose for which is to be able to invariably point towards a coherent comprehensive metaphysical absolute Truth within the phenomenal universe, and time directionality is the indicator employed to convey that progressive evolution.*

The principal historical problem concerning comprehension of the nature of causality with respect to time is that it has become deeply enmeshed with the issue of predeterminism. There are many who have believed in the past, and many who currently believe, that to have a causal relationship implies wholesale determination of outcomes. This construct would introduce severe limitations on the development of creativity and endanger the ultimate aim of causal processes. Of relevance, therefore, is a logical distinction introduced between determinism and predeterminism. The first is a constraint, while the second is a limitation: "Determinism most often refers to the mechanism of decision making, and predeterminism to the outcome of decision making; one is means-related and the other is ends-related; one is influential and the other is controlling."[48]

Causality is deterministic: preceding conditions oftentimes circumscribe the range of available options. This neither suggests nor demands the option chosen is absolutely required. Originality, the introduction of the new, is still feasible. Augustine indirectly referred to this distinction in *On Free Choice of the Will*: "The freedom to choose in a way...not determined by anything outside my control is...*metaphysical* freedom.... We are capable of making a real difference in the world. We do not merely play the parts that the laws of nature and the past history of the universe have written for us; to some extent at least, we write our own scripts."[49]

If Augustine was correct in his ascription of "metaphysical freedom" to acts of human willfulness, then we have at once a second piece of evidence unifying *transcendental* with *continuity* with *self-consciousness* according to Fichte's *Foundations of Transcendental Philosophy*: "Willing concentrates; to this extent it is a *pure* power of the intellect. Our will is also supposed to possess causal power in the world of appearances, however, where it appears as the I's progress through a series of feelings: from the state in which the I began when it willed something, to the state *that* it willed, these constitute a continuous series."[50] Fichte's passage conjoined human will and human self-consciousness to causality. If causality implies continuity originating in the transcendental, then self-consciousness must be associated in the same way.

Hume's formula for causal sequences, in conjunction with the possibility of confusion between determinism and predeterminism, both presuppose a single cause can and must lead to a single effect. G. E. Moore's elaboration of the dilemma in *Principia Ethica* was clearly dependent upon this premise: "It is assumed that absolutely everything that happens has a *cause* in what precedes it.... This is to say that it follows *necessarily* from something that preceded it...so that, if *everything* has a cause, *nothing* ever could have happened except what did happen.... Those who hold...we *have* Free Will, think themselves bound to maintain that acts of will sometimes have *no* cause; and those who hold that everything is caused think that this proves completely that we have not Free Will."[51]

Life is more complex than this sequence suggests. A single cause can lead to a multitude of effects: consider the numerous ramifications of the destruction of the World Trade Center in New York City. On the other hand, a diverse array of causes can, in certain circumstances, often adduce a single resultant: ponder all the different factors culminating with the assassination of Austrian Archduke Francis Ferdinand precipitating the outbreak of World War One. Similar illustrations can be generated from any other arena. Any *single* path chosen in response to the initiating cause(s) can be selected in a completely controlled manner, or by such an arbitrary process as to liken the outcome to a random number generator program. It is this highly fluxional element present in all situations that allows for the introduction of novelty, of creative newness, into the universe.

Truth, self-consciousness, number, space, energy, time, and causality all display transcendental character in their capability for unifying the manifold discreteness readily apparent in manifestation with the continuity that is the necessary essence of metaphysical entities. The danger exists that a critic could claim a plurality is being entertained; that these constructs are discrete and innately separate. What remains to be shown is how all of them are representative of a singular continuity.

The key to the needed demonstration is to consider all of the thematic ideas listed and select, as the one most in need of explanatory harmonization, the notion seemingly most discordant with the others. Energy, space, and time are connected: concepts with practical consequences making physical existence experientially possible. Truth, number, and causality are also kindred: umbrella terms with theoretical implications covering a vast domain making abstract knowledge potentially conceivable. Given experience as both the source and unifier of comprehension—action coupled to thought are the hallmark characteristics of living entities—it is reasonable to suggest that in the phenomenal universe self-consciousness is the connective link, the *metaphysical* agency unifying *ethical-moral* practice with *epistemological* theory for all sentient beings. It is *prima facie* plausible self-consciousness might be the cornerstone to which the rest adhere.

Modern philosophy is frequently alleged to have begun with the writings of René Descartes. The needed demonstration must begin with one of his theories. *Meditations Concerning First Philosophy* was the publication first containing his famous division into mind substance, *res cogitans*, and body substance, *res extensa*:

> From the very fact that I know with certainty that I exist, and that I find that nothing else belongs to my nature or essence except that I am a thinking being, I readily conclude...my essence consists solely in being a body which thinks. And although perhaps, or rather certainly...I have a body with which I am very closely united, nevertheless, since on the one hand I have a clear and very distinct idea of myself insofar as I am only a thinking and not an extended being, and since on the other hand I have a distinct idea of body insofar as it is only an extended being which does not think, it is certain that this "I" is entirely distinct from my body and that it can exist without it.[52]

The problem he instigated, of course, was if they were entirely separate, there could not be any linkage by way of actual experience; hence, there would be no viable means to ascertain causal connectivity for distinguishable events. Self-consciousness must have jurisdiction in both transcendent and immanent domains. This is a declaration previously claimed, but what does it mean? *Self-consciousness is not solely a designator for phenomenal human being; it is the connective element uniting phenomenon with noumenon.* Self-consciousness is an aspect of the wholly transcendent, a subsidiary element of Consciousness. Put another way, "self" in self-consciousness is immanent; "Consciousness" in self-consciousness is transcendent. It is Consciousness that renders the combination "pure, original, and unchangeable," to employ Kant's description.

The phenomenological and linguistic dissection of substantial being into self and Consciousness serves as an explanatory resolution to the semantic difficulty previously articulated by Fiona Steinkamp: "If the transcendental self is the essential self and if it remains constant whereas the self it observes does not, there must be a sharp distinction between the transcendental self and the self that it observes." If the reference is to a singular entity, there can be no "sharp distinction." The union of transcendence with immanence, the constant with the fluxional, is tantamount to acceptance of the significance of the helix within manifest reality. A helix allows for constancy at the root while allowing for diversity in the outward expression. Just so does self-consciousness admit a multiplicity of individual selves simultaneous with the singularity of internal essence. To repeat: "It is the middle way, the path of least resistance, never leaning too far in one direction or the other but always finding perfect balance."

This chapter began with the assertion that human beings are faced with a dichotomy: we seem to experience reality as continuous, yet through analysis of the components seeming to comprise such a reality we encounter limitless discreteness. The thesis of self-consciousness as composed of a discrete self indissolubly linked to a continuous Consciousness would appear to resolve the problem. In contradistinction to Protagoras, *man is not the measure of all things*; *Consciousness is.* We are innately physicochemically constructed to be able to discern only discreteness in natural phenomena, yet we have an unquenchable desire to correlate all such instances in order to conceive of an underlying continuity *because* our discrete self is inalterably connected to the continuity of Consciousness that is not only coextensive with the manifest universe, but is the True transcendental agency serving as origin and support of this physical material plane of existence.

In the previous chapter, the Principle of Psychological Continuity was invoked to counter the claim of Morgan's Canon asserting humanity as the unique species on this planet in possession of mental states whose underlying substrate is self-consciousness. The invocation of some such Principle must be made because an absolute epistemological either/or has been reached: either physical reality is actually innately continuous or it is unbridgeable in its discreteness.

Over a century prior to Lloyd-Morgan's proclamation—and a century before Charles Darwin's theory of evolution—J. B. Robinet advanced what was essentially the notion of the Principle of Psychological Continuity in *De La Nature*: "What continuity can there be between the organic and the inorganic, between the animate and the inanimate, between the rational and the nonrational?... If we wish to allow Nature to pass insensibly from one of her productions to another, without compelling her to make leaps, we must not admit the existence of any inorganic beings, or any inanimate, or any nonrational."[53]

Discreteness, invariably and inevitably, leads to radical solipsism: there is no forest, only trees; no people, only separate arms and legs; no cells, only atoms; no atoms, only subatomic components; no words, only letters; and so on and so on. Radical solipsism, the taking of the discrete as normative, goes counter to a simple unassailable fact: things interact. The universe as we know it is a magnificent exemplar of relationship. Radical skepticism, radical solipsism, and radical discreteness are kin in that none can be consistently maintained. In fact, if Ouspensky was correct that our laws of logic reflect our perceptual ability, the very notion of consistency in thought demands continuity, some indefinable "something" joining one idea to another. We *expect* ideas to be connected because we envision connectedness around us.

Kant accepted the legitimacy of our *a priori* insistence on connectedness within manifestation: "The unity of the universe, in which all phenomena are supposed to be connected, is...a mere deduction of the...principle of the communion of all substances as coexistent."[54] The union of ideas cannot be performed by another idea; in conjunction with a central theme presented in chapter one, the union can only be accomplished by some meta-idea. The ultimate meta-idea must be transcendental in type, for only what is beyond perceptual and conceptual reality can unify what is within reality. Aurobindo conjoined the meta-ideas explored throughout this chapter, and emphasized their transcendental nature through his usage of Capital Letters, in *The Life Divine*: "Consciousness and Force are the twin essential aspects of the pure Power of existence; Knowledge and Will must therefore be the form which that Power takes in creating a world of relations in the extension of Time and Space."[55]

The ultimate meta-idea is Consciousness. All sentient species participate in that meta-idea through the formation of self-consciousness. The Principle of Psychological Continuity acknowledges the relationship, and is subsidiary to a more general construct, the *Principle of Co-Inherence*. The meaning of this term was given by Bernard McGinn in his essay, "Ibn Gabirol: The Sage Among the Schoolmen": "The principle of co-inherence means not only that the forms imitate the *factor primus* in diffusing themselves and their powers insofar as they are able, but also that everything in the lower forms must be pre-contained in the higher, so that lower forms exemplify higher ones."[56]

The second part of this citation—"lower forms exemplify higher ones"—concisely articulates the view that Psychological Continuity—"For any given psychological state, act, or function, P...there is...some psychological state, act, or function, P', and P' is a model of P"—is properly considered a subset of Co-Inherence. The first portion is even more significant in the present context. It offers a mechanistic basis for transitioning from a transcendent source, the "prime factor," to an immanent receiver, and effectively closes any conceivable ontological gap. What is "the *factor primus*"? What are the particular details of "imitation" needed in the expression of a mechanism connecting transcendence with immanence?

For monotheists in Western civilizations, the *factor primus* generically assumes the cognomen God or Allah; for substantial portions of the world's population in the Eastern Hemisphere, it is often designated Brahman or Tao or The Great Ultimate, among numerous other terms. To repeat: "Ideas such as God, Truth, and Good have been proposed as absolutes"; and, "Self-consciousness is...inalienably linked to all three of these constructs." The connection has become transparent: God, Tao, Brahman, Truth, or any other similar designation, *is* Consciousness.[57]

With respect to imitation, the first and best illustration of the Principle of Co-Inherence is self-consciousness. In general, this epistemological concept has received substantial support in the religious literature, which is not surprising since theology is the primary discipline most interested in offering substantiation for any alleged relationship between a transcendental agency and a subordinate, contingent physical world. David Buchman wrote in his introduction to al-Ghazālī's *The Niche of Lights*:

> An important concept...is that all things in the visible world are similitudes for things that exist in the unseen worlds that stand above it in the hierarchical cosmos. There is a qualitative correspondence between all things that exist in these numerous worlds.... The qualities of all things in the visible world provide knowledge of things in the unseen world and *vice versa*. By ascending to the unseen world, one can obtain knowledge of things in the visible world...by meditating on the qualities of things in the visible world, one can gain knowledge of things in the unseen world.[58]

Muhyi al-Din ibn al-'Arabi's *Meccan Openings* mimicked al-Ghazālī's proposition.[59] The central document of Jewish Kabbalah, the *Zohar*, offered a parallel postulate phrased in its own theological language.[60] In *A Treatise on Cosmic Fire*, the mystically-oriented Theosophist Alice Bailey rendered the principle of co-inherence quite simply as the dictum, "that which is above is as that which is below and that which is below is as that which is above."[61]

This is an extraordinarily serviceable epistemological tool; it allows one to claim all sorts of analogies between the immanent and the transcendent for which we can subsequently seek suitable naturally occurring evidence. In this manner, we can come to learn and understand a great deal about that which is beyond our commonplace experience. Moreover, we can establish confidence in the validity of conclusions in seeing a principle of consistent continuity comprehensively employed in development of far-reaching meta-theories.

If self-consciousness is going to be the principal connective link between transcendent Consciousness and an immanent individual self, then extensive investigation into its properties is not merely warranted, but is absolutely necessary. Researchers—primarily philosophers, psychologists, neurologists, and artificial intelligence specialists—tend to divide themselves into three schools of thought. According to John Beloff's "Body and Soul: Another Look at the Mind-Brain Problem," these are: "One can be a materialist and hold that the world consists exclusively of matter in space. One can be an idealist...and hold that mind alone exists. Or one can be...a dualist and acknowledge the existence of both mind and matter."[62]

The first group—inclusive of the majority of contemporary physical and biological scientists as well as artificial intelligence experts—do not believe in the actuality of self-consciousness and consider it either nonexistent in any unique sense, which is the *eliminativist* position, or a derivative of organic brain function, the *epiphenomenalist* approach. The second class of thinkers is comprised of philosophers advocating a no-longer-viable remnant of the scientifically naïve philosophy of Idealism, perhaps best exemplified by the approach of Bishop George Berkeley. The last category of investigators—a distinct minority of physical and biological scientists, but the vast majority of theologians—accepts the independent existence of self-consciousness. This latter group typically—though not always, as has been documented in the last chapter—believe human beings are the elite class of entities exhibiting self-consciousness.

An example of materialist thought can be found in Güven Güzeldere's introduction to the compendium of papers, *The Nature of Consciousness: Philosophical Debates*. Güzeldere credited biologist Thomas Henry Huxley as originator of the doctrine that came to be called *epiphenomenalism* or *reductionism*: "In Huxley, consciousness plays no contributory role in the causal chains that take place in the nervous systems that totally determine the behavior of an organism; it only gets affected by the neural interactions."[63] Berkeley's *Three Dialogues Between Hylas and Philonius* presented a clear statement of the idealist view in the second dialogue: "Sensible things cannot exist otherwise than in a mind or spirit. Whence I conclude, not that they have no real existence, but that seeing they depend not on my thought, and have an existence distinct from being perceived by me, *there must be some other mind wherein they exist*."[64] Samuel Angus revealed in *The Mystery Religions*: "Dualism was introduced into Greek thought by Anaxagoras, who, in opposition to the preceding materialistic monism, segregated spirit and matter."[65] The two best-known champions of dualist thought in the history of philosophy were Plato and Descartes.

The fact of survival into modern times for two of these three alternate ways to conceive of the essential nature of self-consciousness, the materialist and the dualist, suggests resolution will not be a trivial issue. Yet, resolution can be had if these approaches are viewed in the context of discreteness *versus* continuity. The doctrine of epiphenomenalism, and the more extreme thesis of eliminativism, cannot evoke evidence for a plausible justification of connectivity. Being confined to a material outlook relegates such proponents to a Whiteheadian atomism where everything is a particular thing with no viable medium for bridging individual, independent units into a continuum. As has been emphasized throughout, if one starts with discreteness, one can only generate discreteness.

The materialist argument is completely subverted by adoption of notions like energy and force, neither of which is literally perceptible or comprised of particulate matter, although they clearly contribute to the nature of, and have effects on, physical entities.[66] Dualism also fails as a rationale for a necessary mechanism of connectivity: the premise of Cartesian dualism is separation, and one cannot unify separate things, whether wholly or partially material, through divergent concepts, namely spirit and matter. Materialism and dualism are doctrines constructed from the presumption of discreteness as basic. It would be logically impossible to reconcile either of the two with manifest reality when neither recognizes the legitimacy of continuity as the preferred orientation.

This analysis leads to a simple conclusion: *continuity is possible only if it is the fundamental characteristic of ontological origin.* Acceptance of this claim accounts for the potency of Ken Wilber's assertion in *The Spectrum of Consciousness*: "But since consciousness is Reality...it would be much more accurate to view consciousness not as relative subject confronting objects but as Absolute Subjectivity above the dualism of subject and object."[67] As will be repeatedly professed, self-consciousness is only the immanent analog to transcendental Consciousness, an approach allowing for both continuity and discreteness: a common universality among all species embedded within an acceptance for distinctions differentiating each type of creature on the level of the individual. Hegel posited this condition as fact in *Phenomenology of Spirit*:

> For that universality which does not let itself advance to the reality of an organic articulation, and whose aim is to maintain itself in an unbroken continuity...creates a distinction within itself, because it is movement or consciousness in general. And, moreover, by virtue of its own abstraction, it divides itself into extremes equally abstract, into a simple, inflexible cold universality, and into the discrete, absolute hard rigidity and self-willed atomism of actual self-consciousness.[68]

Our bodies are *a priori* structurally conditioned to degrade the continuity of electromagnetic radiation around us into discrete units of sight and sound. Other species break up this continuity differently, which explains why dogs can hear pitches of sound inaudible to us, or why bats employ sonar instead of vision as their principal means for direction-finding. Self-consciousness of the human variety is a discrete form of the continuity of Consciousness. A different "cut" of the continuity of universal Consciousness enables other species to possess a functionally different form of self-consciousness. The integration of self-consciousness over all physical space *is* that universal Consciousness. Using a mathematical formalism seen before:

$$\int_{\text{all sentient life}}^{\text{3-D space}} \text{self-consciousness} = \text{Consciousness} = \text{God}$$

This equation makes plausible the attribution of omniscience. If God is the same as universal Consciousness, then it necessarily contains complete awareness of all possible information over the entire sensory range of all potential experience. When we have a thought concerning some phenomenal condition, it is because we have taken a slice of continuous information flow into our discrete brain states. Alice Bailey assigned the "omniscient soul" as the source of all intuition in *From Intellect to Intuition*:

> What is this mysterious thing we call the intuition?... We might define it as direct apprehension of truth, apart from the reasoning faculty or from any process of intellection. It is the emergence into the consciousness of some truth or beauty never before sensed. It does not emerge from the subconscious, or from the stored-up memory...but drops into the mind directly from the superconscious, or from the omniscient soul. It is immediately recognized as infallibly true and arouses no questioning.[69]

Bailey's characterization of intuition as both a specific product of, as well as a process distinct from, typical rational thought reaffirms the thought of Descartes in *Rules for the Direction of Mind*: "Two things are requisite for intuition: namely, that the proposition be understood clearly and distinctly, and...be perceived at one glance and not successively.... Deduction...does not seem to be accomplished all at once, but...involves a certain movement of our mind in inferring one thing from another."[70] Bailey's error was her belief that intuition occurs "apart from the reasoning faculty." If this were true, then there would be a discontinuity in process. If continuous, however, then our intellectual faculty must accept and employ the introduction of an intuition at some point. Duns Scotus "identifies a second act of cognition, belonging both to the senses and to the intellect. This act is immediate, with no representational species to mediate the mind's encounter with the object. Scotus calls it intuitive cognition, and explains that it is...an immediate awareness of an object in its entirety as present and existing."[71]

If it is the case that Consciousness is the ontological origin of continuity, how does this construct unify the other transcendent/immanent pair notions already encountered—Truth/truth, Number/number, Space/space, Time/time, Energy/energy, and Causality/causality? Consciousness *is* Space, not merely coextensive with it, while physical space is its existential analog. Time is that which happens to Space on the non-manifest level of reality, while time is, according to construal of the principal of co-inherence, the means employed to relate the progressive unfolding of events in space. The radiant energy permeating the universe, even that inhering in substantial bodies, is a this-worldly image of transcendental Energy, which is still another term kin to God or Space in being understood as absolutely identical to Consciousness. Causality and number within the existential dimension serve as means to collate one self-consciousness event to another, just as their transcendental counterparts do the same in the metaphysical realm. The accumulation of all such events within the phenomenal universe is regarded as truth, while Truth denotes the cosmogonic coherency of ontological and teleological events.

An optimally coherent epistemological system must be a metatheory; it should unify noumenon with phenomenon consistently and comprehensively. To accomplish this, bridging principles must be defined and characterized. One would like to incorporate these principles as much as possible into quantifiable laws, since mathematics has its roots in both transcendental and immanent realms. Over the last century and a half, *psychophysics* has been used to connote the merger of mental properties with physical properties. Invented by Gustav Theodor Fechner, his *Elements of Psychophysics* serves as the forerunner for any modern attempt at such formulations: "The physical world operates under one law and we must assume that the spiritual world is no less so. There must be...*a priori*, some exact mathematical relationship between the physical and the psychical, some law of concomitant variations, for all that is psychic is but the self-appearance of the physical; a material process runs parallel to every conscious process."[72]

In *The Conscious Mind*, David Chalmers explained the boundaries of the relationship hypothetically existing among phenomena, noumena, and the laws of operation for both: "Certain phenomenal properties will...be taken as *basic* properties.... Perhaps there is some...class of...fundamental properties from which phenomenal properties are derived.... Where we have new fundamental properties, we also have new fundamental laws.... The...laws will be *psychophysical* laws."[73]

Within chemistry, the interactive subdisciplines of thermodynamics and kinetics serve as an appropriate scientific model for the role envisaged for the elaboration of psychophysical laws. Kinetic theory embodies continuity through its postulation of a dynamically continuous exchange of matter and energy over the coordinates of spacetime among a plurality of molecules in a system. Meanwhile, thermodynamics employs the notion of discreteness as applied to spatiotemporal energetic states of being through its focus on a unified system establishing a stable equilibrium state. Where kinetics treats the many by appeal to continuous, unified change, thermodynamics applies to the ensemble collection of molecules through employment of discrete parameters characterizing the system as a whole.

This methodology provides another illustration of the construct of "each inhering in each": the multitude relates to continuity, typically conceived as the realm of singularity; the collective system is treated as discrete, contrary to its usual mode of conceptualization. Despite an apparent divergence of approach, the two disciplines converge: the arithmetic ratio of forward and reverse reaction kinetic rate constants equals the thermodynamic equilibrium constant. In formulaic symbolism: $K_{eq} = k_{forward}/k_{reverse}$. It is possible suitable psychophysical laws will display analogously similar capabilities in unifying the transcendent with the immanent.

Chapter Four: Metaphysics

> We must first form a conception of the existence of the Creator according to our capabilities; that is, we must have a knowledge of Metaphysics. But this discipline can only be approached after the study of Physics; for the science of Physics borders on Metaphysics, and must precede it in the course of our studies, as is clear to all who are familiar with these questions.
>
> —Moses Maimonides, *The Guide for the Perplexed*

Maimonides was right in two respects: first, metaphysics and physics are contiguous disciplines for investigation just as their nomenclature suggests; second, in the acquisition of information leading to knowledge it is more efficient to obtain sufficient accumulations of physical facts about the world, both with respect to ourselves and our environment, before attempting to develop any meaningful insight into metaphysical actuality.[1] Indeed, it might be convincingly argued that *the* principal weakness in the thoughts of those engaged in metaphysical speculation has been that most of the speculators were inadequately versed in physics. In part, this situation was circumstantial in nature. Until recently—roughly commencing with the second half of the twentieth century—the state of physics was such that experimentalists and theoreticians had neither the instrumental nor mathematical tools available to render them able to compile and analyze sufficient data about the universe and its inhabitants to construct coherent theories.

Of equal significance in the failure of most metaphysical speculations to be thoroughly convincing, even into the present era, was the historical irony that the birth of the scientific revolution also served to introduce the age of specialization: philosophy and science underwent a trial separation lasting over three centuries prior to beginning to take steps in reconciliation. Hence, metaphysicians were generally unaware of the advances in physics and knew nothing about how to utilize for their own ends the knowledge gained. In sum, before the seventeenth century, students of physics and metaphysics had inadequate information; since that time, physicists have established the information base, but metaphysicians have not been paying attention.

Nonetheless, epistemology must not be confused with ontology. To do so is to employ the wholly erroneous anthropocentric presupposition. What we learn first should not be thought to be what comes first. What comes first is ontologically prior to, and absolutely independent of, the epistemological process by which we arrive at comprehension. What we learn first, we learn first relative to its practical convenience for us in daily living. Consequently, though physics truly precedes metaphysics "in the course of our studies," any exposition of a system of thought begins with metaphysics precisely because it is an ontological *a priori* with respect to physics. Leibniz admitted in his *Discourse on Metaphysics*: "It appears more and more clear that although all the...phenomena of nature can be explained mathematically or mechanically by those who understand them...the general principles of corporeal nature and even of mechanics are metaphysical rather than geometric, and belong rather to certain indivisible forms or natures as the causes of the appearances, than to the corporeal mass or to extension."[2]

Why should this be the case? The previous chapter demonstrated several prominent physical concepts—among them space, time, energy, force—that are incapable of self-generating the attribute of continuity. They derive their character of infinite application exclusively from the transcendental realm. Hence, the notion that metaphysics, exploration of the transcendental, must ontologically precede physics, investigation of the immanent, is tantamount to the allegation of continuity as an ontological precursor to discreteness. This is a *necessary* Truth: the accumulation of discrete entities never suffices to produce a continuous entity, whereas the parsing of a whole into parts is both theoretically and practically unproblematic.

James Wayne Dye was correct in his assessment of the consequences for philosophy in recognizing metaphysics as central to all forms of theorizing: "Metaphysical systems disagree about how the most general kinds of things are to be ordered with respect to each other; they are...disagreements about everything."[3] Given this, it should come as no surprise that the system to be promulgated is different from others proposed throughout the lengthy history of philosophy.

It is impossible to claim any particular speculative metaphysical system as capable of being *proven* to be True given humanity's current depth and breadth of knowledge. This error of attribution was a problem adhering to most previous philosophic expositions having a metaphysical focus. Each developer boldly proclaimed his version of ontology and teleology—whether consisting of an elaboration of a partisan experience-based religious platform or deriving from a neutral abstract philosophical vision—was the absolute Truth about the nature of reality without likelihood of error.

All metaphysical propositions were and are innately speculative, as is the one promulgated in these pages, for the simple fact that we are incapable of perceiving, even with the aid of instruments, what occurred prior to our existence in manifestation. Nor are we capable of comprehending the nature of transcendent reality in all aspects given our evolutionary developmental status. Nor can we accurately and precisely predict what teleological path will unfold in the future either cosmologically within physical spacetime or cosmogonically in some postulated transcendent dimension of existence. The best that can be claimed for any metaphysical system is its internal coherence and compatibility with all known facts, even those expressly contingent. As Saadia Gaon accurately stated in *The Book of Beliefs and Opinions*:

> If we seek to establish the truth in the domain of knowledge obtained by logical inference, we must...make certain (a) that there is no other means than the theory in question of sustaining the truth of what is perceived with the senses, nor (b) any other method of upholding what is intuitively apprehended by reason. Furthermore (c) it must not invalidate any other accepted fact, nor (d) must one part of it contradict another, let alone (e) that a theory be adopted that is worse than the one that has been rejected.[4]

Despite the lack of guarantee for absolute invariant Truth content, and consistent with the doctrine of syncretism—which acknowledges value in abstracting viable elements from otherwise disjunctive theses and coalescing them into a coherent self-consistent model—most aspects of the speculation contained herein will be echoed, each in its own unique manner, by virtually all previous metaphysical expositions. As mentioned in the preface to this volume, it is grossly inefficient to restate in exhaustive detail the wealth of documentation and subsequent conclusions contained in *The Other Perennial Philosophy: A Metaphysical Dialectic* (*TOPP*). Enough of a review must be presented, however, both to set the groundwork for forthcoming chapters, and so that only the deeply committed reader need consult the previous book for supporting evidence. Hence, what follows is a reconstruction of only the minimal basis set of data needed to proceed supplemented by an analytical enhancement of the justification underlying the developed theory.

A detailed analysis of skepticism was made in the first chapter. It was concluded that radical skepticism is impossible in execution however much it might be idealized as a potent philosophical technique. It was acknowledged that presuppositions must be initially employed to begin any discussion on a theme, since a beginning point and the grounds necessary for starting with that particular postulate are always arbitrarily determined. This is not to say either the presuppositions or the point of entry are immune to investigation. On the contrary, they must themselves be analyzed using other elements of the field of enquiry in order to unequivocally establish self-consistency of argument. If a discussion involves aspects 'A,' 'B,' and 'C,' a proponent of any thesis must eventually relate each to each regardless of where one starts.

Philosophy, particularly metaphysical speculation, contains examples of both inappropriate extremes. There were thinkers who believed the starting point for investigation must begin from rigorously provable and absolutely unimpeachable first principles: philosophers such as René Descartes who insisted everything must be "clear and distinct." There were others who believed their starting point, though innately relativistic and arbitrary, could not or should not be questioned, such as Edmund Husserl, who asserted without adequate justification in *Logical Investigations*: "The empirical ego is intentionally constituted out of the phenomenological ego."[5]

In the present context, the propriety of logic will be assumed as the operative background for discussion. This appears justifiable: for humans, knowledge requires rational thought connecting discrete pieces of data and employs a human-derived logic as mechanism of process in order to arrive at proper means of unification.[6] There is no stipulation here that what is true for us must be true for all forms of life, hence the qualifications, "for humans" and "human-derived logic." The anthropocentric presupposition is rejected outright. It is easily possible to imagine other intelligent species learn—and learn to a degree of comprehensiveness and self-consistent validity equal or superior to ours—without undergoing modes of thought *we* characterize as "logical." Chapter three offered numerous examples in which discreteness was transformable into continuity through application of some transcendental principle serving as the effective agent. It is proposed that logic is one of these meta-principles. Human logic may be peculiar in its methodological specifics to *Homo sapiens*; logic in the abstract is a transcendental property of Truth.

Using this powerful tool allows us to start with some characterization of the ontological source of the universe. The first Kantian Antinomy contained in the *Critique of Pure Reason* is famous for expressing the initial either/or many metaphysical theorists have decided must be the starting point for all speculative discussion: "Thesis: The world has a beginning in time, and is limited also with regard to space; Antithesis: The world has no beginning and no limits in space, but is infinite, in respect both to time and space."[7] For the sake of convenience, an interpolation will be made: by "world" was not meant "the planet Earth," but rather "the universe" or "the cosmos in its entirety," or "the physical existential dimension." Severe restriction to the literalness of "world" would result in a trivial, unambiguous resolution: Earth has a definite origin in time and space.

Given the expansion in interpretation, Kant was wrong in two respects: first, he made an illogical leap from the outset in assuming the properties of space and time were identical and had to be treated as an inseparable pair; second, he unnecessarily conjoined origins with limits. If space and time are uncoupled and considered separately, possibilities emerge other than those considered by Kant and most other metaphysicians.

What would it mean for space to have limits? Consider the vastness of space: the galaxies astronomers have catalogued; those not charted because our instrumentation is not sufficiently powerful; those not even in principle observable because they are farther away than the distance light can travel in the relative vacuum of space in a given period of time; and, those having dimensionality other than the three we are equipped to detect. The almost incomprehensible size of the universe has led astronomers and cosmologists to articulate the *horizon problem*: "As we look farther and farther away...we reach distances for which the look-back time is comparable to the age of the universe, fifteen billion years. Beyond such distances...we clearly cannot see anything. These distances establish our cosmological horizon. This is not to say that regions beyond the horizon do not exist."[8]

What would, or could, be outside space such that a distinct boundary is a conceptual possibility? To say that beyond the limit is something outside our conception is to say nothing intelligible. To say it is nothingness also evades the issue, for even the philosophers of religion, whether Eastern or Western in outlook, admit the ultimate Void, *Śūnyā*, has a direct connection to Being, as has been mentioned in a prior chapter.[9] It is an inescapable conclusion: a limit to space in its entirety would demand something was definable as being external to space. Would not that something also occupy an extended space, or be imagined co-extensive with that space? Kant was right to regard space as an epistemological *a priori* for the human reasoning process. That same inherent *a priori*, however, means it would be illogical, literally against the human definition of logic, to consider universal space as limited.[10] *Infinite* space automatically alludes to the transcendental, and this fact constitutes yet another justification for metaphysics to ontologically precede physics.

This conclusion is *not contrary* to astrophysical claims that our universe is currently undergoing a cycle of expansion for two divergent reasons. First, what is generally meant by the notion of an "expanding universe" is that the density of objects is decreasing within the boundlessness of space, not that the absolute size of the universe is growing. In a limitless space, objects can move further away from each other. Since object density is determined by extrapolation from the ratio of the number of objects in a defined region of space to the expanse of that space, the density will decrease as stellar objects recede from each other.

Second, a universe capable of expanding in its dimensional vastness does not overturn the notion of spatial infinitude. To think otherwise is to assume the universe we inhabit is the *only* universe—the anthropocentric premise in disguise. There is no contradiction in conceiving our universe as expanding in concordance with another universe undergoing a simultaneous contraction equal in magnitude resulting in no overall change in total spatial expanse. These mathematical relationships can hold with consistency: (a) rate universe 'A' expands = rate universe 'B' contracts, or equivalently, rate of universe

'A' expansion = – rate of universe 'B' expansion; and, (b) universe 'A' + universe 'B' are cumulatively unbounded.

The number of universes can be limitlessly multiplied and the concept of spatial infinitude is not undercut. While the precise nature of the differences comprising these universes so as to keep them distinct and separate need not concern us yet, it is reasonable to speculate they would operate according to different physical laws. This distinction does not lead to an overall lack of continuity on a transcendental level for the supposition of different physical universes does not also demand supposition of unique transcendent domains as progenitors. What is necessary is each physical universe somehow serves as a mirror for the transcendental according to the "law" of similitudes.

Thinking about time does not yield the same conclusion with regard to its essential nature. Space can be imagined to be devoid of objects within the space; the absence of objects does not imply the absence of space. Time, however, does require objects for its intrinsic meaning. Time is not a stand-alone concept, it is a measure of change, a "from–to" alteration of something. This is a main reason space and time must be pondered as segregated ideas: the intrinsic absolute and independent nature of space is completely different from the innate relative and relational nature of time.

In consequence, both variations—time as eternal; time as possessing a specific beginning and ending—are imaginable: (a) whatever time measures now is in some manner related to whatever time previously referenced, and is what time will in the future continue to allude to in some manner, with this relational process eternal in its unfoldment; (b) time began, in the sense that everything before—not in the temporal sense of before, but used solely as a language descriptor needed for our innate tendency to use linear sequences to describe conditions; that is, an illustration of the Kantian *a priori* template effect—the beginning of time was static without undergoing a change of any type, and will cease its adjustments at the end of time, after which everything within phenomenal existence will again undergo a transformation to absolute stillness with no further alterations.

If space and time cannot be treated as an aggregate pair, then it becomes necessary to alter the first Kantian Antinomy to reflect this fact. Potentially feasible optional candidates are: (a) Space is unbounded, has always had, and will always have, objects within it undergoing ceaseless transformations, and therefore time and the objects it references are eternally coexistent within space; (b) Space is unbounded, originally had no objects within it, at some definite moment a unique event occurred generating objects and marking the beginning of time, which will now continue forevermore as a measure of eternal transitions in the status of objects in space; (c) Space is unbounded, originally had no objects within it, at some definite moment a unique event occurred generating objects and marking the beginning of time, and these objects will eventually depart from manifestation, leading to the end of time as a characteristic within a restored objectless space; (d) Space is unbounded, originally had no objects within it, at some definite moment a unique event occurred generating objects and marking the beginning of time, and these objects will eventually cease their dynamic alterations, resulting in the end of time as a characteristic within a space still filled with objects.

It is important to comprehend that the notions of ontology and teleology, philosophical concepts pertaining to those items having a beginning and an ending, here can refer only to time and individual objects within space. To claim universal space of a numerically fixed dimensionality—which can only be unbounded because it is impossible to conceive limits on extension—can have a beginning or ending is to confuse domains. It is also unreasonable to hypothesize the ability of a universe to alter its physical dimensionality. How could a two-dimensional universe transform into a three-dimensional one, for example? Invariant dimensionality must be an inherent property of any given existent physical space even if a transcendent domain can produce universes of different dimensional character according to the principle of co-inherence.

While any given object *within* space can have a point of both spatial and temporal origin, as well as of both spatial and temporal termination, the first possibility *declares any discussion about ontology and teleology with respect to space and time in the universal sense is moot,* a result discordant with the nature of metaphysical philosophy in general, since that discipline's entire domain of inquiry is circumscribed by those two terms.[11] Choice number two *admits an ontology for time, a singular starting point, but denies teleological closure*, while the third variant *allows for both an ontology and teleology of time as a whole*. The last option *agrees with the third in its acceptance of a temporal ontology and teleology*, but provides a different reason.

Designated variant (c) includes a temporal teleology due to permanent disappearance of objects for which time is the measure in characterizing their successive transformations; choice (d) eliminates time due to the absence of any need for measurement of alterations of the properties of objects, which have themselves not ceased to exist. Another way to phrase the variations is: the first refers to an eternal dynamic universe filled with objects; the second differs from the first in that the objects originate within existence; the third and forth reference a universe containing objects the composition of which both originated in the past and terminates, for one reason or another, in the future.

The preceding treatment is based upon an idea of physical existential space as absolute subject that cannot be removed with respect to time acting as relative measure of all objects within that space. This perspective can be designated a *two-level speculation*. It depicts the metaphysics of *materialism*. There is a corresponding *three-level* approach conveying the metaphysical scenario described by all theories of *representative dualism*: consideration of manifest universal space as an object for a transcendental nonphysical agent now conceived as the new absolute subject serving as the ontological frame of reference.

This perspective regards physical space as, simultaneously, subject and object. It is still possible for space to contain other objects within its expanse. As an object, it is now just as measurable by time as the objects within space are; space itself can have a possible beginning and ending. The main way in which space can change is not through the property of extension—space, whether absolute subject or relative simultaneous subject/object, must always be unbounded. It can change through its actual coming into manifestation in three dimensions. Other universes could come into manifestation exhibiting different dimensional properties. If guided by the principle of co-inherence, an occurrence of this kind is not impossible for one or more transcendental domains acting as agents for the transformation.

As a subject/object, the first transformation physical space can undergo is its own appearance as an object for a transcendental subject. Ontologically, this means space and time are inseparable in the sense assumed by Kant in his First Antinomy: time began when space came into being. More precisely, time is an intrinsic measure for the initial appearance of physical space. With this condition, a specific definable teleology becomes recognizable: physical space ceases to exist, and time disappears with it. This is not, however, the only possible outcome. Another is the continued eternal existence of space, and of time as its measure, while any actual objects within space continue to undergo unceasing transformation.

From this analysis, Kant was in error when he thought the notion of a transcendental absolute subject was hypothetical: "You never come face to face with anything unconditioned...to determine whether the unconditioned exists in an absolute beginning of the synthesis, or in an absolute totality of the series without any beginning."[12] He failed to understand the attribute of extension does not pertain to space in the same manner as the property of limitation may refer to time. The formulation in his First Antinomy *demands* prior institution of a three-level speculation where the origination of physical space becomes optional. The transcendental becomes an epistemological *a priori* for his consideration of an ontological beginning for space.

Kant's second misjudgment can be elucidated from another statement in the *Critique of Pure Reason* concerning the alleged absence of certainty to be attached to any proposition incorporating a transcendental subject:

> If the highest Being should stand itself in that chain of conditions, it would be a link in the series, and would, exactly like the lower links...require further investigation with regard to its own still higher cause. If, on the contrary, we mean to separate it from that chain, and as a purely intelligible Being, not comprehend it in the series of natural causes, what bridge is then open for reason to reach it, considering that all rules determining the transition from effect to cause...refer to nothing but possible experience, and...to objects of the world of sense...and are valid nowhere else?[13]

His presupposition that a transcendental source to phenomenal spacetime would itself require inquiry as to its origin—"highest Being...would, exactly like the lower links...require further investigation with regard to its own still higher cause"—was derived from an ignorance of mathematics. *Existence of an infinite series of terms does not preclude an ability to denominate a first term for the sequence.* Ontology and teleology need not be symmetrical. It is a confusion to think of an infinite series as a one-dimensional line extending endlessly such that neither first nor last terms can be precisely determined. More accurate is cognition of the series as beginning with a zero-dimensional point and extending limitlessly in a single direction. *Inclusion of a point of origin for countable terms represents the directionality in time used as the measure for the appearance of those terms within manifest reality.* The zero-dimensional point is by definition transcendental because, possessing zero-dimensions, it is a no-thing.

The teleological options in three-tiered metaphysical speculations are not confined to those cited by Kant in his Antinomy. The remaining outcomes outlined for a two-level speculation are still conceivable: (a) space does not disappear, although the objects within it do, causing space to become static and time to have a definite endpoint; (b) space does not disappear, neither do its objective contents, but those items fail to undergo further transformations resulting in time ceasing to serve as a meaningful measure and coming to an end.

There are significant similarities as well as distinctions between the two-level and three-level alternatives. The schemes have in common: (a) physical space is unbounded; (b) time measures changes in the nature or properties of objects; (c) time has a comprehensible ontology; (d) four mutually exclusive teleological choices are cognizable with respect to time and objects in space. As distinctions: (e) absolute transcendental subject(s) is/are posited only in the three-level case; (f) existential space becomes subject/object rather than subject only; (g) space has an ontology, and may have a teleology, in one depiction, but not in the other.

A choice between the two-level and three-level approaches can be made if we consider the influence of two postulates developed in the last chapter. First, the transition from discreteness to continuity for objects and processes within manifestation requires the introduction of a transcendental construct unifying the manifold from a meta-perspective. As Jean-Paul Sartre wrote in *Being and Nothingness*: "The *unity* of change and the permanent is necessary for the constitution of change... A unity of being amounts to requiring that the permanent *be* that which changes."[14] Second, noumena are the model for phenomena, or as the *Zohar* succinctly expressed the notion: "The process of creation, too, has taken place on two planes, one above and one below, and for this reason the Torah begins with the letter 'Beth,' the numerical value of which is two."[15]

In combination, these factors suggest we opt for the three-level approach in which some conception of transcendental Space and Time serves as the source from which immanent space and time acquire analogous attributes. Since time is a relative measure of the transformations of objects irrespective of whether the object is space or discernible things in space, it is concluded Time is relative to absolute Space on the transcendental level also. How can Space, unbounded as much as its existential analog is, be characterized? What would be the nature of its Being?

Three ideas contained in the third chapter can be used to answer these questions definitively. First, self-consciousness was demonstrated to be able to unify concepts of space, time, energy, number, truth, and causality through its ability to express the relationship between abstract thought and practical experience. Second, the needed transcendent precursor to self-consciousness is Consciousness, which bridges the manifold of individual selves and turns them into a transcendentally originated continuity. Third, the aforementioned relationship between noumenon and phenomenon allows establishment of a reliable metaphysical and epistemological model built on the premise of "as above, so below." If Ockham's razor is applied, it is *prima facie* reasonable to posit Consciousness as the agency collectively unifying the transcendental analogs. In other words, Consciousness unifies within itself the notions of Space, Time, Energy, Number, Truth, and Causality.

This idea receives partial support through reinterpretation of a forceful declarative expression offered by Rudolf Otto in *Mysticism East and West*: "Being...has its existence, and as existing it is consciousness. Consciousness is at the same time the only possible form and mode of existence of Being."[16] Although Otto may have meant self-consciousness rather than Consciousness because "existence" is often equated in meaning with "manifestation," his statement—especially with the capitalized "Being"—is more appropriately susceptible to a *supra*natural interpretation. Further evidential support for the conception of Consciousness as appropriately identifiable with transcendent Being can be uncovered from an array of philosophic and religious sources. One is Julius Lipner's recounting in *The Face of Truth* of the philosophy of

Rāmanujā, who avowed: "From the transcendent point of view, the absolute Self...*is* Consciousness, and the distinction between individual knower, object known, and act of knowing ceases to apply."[17]

Both Eastern and Western metaphysical philosophy have depicted an unlimited transcendental Being as God although the specific word employed as the referent may be unique to each language and culture. An example of this tendency to merge the notions of God, Being, and Consciousness can be found in Rajmani Tigunait's *Seven Systems of Indian Philosophy*: "Vedāntins view reality as Supreme Consciousness, which is simultaneously involved in the phenomenal world and transcendent. To avoid confusion, Vedāntins use the term 'Brahman' to denote the transcendent aspect, and the word 'Isvara' to denote that aspect of...Reality that oversees the phenomenal world."[18] This statement's next sentence is especially significant: "All the schools of Vedānta philosophy agree that the existence of God cannot be proved or known by logic alone; God's existence can be established only through direct experience or by testimony."

Various thinkers have argued for God's existence as a necessary reality, while others have contended for an inherent inability to prove or disprove the Truth value of such a proposition.[19] Thus, the perspective of the Vedāntins offsets contrary propositions, like the one uttered by Hans Küng in *Does God Exist?*: "Scientific proofs of God...are not possible. Since they are dependent on intuition, no scientific judgment can be made about God, who is not in space and time and thus not an object of intuition."[20] Vedāntins and other theorists argue in rebuttal to Küng that profession of a mystical connection to a transcendental agency, independent of that agency's description as personal or impersonal, cannot be discounted through stipulation of the unconvincing nature of an experience-based report, or rejected automatically as *prima facie* impossible.

Since some concepts of transcendental character are necessary, one of these is Space, and Space must be unbounded, it is reasonable to allege as a first approximation that God is to be identified with Space. Under the logical and mathematical rule, "things equal to the same thing are equal to each other," the conclusion must be that Consciousness, the agency proposed to unify the other transcendental constructs, is identical to that idea denoted as God. Thus, *God is Consciousness, the agency in which Space, Time, Energy, Number, Truth, and Causality all coalesce.*

This definition of God might be considered heretical by the three main monotheistic religious traditions because Western theologians are faced with a dilemma pertaining to God: Supreme Being as transcendent absolute *versus* deity as immanent relative. Judaism, Catholicism, and Islam are unanimous in representing God as Creator, necessarily relative to that made manifest, while simultaneously adhering to the dogma that God is intrinsically Other than creation, absolute beyond comparison.

By way of contrast, in Eastern religious philosophy God is frequently referenced as Tao, Brahman, or the Great Ultimate, an impersonal absolute without a direct role in the formation of phenomenal reality. While these terms are united in describing the absolute as the foundation of that which exists, they are equally committed to the prospect of needing an intermediary to connect that ontologically distant absolute with the conditional production of physically detectable objects: the position occupied by Brahma or Isvara within Indian religious doctrine, for example.[21]

Western monotheism has acknowledged the problem by its occasional use of the term "Godhead" as the absolute source from which God is the link to manifest reality. In such situations, theology in the West has followed its Eastern counterpart. Nikolai Berdyaev offered a plausible though not entirely accurate description of the conceptual relationship between Godhead and God in *The Beginning and the End*:

> The Absolute is that which is separate and self-sufficient; there is in the Absolute no relation to any other. In this sense God is not the Absolute, the Absolute cannot be the Creator, and knows no relation to anything else.... God is the Relative, because God has a relation to his other, that is to say to man and to the world.... The Absolute is the unique, and the thinking mind can assert this of the Godhead, though it says it very poorly.[22]

Berdyaev's deficiency follows from failure to resolve the issue through his insistence that "there is in the Absolute no relation to any other." The construct of "absolute" does not demand complete isolation from everything else. If Berdyaev were correct, there would be an ontological gap between Godhead and God. More precisely stated, an *absolute* is the starting point through which all relation is introduced. With this emendation in place, it is noteworthy that others have echoed Berdyaev in expressing this particular solution to the problem of God's transcendental nature.[23]

If Godhead is accepted as the absolute, what is the role of God in the proposed metaphysical scheme? God *is* the universe: "God does not operate on the world from above or from outside as an unmoved mover, but from within as the dynamic, most real reality in the process of the evolution of the world, which He makes possible, directs, and completes.... He is source, center, and goal of the world process."[24] *TOPP* made a similar statement in the form of an aphorism: "God is Consciousness within existence."[25] If God is Creator, then the term "God" serves best as an existential concept unifying the manifold diversity of creation into a unity in precisely the same manner as self-consciousness conjoins the otherwise discrete components of a single entity into a functional whole. Sarvepelli Radhakrishnan, in *An Idealist View of Life*, supported the idea of God as a link in the chain of Being connecting the immanent to the wholly transcendental:

> If God is bound up with the world, subject to the category of time, if his work is limited by the freedom of man and the conditions of existence, however infinite he may be in the quality of his life, in power, knowledge, and righteousness, he is but an expression of the Absolute.... While the Absolute is pure consciousness and pure freedom and infinite possibility, it appears to be God from the point of view of the one specific possibility, which has become actualized.[26]

Acceptance of this formulation implies the transcendental equation must be modified slightly: *Godhead* is Consciousness, the agency in which Space, Time, Energy, Number, Truth, and Causality all coalesce. Consistent with this formula, *Consciousness* can best be likened to *a homogeneous, omnidirectional field of pure Energy coincident with its Space*. This Energy has an immanent analog, energy, when transformed into the manifest physical reality we know as defined by time and space. As long as bounded finitude is not automatically understood to be an appropriate characterization of Space, Dimensionality can be considered one of its core traits. The Dimensional Space of transcendental Consciousness is not contained within a grid system possessing 'X' Coordinates corresponding to 'X' Dimensions. This means Consciousness is not embedded *within* some Space, it *is* the Space; it is the *sentient* Space of the Godhead.

Consciousness is not engaged in physical manifestation of the universe. Since self-consciousness is transparently so involved, there must be some intermediary connecting the two notions. This role is served by the construct of *global consciousness*. In any meaningful interpretive philosophic sense, global consciousness is identical to *both* transcendental Consciousness *and* personal self-consciousness. Global consciousness conforms to the notion of an *imaginal* reality as defined by William Chittick in *Imaginal Worlds*: "An imaginal—not 'imaginary'—reality is one that dwells in an intermediate domain between two other realities and shares in the attributes of both sides. An imaginal thing is both the same as and different from each of the two sides that define it.... A commonly given example of an imaginal reality is a mirror image, which acts as a bridge...between the reflected object and the mirror."[27]

The theoretical "difference" between transcendent Consciousness and global consciousness is that where the first is a transcendental Space beyond physical manifestation, the second is space coextensive with the existential universe. On the other hand, the "distinction" between global consciousness and individual self-consciousness should be obvious from the terminology:

the former refers to the universal Being of Space *as* the form of space, while the latter is singularly unique to each sentient being *within* the universal space. In conjunction with the declarative aphorism, "God is Consciousness within existence," the relationship between transcendental Consciousness, global consciousness, and self-consciousness inextricably links Godhead, or the first-named essence, with God—the middle term, the imaginal reality—and individual beings.

Consciousness can thus be said to come in three modes—transcendental Consciousness or Godhead, global consciousness or God, and individualized self-consciousness—though this division is fundamentally a pedagogical tool and not a real distinction with inherent differentiable qualities. Therefore, one must appreciate that, in essence, transcendental Consciousness *is* global consciousness *is* self-consciousness. Distinct terminology is used solely to facilitate discussion on the various roles Consciousness takes on in a reality inclusive of realms both material and immaterial in form.

Unaware of doing so, in the space of a few lines, Aurobindo referred to all three modes of Consciousness in *The Life Divine*: "Consciousness can exist where there are no...operations, no signs revealing it, even where it is withdrawn from objects and absorbed in pure existence or involved in the appearance of non-existence.... Where consciousness is one with being and immobile, it is not a separate reality, but...the self-awareness inherent in existence."[28] The transcendental mode of Consciousness is conveyed by the first portion, with the relevant indicators being "no signs revealing it" and "absorbed in pure existence." Global consciousness is referent for the phrase, "one with being and immobile." An accurate depiction of a primary role for self-consciousness is suggested by "self-awareness inherent in existence."

In a single stroke, all three options conceived by Charles Hartshorne in *The Divine Relativity*—"There are...the three views: (1) God is merely the cosmos, in all aspects inseparable from the sum or system of dependent things or effects; (2) He is both this system and something independent of it; (3) He is not the system, but is in all respects independent"—are favorably reconciled and transformed from a seeming incompatibility of choices to a harmonious relationship dependent only upon the perspective adopted from which to view the metaphysical-physical interaction.[29]

Metaphysicians, those with a religious orientation and those without, have almost always assumed the unceasing quest for unity among discrete events leads to the conclusion there must be a single transcendental agent responsible for manifestation. *TOPP* rejected this position as illogical. First, if this were a valid description of metaphysical reality, there would be no origin to the utility of the concept of number *within* manifestation. *TOPP* went into extensive detail in demonstrating that unity can only beget unity.

Chapter three proved numerical discreteness, as exemplified by counting integers—as well as the infinitude represented by proper fractions and non-terminating decimals—could be unified into a continuum through inclusion of Number as a transcendental idea. Relational bridges within manifestation necessarily demand prior existence of two "somethings" to be conjoined. Consequently, number must commence from the existence of duality in the form of "zero" and "one." As the source of origination for number within the universe, duality with respect to Number must extend to the metaphysical Dimension *necessarily*.

Second, if it were the case that a sole transcendental agent causing the production of manifestation inherently harmonized all potential singularities into a continuum, why produce a physical reality ruled by discreteness and ephemeral relationships between objects? As Aurobindo commented in *The Life Divine*: "There must first be some obstruction in the flow of the Force ocean, some contraction and expansion, some interplay of vibrations, some impinging of force upon force so as to create a beginning of fixed relations and mutual effects."[30] Number is incorporated directly through the notion of Dimensionality. If Consciousness is depicted as an unbounded Dimensional Space of sentience, the positing of two or more such unbounded Dimensional Spaces of Consciousness is viable *if* multiple unboundedness can be shown to not be self-contradictory.

In support of this notion, *TOPP* again went into excessive detail. In brief: an infinite number of zero-dimensional points, one-dimensional lines, two-dimensional planes, and three-dimensional regular geometric shapes—cube, tetrahedron, octahedron, dodecahedron, icosahedron—can border each other without overlap.[31] This suggests, with respect to properties attributable to Conscious Spaces, they can exhibit Dimensionality such that they can border each other yet still be unbounded. To re-emphasize: Dimensionality is *not* a synonym for restriction to a coordinate grid system; it is a term expressing a category difference between two or more equally unbounded, infinite Spaces.

Aurobindo's mention of Force in the cited passage leads to articulation and proper comprehension of the next element in the metaphysical puzzle. Although contemporary astrophysicists claim the *Big Bang*, or the derivative *Big Bang plus Inflation* scenario, as the source origin of space and time, what cosmologists leave unstated is the precise cause and nature of that energetic burst. *TOPP* posited the universe we live in was formed from the intersection of dual Conscious Spaces in a process called *Interpassage*. The fit is exact: the cosmologic Big Bang, mirroring precisely cosmogonic Interpassage, was the outward expulsion of enormous amounts of energy in the form of light and heat from an infinitesimal point, with the result that physical space and time came into manifest existence. On the transcendental level, Interpassage was the onset of a Causal Process in which there was an exchange of Energy between Dual Spaces.

Here is the origin of the innate characteristic of "each inhering in each" for Causality, or for elements found in a causal chain on the immanent level of existence. The intrusion of Consciousness from one Dimensionality into another represented a mutual movement of Conscious Energy, and was *a novel occurrence marked by the introduction of Time serving as reference for this unique transition*. Hence, Time correlates with Energy transition and event occurrence in the same manner as time does. This speculative outline completely, consistently, and comprehensively explains how Consciousness unifies Space, Time, Energy, Number, and Causality, with coherency of the whole being denoted as Truth.

F. W. J. Schelling proposed essentially this exact scenario in *Ages of the World*, although he made the conflict internal to an absolute transcendental unity and imaged it in terms of a battle between willful elements:

> The relationship is the following: one will wants the essence (what-is and being) to remain closed and hence concealed, just as it was before. The other will wants the essence to open up and abandon its concealment. How is a decision possible here? Perhaps someone might say...one of the wills is by nature subservient to the other, in which case it would necessarily be overcome by the other will, and this other would then be the victor. But this presupposition is false. Both wills are by nature equally important, each has the same right to be active, and it is necessarily true that neither retreats before the other.[32]

As is unavoidably the case for all previous, and all subsequent, attempts to endow a single transcendent unity with contrary properties of being both one and many, exhibiting continuity and discreteness, Schelling's proposal left an unanswered question: Why have two conflicting wills?[33] He tried to explain it away as a consequence of the innate distinction exhibited by the transcendental in contrast to the material: "If, in the expressible, the relation between the opposites is characterized by an inner necessity compelling them to be one (because both are equally necessary to the whole)—then, in the expressing, the relation...is characterized by an inner freedom not to be one, but rather for each to be for itself. Each of the wills is individual and self-sufficient, and each has the complete freedom to posit itself and to negate the other."[34]

His approach failed in its attempt to resolve the apparent difficulty in comprehending any necessity to postulate an ontological source with divided willfulness: "Since there was no time when God could have deliberated with himself and acted, and since only the supreme free will could have decided, it follows that the conflicting wills...must have been united through an instantaneous, all-illuminating action."[35] Aside from the ambiguity embedded within the notion of "no time"—Was this a function of the lack of existence of Time, or was the reference meant to indicate activity proceeded so quickly that no discernible, measurable amount of time passed?—the passage can be interpreted to imply that the "conflicting wills" were independent of, and yet somehow subordinate to, "the supreme free will." If this were the case, it is unclear why either will should have any potency whatsoever, or how they could forcibly induce manifestation.

An Absolute composed of two parts, two willful elements in Schelling's scenario, can never be reconciled, but must forevermore remain a duality, in precisely the same manner in which two immiscible chemicals cannot be combined into a homogeneous solution unless a third medium capable of homogenizing the two were introduced. Yet, once added to the mixture, Schelling's "supreme will" could not then be overturned in its effect except through inclusion of still another component with the capability of reversing the homogenizing process. Schelling's explanation does not provide for this fourth element. Hence, only an external dualism can metaphysically account for both continuity and discreteness, for both Consciousness as un-manifest and self-consciousness as manifest. Extending the metaphor of archetypal principles and chemical compounds, the twin constructs of dual Conscious Spaces and Interpassage allows for Separation and Union in the same manner as if two miscible chemicals were stored in individual containers or mixed together into a uniform solution and stored in that form.

The standard rebuttal to any proposition of primordial dualism as the foundation for metaphysics is to assert multiple absolutes are impossible in concept, to claim "absolute" covers all possible diversity.[36] This contention assumes precisely what is at issue, namely the twin theses: (a) absolute necessarily implies singularity; (b) a singularity must be ontologically prior to discreteness and multiplicity.

The previous chapter presented numerous ways in which discreteness is, literally, natural to manifest existence—as a fundamental aspect of physical reality. It was emphasized that unity can only accrue from the imposition of a meta-perspective, from a transcendental construct. However, as was explored at length in *TOPP*, it is not possible to produce the many from an absolute One: there must be original division because "from what is one and simple only what is one and simple may come to be."[37] Discreteness in phenomenal reality must derive from multiplicity in Being.

As an additional piece of supporting evidence for the thesis of inherent duality, consider the issue from the perspective of information theory. To repeat a statement from chapter three: "The amount of information conveyed by the message increases as the amount of uncertainty as to what message actually will be produced becomes greater." An infinitely long message of all "ones"—"on" states in terms of a binary digital computer—or all "zeros"—"off" states—conveys no information simply because there is no uncertainty as to what the message may be. A monolithic, inalterable "one" does not mean *any*thing precisely because it stands for *every*thing. It may indeed be representative of an absolute Conscious Space, conceived as Godhead, as such an agency existed before commencement of Interpassage when it had no objects with which to be occupied and no information to express.

If, however, such an inexhaustible continuity is transformed into discrete units by inserting a stop signal at random through the addition of intermittent "zeros," information content is possible with respect to its communicability. Such a condition would describe either a transcendental Godhead after the onset of Interpassage—since an absolute Other becomes an object *in* each Space and represents a "stop" signal in continuity *for* each Space—or is appropriately descriptive of manifest global consciousness, God, with an entire universe of physical objects worthy of description. In either case, information requires distinction in context, meaning at least two characters must be used in any information transfer operation.

Similarly, multiplicity with respect to transcendental Dimensions must be compatible with a singular character for any particular unique Dimension. Sufficient disproof of the notion that absolute implies singularity comes from consideration of the following question: Would an absolute scale of Beauty undermine an absolute scale of Justice? The response must be "no." The two ideas occupy different domains, and the mere assertion of the existence of either does not impede on the propriety of adopting the other as a reference frame for the coordinate concepts it includes under its particular umbrella.

Nor is the criticism potentially levied against the proposed speculations concerning the implausibility of coexistent Conscious Spaces adjoining each other without possibility of merger prior to the onset of Interpassage a valid argument. First, the contention assumes the existence of Time as a parameter of pertinence for the existence of Space, as indicated by the phrase "prior to the onset of Interpassage." It has already been pointed out that Time and Space must be treated separately in any discussion of metaphysical ontology and teleology. If it is posited, as this scenario claims, that Time *began* with Interpassage, there cannot be any suggestion of a "prior to."

Second, the seeming self-contradiction of "adjoining Spaces" that are simultaneously conceived as infinite in scope can be dispelled by application of examples from geometry. This problem was addressed with discussion of Number, where known shapes can be configured adjacently without overlap when extended to infinite dimensions. There is also, with respect to defining borders where the dual Spaces connect, no difficulty in citing precedents from geometry.

One can easily construct a two-dimensional rectangular plane of finite length and width tangent to a two-dimensional circle of finite circumference meeting at a single point without the two figures merging into a continuum. It is likewise possible to construct a three-dimensional rectangular solid of finite length, width and depth tangent to a three-dimensional sphere of finite circumference meeting at a single point without the two figures merging into a continuum. Existence in a three-dimensional universe bars physically drawing an infinite-dimensional object of any kind; yet, it is hypothetically possible to construct a three-dimensional rectangular solid of *infinite* length, width and depth that is also tangent to a three-dimensional sphere of *infinite* circumference meeting at a single point without the two figures merging into a continuum.

The principal problematic in this scheme does not reside in the postulate of dual Conscious Spaces as the ontological source of physical existence. It consists in justifying the commencement of Interpassage, the onset of merger into a singular Space with concomitant creation of manifest reality. There are two mechanisms, *intentionality* and *accident*, capable of accounting for this event. As Ilham Dilman remarked in *Free Will*: "A thing is necessary...when it cannot be otherwise, and...when it follows from a...sufficient ground.... The antithesis of necessary is accidental; but what is accidental is...relative to my intention."[38] Intentionality and accident are associated with a collection of precise technical philosophical meanings, and both evoke certain fixed repercussions as instrumental for Causality.

Actions are considered to be intentional if: (a) they arise from an agent having pre-existent knowledge of a specific situation; (b) this knowledge serves as a motivational source for the desire to commit a certain act; (c) the willfulness associated with desire is equivalent to taking the pragmatic steps necessary to carry out the intention; (d) the agent accepts responsibility for the consequences of the activity. Recognition of several of these aspects of intentionality allowed Carolyn Ristau to claim, "an intentional creature will have beliefs and knowledge, and it will act in accordance with them," when discussing the behavior of birds.[39] This model of intentional activity permitted Joseph Goldstein and Jack Kornfield to assert, "inherent in each intention in the mind is an energy powerful enough to bring about subsequent results," in the context of an investigation of Buddhist meditation practices.[40] Other expressions found in diverse literature sources also support this dissection of intentionality into component elements.[41]

The idea of intentionality, as utilized here, is not limited to its meaning within the methodology applicable to phenomenology, where it is considered to be a specific object reference for a subject's thought independent of the potential external actuality of that object.[42] Under the present circumstances, intentionality is not merely a mental act; it is a direct relationship between thought and experiential activity. Although these characteristics are derived from the idea of intentionality as it applies to the life experiences of sentient creatures, it is reasonable to employ them in the service of transcendental agents. Thus, Conscious Spaces may also be said to act with intentionality—given the freedom embodied in the epistemological operating principle, "as above, so below; as below, so above"—if the same attributes can be properly said to be involved.

With specific application to Interpassage, intentionality can be in force if and only if there was mutual intent on the part of both Spaces. Any one-sided activity would imply the acting agent had superior pre-existent knowledge, and a greater ability to act on the consequent willfulness instigated by that knowledge, than the Other with respect to the potential conditional nature of the interaction. This precept violates the notion of equipotent Spaces. Hence, the dual Conscious Spaces had to simultaneously perceive the reality of the Other, and mutually choose to interact Dimensionally with the Other, if the concept of Interpassage is to have a consistent foundation.

In this statement of requirements, a tendency to interpret "simultaneous" as a latent admission Time was already operational must be rejected. Human language has the characteristic of primarily serving human interests and is not always literally applicable to metaphysical subjects. This difficulty has been highlighted in the earlier discussion about the anthropocentric premise. That *caveat* notwithstanding, this proposal does make coherent the origin of transcendental Time as indicative of the onset of Interpassage. Problematic is a demand for Time's commencement characterizing transcendental activity "prior" to the onset of Interpassage, as the use of "simultaneous" or 'prior" with respect to the mere perception of the Other might hint.

The principal consequence of the postulation of mutual intentionality by dual Conscious Spaces as the cause of Interpassage is teleological. If both Spaces expressed mutual intent in undergoing Interpassage leading to the creation of physical spacetime, the only self-consistent teleology would be the eternal continuance of that condition. In terms of Kant's Antinomy, both alternatives would be eliminated: it would be false to allege space and time are eternal, and it would be false to claim space and time had a defined beginning and will have a definite ending. What would accrue would be a combination of the two: space and time had a definite beginning, but will ever after that moment of initiation, endure eternally. In other words, *mutual intentionality demands fulfillment of one of the four original teleological scenarios previously outlined for a three-level speculation.*

The rationale for such a conclusion is obvious: if Consciousness acts with intent, it would be internally inconsistent to posit that it somehow came to later realize the resultant actions were a mistake and seek to undo the results of initial movement. Such an allegation would imply Consciousness could be in error due to ignorance, meaning deprived of sufficient knowledge of conditions and their repercussions, which would be a clear repudiation of omniscience as applicable to transcendent agencies. We must willfully reject the anthropocentric premise alleging what is possible for us, a deficiency of reliable data, must necessarily be possible for a *supra*physical agent.

What about Interpassage as the outgrowth of an accidental intermingling of Conscious Spaces? Accident characterizes an event neither necessary nor typical in its occurrence according to Aristotle's *Metaphysics*: "Some things, then, are always and of necessity as they are...in the sense in which we say that they cannot be otherwise; others are so, not of necessity or not always, but usually; and this is the principle or ground by which the accidental is distinguished, for what is...neither always nor usually...is accidental."[43] He added an addendum to the notion of accident consistent with the advocated thesis: "Moreover, since nothing accidental is prior to the essential, neither are accidental factors prior.... If chance or spontaneity is a fundamental factor of the universe, mind and nature are factors prior to it."[44] Hence, the accidental coming into being of the universe, as posited by cosmologists, must be preceded by the essential existence of Consciousness.

Interpassage could reasonably be defined as accidental: nothing about the event suggests it was necessary in the sense of being unavoidable. Even less plausibly could it be said to have been typical, since it is specifically posited as a unique event with no precedence. In favor of accident as the ultimate ontological justification for Interpassage and manifestation is that all possible teleological variations previously explicated are retained, in contrast to the single option emanating from mutual intentionality: (a) space and time could both continue eternally once begun; (b) space could continue and time could be terminated; (c) both space and time could eventually cease to exist. On the cosmogonic level of reality, Time could terminate upon Separation of Conscious Spaces or continue indefinitely into a future with ongoing Union. In either instance, Space could not, of course, ever cease to be.

In the absence of human certainty about the True cause for formation of the universe, it is reasonable to maintain under consideration all teleological options without prematurely foreclosing any. This rationale leads directly to the thematic conclusion to be found in *Going Against the Flow: An Exercise in Ethical Syncretism* (*GATF*), in which a critical analysis of the foundations of frequently enunciated ethical systems was produced. Coverage included: (a) virtue ethics; (b) obedience to God's Will; (c) *eudaemonism* (happiness); (d) hedonism (pleasure); (e) *apatheia* (dispassion); (f) *ataraxia* (tranquility); (g) Kantian duty; (h) utilitarianism; (i) consequentialism.

A primary outcome of significance was the finding that all of the systems taken individually were flawed, but when aspects of several were combined, a powerful product resulted consistent with the metaphysics of *TOPP*. It was found that, to be of ultimate value, ethics must be placed in the service of epistemology, of the attainment of knowledge by all sentient creatures for the purpose of their intentional participation in resolution of the cosmogonic-cosmologic crisis initiated by the onset of Interpassage.

What crisis? There is presently an unanswered cosmological question as to whether the universe is *open* or *closed* spatially. If physical spacetime is closed, Interpassage will be reversed. A closed spacetime would mean the universe would begin to contract, and eventually shrink to the dimensionless point from which it came, before disappearing from the manifest dimension altogether. Astrophysicists refer to this process as the *Big Crunch* in direct analogy to the Big Bang from which it originated. On a transcendental level, it would mean dual Conscious Spaces would Separate, disentangle from each other.

If, as the logical alternative, physical spacetime is open, the universe will continue to expand with no foreseeable terminal point. As was mentioned previously, "expansion" is a misrepresentation, since there is nothing devoid of space to which space could expand into. Physically, objects contained in space would move further apart. Union would be the necessary implication of a perpetually expanding universe: Interpassage would continue unimpeded until complete Dimensional overlap of Consciousness accrues.

The determination of cosmological openness or closure is hypothesized to depend upon the mass-energy balance in physical spacetime. Insufficient mass is indicative of permanent universal openness because the force of gravitational attraction would be unable to reverse the current expansionary trend; a sufficiency of mass would allow for a dominance of the gravitational force of attraction. The amount of mass in a given volume of space needed to balance the two tendencies is the *critical density*, which has been calculated to be $\sim 10^{-29}$ grams/cubic centimeter.

Using the dispersal of galactic clusters as a means of measurement, the universe is expanding at an accelerated rate over time, meaning Interpassage continues unopposed by any countervailing tendency. The density of matter from recognized sources—regular matter, *dark matter*, interstellar gases—is close enough—estimated at 4×10^{-31} grams/cubic centimeter—to the critical density to allow the fate of the cosmos to remain uncertain. In a paper written for *The Irish Astronomical Journal*, B. J. Carr attested to these facts: "It should be stressed that the really surprising point is that the density of the universe is so close to critical at all.... The fact that the density is within a factor of ten of the critical value now implies that its deviation from the critical value must have been even smaller at earlier times."[45]

It is the *mass-energy* balance that is the crucial determinant, and not just the *mass-volume* ratio. Why? Mass and energy are inconvertible when matter moves at velocities approaching the speed of light in the near-vacuum of space according to the relationship, $E = mc^2$, derived from Einstein's Special Theory of Relativity. Metaphysically, this allows for involvement of sentient species in the determination of cosmogonic and cosmologic teleology. All lifeforms have an associated self-consciousness both identical to and distinct from, global consciousness, the manifest version of Interpassage, as well as Consciousness, the transcendent field of sentient Energy. Self-consciousness, if properly directed, could *create* the mass necessary for universal closure—or conversely, prevent that mass from ever being created—if physical beings acted in consonance with each other.

How should all self-conscious entities orient their involvement in this cosmological drama? *GATF* proposed evolution to a sufficient level so as to be capable of responding in the most preferable manner was the ultimate aim of individual ethical development, with "individual" not being restricted to *Homo sapiens*. To facilitate the solution of a metaphysical conundrum, ethics posits epistemology as the ethical *raison d'être*: to procure wisdom and self-development such that intelligent beings could be capable of making the best choice *vis à vis* allowing Interpassage to continue unabated and permanently, or initiating steps to reverse the process and return to a state where the dual Spaces were separate and non-interactive.

This lofty aim was suggested as most reasonably accomplished through a process of unending growth in knowledge and virtue. The guiding rule for ethical theory would seem to be promotion of comprehension of the ways in which the universe operates on all levels, from the most microscopic to the most macroscopic, coupled to cultivation of those qualities deemed to be virtuous by each being. Why virtue? Wisdom, acting in the most appropriate manner in any given situation, is generically accepted as a virtuous quality, and its full development could only be attained through inclusion of all other similarly enumerated and defined traits.

In this respect, Nikolai Berdyaev's remark in *The Beginning and the End* was highly insightful, not ludicrous: "The fact...the world exists in time and not only in space means...the world is not completed...its creation has not yet reached its crowning consummation, that it continues to be created. If the creation of the world were closed, newness would not be possible. Finished and closed reality does not exist."[46] "Creation" could suggest formation of additional mass from Conscious energy; "newness" could refer to intentional use of self-consciousness towards some preferred cosmogonic-cosmologic solution.

These speculations make rational sense. On the *supra*physical level, the two Spaces are equipotent. *If* the Cause inducing the start of Interpassage were accidental, only application of externally based Energy could invoke, and enforce, a decision to continue or discontinue merger, to compel Union or Separation. Analogously, only externally based energy can command the determination of universal expansion or closure. *Teleological uncertainty in the manifest dimension suggests a similar lack of resolution cosmogonically.* This is to argue that mutual intentionality by Conscious Spaces to undergo Interpassage leading to a predetermined teleology of Union is most likely incorrect. Given the principle of similitudes, immanent reality takes its cue from transcendental reality, there is no obvious rationale for the existence of cosmological uncertainty in the face of cosmogonic certainty, and the former is an established fact.

On both levels, there is currently lack of resolution for the same reason: in one situation, the balance of power between the two Spaces does not admit decision-making; in the other, the universal density of objects is sufficiently near the critical level for the decision to swing either way. In the context of this speculative metaphysical model, justification for feasibility of the stated potential epistemological/teleological goal could do worse than adopt Pitirim Sorokin's remarks from *Social and Cultural Dynamics*:

> Bearing the seeds of its change in itself, any sociocultural system bears also in itself the power of molding its own destiny or life career.... This destiny, or the system's subsequent life career, represents mainly an unfolding of the immanent possibilities of the system... The process of unfolding the immanent potentialities...is somewhat predetermined by the system, but this predetermination leaves a considerable margin for variations.... Only the main direction and the main phases of the unfolding are predetermined; the rest...are "free" and become an unforeseen and unpredictable matter of chance, environment, and free choice of the system.[47]

TOPP included numerous statements from religious writings supporting the notion of humanity—by extrapolation other sentient lifeforms also—as having not merely a contributory role to teleology, but a decisive one. This doctrine is found especially frequently in Judaism due to the Bible-inspired belief they were the "chosen people." The remarks by Adin Steinsaltz in *The Thirteen Petalled Rose* can be perceived not as hyperbole, but as a prescient vision of possibility:

> In Judaism man is conceived, in all the power of his body and soul, as the central agent, the chief actor on a cosmic stage... Everything he does constitutes an act of creation, both in his own life and in other worlds hidden from his sight. Every single

> particle of his body and every nuance of his thought and feeling are connected with forces of all kinds in the cosmos.[48]

In a similar vein, Isaiah Tishby asserted, "The preservation of harmony in the life of the Godhead depends upon human actions," and clarified his meaning in "The Doctrine of Man in the Zohar": "The relationship between man and God is not confined to man's dependence upon God's mercy or his receipt of benign influence from above, but is one of mutual influence. They afford one another reciprocal aid. Man lives and has his being because of the flow of influence that originates from a stimulus in the upper world, but he in his turn can use his power to influence that world."[49] Other literature sources, philosophical and theological, can be interpreted as containing veiled hints leading to the same conclusion.[50]

Sorokin stated, while Steinsaltz and Tishby merely implied, influence by intelligent creatures on cosmologic or cosmogonic scales necessarily requires such beings be capable of independent action through possession of free will. In another illustration of the utility of similitudes, in *The Divine Relativity* Charles Hartshorne affirmed an inherently reciprocal relationship pertaining to free will involved in the interactions conceivable between immanence and transcendence: "We influence God by our experiences but do not...deprive Him of freedom in His response to us. This divine response, becoming our object, by the same principle in turn influences us...without removing all freedom."[51]

Granted these assessments reek heavily of anthropocentrism, a belief the universe turns on a human axis. As evaluations of a metaphysical state of affairs, they need not be interpreted so exclusively. If animal, and even plant, self-consciousness is a reality, then they too can contribute to resolution of the situation initiated by Interpassage. Moreover, the issue of intelligent life elsewhere in the expanse of space is certainly within the realm of feasibility.

The biggest obstacle to the original Big Bang theory's acceptance on its own was the uniformity of distribution of energy throughout physical space.[52] If one were to detonate a grenade, the resultant distribution of matter and energy expelled by the force of detonation would be random, but completely determinable through calculation of the physics associated with projectile motion: there would be regions of the surrounding space towards which more matter would have been propelled than others; the shock wave energy would not be equally arranged in all directions. Yet, on a cosmological scale, the universe appears to have a uniform distribution of energy. In 1965, satellite data revealed there is an *isotropic*, same in all directions, and *isothermal*, same temperature, background radiation with an energetic equivalent of approximately three degrees above absolute zero in temperature. It was to account for this improbable, unanticipated symmetry of energy distribution that provoked Alan Guth to postulate the concept of *Inflation.*

Paul Davies explained Guth's proposal in "What Caused the Big Bang?": "In the beginning the universe found itself in an excited vacuum state... In this state the universe would be subject to a cosmic repulsion force of such magnitude...it would cause headlong expansion at a huge rate.... This means that every 10^{-34} seconds, every region of the universe doubles its size, then goes on doubling again and again in a progression."[53] Inflation accounts for the isotropic distribution of background energy because it asserts universal expansion was initially so swift that, after billions of years, the cooling of the now-immensely large universe proceeded everywhere at the same rate.

This elegant solution introduced a new problem. If inflation accounted for equal distribution of energy along cosmic distances, how is it matter is not as uniformly arranged? The fact is, objects in space are grouped non-homogeneously; that is, density is not uniform throughout space. If the initial energy latent in the universe was evenly distributed in all directions, and if energy is simply another form of matter completely interconvertible with it according to special relativity theory, why is the mass of matter *anisotropic*?

The usual explanation is framed negatively: if there were no asymmetry, there could be no creation of living beings.[54] There is a preferred explanation based upon a recounting of cosmogonic events according to the theory being offered: the dual Spaces began merging asymmetrically from a single point of Dimensional overlap and this interaction subsequently spread in a non-uniform manner. The cosmologically heterogeneous distribution of matter is a visible physical representation of such an unfolding dynamic process.

Analysis of the ontology of manifestation has revealed three conditions: universal openness, closure, or criticality. If symmetry of process suggested by the idea of similitudes is to remain in force, they must have corresponding cosmogonic counterparts. In *TOPP*, these were the aforementioned Union and Separation, plus the addition of what was designated *Conflict.* Conflict is the most theoretically intriguing of the three, for it proclaims the Conscious Spaces do not possess equivalent intentionality with respect to continuance of the process of Interpassage. Having begun accidentally, both Union and Separation must be intentional with respect to both Spaces. Any discordance culminating in Conflict, however temporally transitory such a state might be, would then be conceivable on the basis of a posited Energy incompatibility.

This thought leads to consideration of an option not previously explored. It was *assumed* the two Spaces are best characterized as Being. Perhaps this is a flawed premise; perhaps the distinction between Consciousness fields is the difference between Being and Nonbeing. This notion begins to achieve plausibility when we recall the statement by Masao Abe in which Buddhist philosophy posits a distinction between *U*, Being, and *Mu*, Nonbeing, which is also an identity.[55]

Further, it is consistent with data obtained from physics and chemistry, where experiments demonstrating the existence of matter particles—protons, neutrons, electrons—are countered by observations confirming the identities of antimatter particles—antiprotons, antineutrons, positrons. Stephen Mason explained the relationship in *Chemical Evolution: Origin of the Elements, Molecules, and Living Systems*: "According to the principle of charge-parity-time conservation, the negatively charged electron and the positively charged positron are charge-polarity mirror-image forms, and the hydrogen atom, composed of an electron and a proton, has a charge-parity enantiomer made up of a positron and an antiproton."[56] An *enantiomer* is any object whose mirror image does not precisely spatially overlap the original. The classic illustration of enantiomers is right and left hands: they are mirror images, but if you try to slide one image over the other laterally, the fingers will not be aligned because thumb will overlap with pinky, and ring finger with index finger.

It does not require a large stretch of imagination to make a jump from the known existence of matter and antimatter to attribution of *U* and *Mu* as apt metaphoric descriptors of a difference in Conscious Energy comprising dual transcendent Spaces, provided: (a) it can be considered plausible their innate difference in Dimensionality is an adequate label for distinctions in Energy even if vague, and (b) this feature allowed them to border each other without coalescence prior to the onset of Interpassage. Hence, characterization of the Conscious Spaces as enantiomers is viable: they would not Spatially overlap with precision because they exhibit different Dimensionality. Of immense interest in this context is the fact that matter and antimatter undergo mutual annihilation with the concomitant release of a tremendous amount of energy when they come in contact, which would be a viable interpretation of the cosmogonic Cause for the Big Bang plus Inflation cosmological theory.

The mention of "the principle of charge-parity-time conservation" leads immediately to solidification of another aspect of this metaphysical thesis. The theory first expressed in *TOPP*, and summarily reiterated in these pages, suggests the universe we know is not intrinsically symmetrical, but exhibits an asymmetry observable in its natural operative flow. Recall chapter three's discussion of time directionality. As experienced, time invariably moves in a single direction from past to future; we do not detect our lives, or any extant natural process, running backward. There is, however, an apparent breach between the qualitative characterization of this process and its quantitative element. All of Newtonian physics employs mathematical laws founded on time reversibility. Appeals made to relativity or quantum mechanical theories fail to improve the situation: both are overtly time symmetric in concept and use.

The theory of Interpassage contains the remarkable claim that there is a symmetry evoked, but it is a spatiotemporal symmetry only in the sense that it occurs in conjunction with the "parallel" universe created by the merging of Conscious fields of Energy. This is the point at which the *Principle of Charge-Parity-Time Conservation* becomes significant. The first property is *charge conjugation*, the symmetry involved in the relationship of matter to antimatter exemplified by the reciprocality between a small mass, negatively charged electron orbiting around the nucleus of an atom and a small mass, positively charged positron orbiting around the nucleus of an *anti-atom*. The second component is *parity inversion*, whereby every point within a three-dimensional coordinate system marking position in space has a mirror image coordinate system producing an *anti-image in anti-space*. The third factor refers to *time reversal* in the ordinary sense, where a directional flow from the past to the future is matched with an *anti-time* flow from the future to the past.

What the CPT theorem says is if one reverses all three parameters, the resultant physical laws are identical to what they were before reversal. *This is what Interpassage would anticipate*: the merger of Spaces was asymmetric when looked at from one perspective. This joining allowed for creation of the manifest universe we inhabit, a universe in which matter predominates over antimatter, and where temporal progression is asymmetric in its flow in one phenomenological direction. Time also takes on the appearance of flowing unidirectionally from the vantage point of either Conscious Space because time has Time as its transcendental analog. However, *merger was symmetric when looked at from the perspective of both Spaces jointly*. In order to maintain overall symmetry, the parallel universe postulated to emanate from the Energy of Nonbeing should possess as attributes a predominance of antimatter over matter and time unfolding in the opposite direction.

In *The Arrow of Time*, Peter Coveney and Roger Highfield stated: "The most important facet of the CPT theorem is the statement that if the CP part is symmetrical then the process under discussion must also possess time-reversal symmetry.... This appears to offer a theoretical foundation for time's arrow."[57] From the citation, one can infer the logical contrapositive must hold: if time reversal is not symmetric, the combination of charge and parity is not symmetric. *To retain overall CPT symmetry, the charge-parity combination must jointly display the same symmetry characteristics as time alone.*

Charge conjugation is known to be asymmetric with matter predominant over antimatter in our universe. Space expresses parity as symmetric in that it is isotropic, hence a parallel universe created as a result of Interpassage must display an isotropic radiant energy pattern in order to be parallel. Though comprised of different kinds of Energy—the difference between Being and Nonbeing—there would of necessity result a point-for-point correspondence in Energy distribution between the two universes, which would be fulfillment of the criterion of an "anti-image in anti-space." Charge and spatial parity would collectively demonstrate combinatorial *asymmetry*, and the passage of time in both universes *must* be fundamentally asymmetric in accordance with the CPT Theorem.

What remains to complete this metaphysical theory is elucidation of the mechanism for coordination of Consciousness, global consciousness, and the individual self-consciousness associated with every phenomenal being. It has been claimed global consciousness is commensurate with three-dimensional space and its background energy, detectable and undetectable, in the physical universe. Global consciousness as Conscious Spatial Energy generated from Interpassage of two transcendent absolutes was also proposed. There must be a means to bridge the two forms in an Energetic/energetic and Spatial/spatial sense. In addition, there must be some procedural mechanism to relate global consciousness to self-consciousness such that transfer of knowledge in the form of information is feasible and overtly accessible.

Both demands are met by incorporation of a single concept: the notion of *theological space* first conceived by Paul Weiss. He described his postulation of theological space and its significance in "God and the World":

> We know...we long for God and...that longing attains its end...adequately. If we did not...also know...God was searching for and finding us, we would have no direct knowledge...He was in any way concerned with us.... They...constitute a "theological space" in which they "move" with respect to one another.... Since everything longs and is answered to some degree, and since God needs all things and feels the effect of every longing, theological space is the region which unites nature and God so that together they form a single cosmos.[58]

Aside from the personification of God, a difficulty pervading all Western monotheistic theology, there are two highly controversial assumptions built into this idea: (a) our yearning for God is satisfied "adequately," and (b) God reciprocates this yearning by "searching for and finding us." To these, Weiss had a ready answer:

> This is a conclusion against which most theologians would rebel, though it follows out of the premises they...would acknowledge. They admit the universe has its own laws and man a free will...God is perfect and self-contained...man longs for and reaches to God...God is concerned with and infects the nature of things.... The theologians shy away from the idea of a theological space because they think...on the one hand, nothing could encompass both God and the world without God being

> reduced to the status of a being dependent on and subsumed under it as an inferior being, and on the other...no one could bridge the gap between Him and us but God Himself.... Nor does the space include God within it; it terminates in God, but God also lives outside it. That is why He can be said to have an independent, immutable being. Nor is it true that the only way of bridging the gap which exists between God and the world is by God coming down to that world.... Reflection shows that God reaches towards us; experience shows that we reach towards Him as well.[59]

Theological space is a partially immanent dimensional analog of what occurs in the transcendental Dimension alone. If Interpassage is defined as a process resulting in formation of global consciousness from Consciousness as a consequence of the unification of two distinct Conscious Spaces—the derivation of God from Godhead on a transcendental level—then, theological space is this same process occurring within manifestation allowing the union of transcendence with immanence, global consciousness with individual self-consciousness, into a single whole. Continuity allows for the discrete.

Picture two circles: one is global consciousness, the other is the natural physical universe. Their area of intersection is labeled "theological space." A capacity for transmission of information from global consciousness to the natural world exists by way of miraculous acts of creation precisely because global consciousness *is* the natural universe in its spatial extent. There is also a capability for transmission through acts of mystical experience and prayer from beings within nature to God—if what is sought is sought in the form of ideal personalized deity—or to global consciousness—if what is sought is sought in the form of impersonal agency.

Within God/global consciousness, an image of natural existence forms, yet since the image is wholly contained within transcendent Consciousness, it is an image of God conceived as particular things. Global consciousness has knowledge of the world without physically experiencing it, a state which when transmitted as information allows for the production of the doctrine of God as transcendent, immutable, independent, uninvolved in the world of time and space, with no feelings or thought—the image of a pure supreme Godhead.

Within the sphere of the universe, an image of God is created, yet since the image is wholly contained within sentient creatures, it is an image of the properties of existence conceived as transcendent divinity. Inhabitants of the universe know global consciousness as it is abstractly conceptualized without comprehension of its essence, a state which when transmitted as information allows creatures to portray global consciousness as possessing properties belonging to phenomenal reality: immanent, mutable, relationally dependent, thoroughly involved in the world of time and space, with deep concern or love for the world—a personal and personable God of Western religion. The world can conceive of a remote Godhead because there is the conduit for information, namely theological space itself.

Ervin Laszlo, based on principles of physics and some accumulated data, postulated the existence of the *quantum vacuum field.* In *The Interconnected Universe: Conceptual Foundations of Transdisciplinary Unified Theory*, this field was described in terms sounding much like those used here to define and characterize theological space: "The quantum vacuum exhibits...zero-point energies—energies that persist even at temperatures close to absolute zero, when all other forces vanish. Hence the vacuum is not empty space, but a fluctuating field filled with zero-point energies.... Some physicists...now view the vacuum as a physical expression of spacetime."[60] The addendum to Laszlo's book contained a contribution from Mauro Ceruti, "Philosophical Perspectives on Quantum Vacuum Interaction Cosmology," summarizing the features of this quantum vacuum field. *Its properties, as well as its potential influence on cosmology, mimic those postulated for theological space*:

> The quantum vacuum possesses more than its own reality: it possesses a reality that precedes matter and radiation in the cosmos. The vacuum plays a decisive role in the birth of our universe, and it continues to exercise a decisive role in its history and evolution, interacting with energy...and producing a primary influence on the explorations and innovations of evolution. The vacuum field...has a different, and in some sense more profound, actuality than matter and radiation.[61]

Theological space serves as the pathway through which information—literally in-form-action, the Energy of Consciousness acting as the form of physical energy—can be transferred between global consciousness and self-consciousness, or between two or more beings, so as to provide the potential for knowledge when each recipient assimilates and associates the data bits registering within self-consciousness. The mind-brain connection also uses theological space, coordinating mental states receiving information with the material brain that processes such data and transforms it into action.

Chapter Five: Physics

> We are proposing that in the formulation of the laws of physics, primary relevance is to be given to the implicate order, while the explicate order is to have a secondary level of significance.
>
> —David Bohm, *Wholeness and the Implicate Order*

Using Bohm's suggestion as a guide, the thematic focus of this chapter will be the relationship between theological space and accepted theories and laws grounding physics. Previous chapters have demonstrated discreteness is always unified through operation of some categorical transcendental notion. This applies to the physics of fundamental forces and material constituents also. A necessary deduction from the claim of transcendental Consciousness as Space is recognition that conventional physical concepts—those involving energy, matter, force, distance, time, electrical charge, spin momentum, and their various interactions—derive entirely from this source through processes conceived as emanating initially through theological space, the unseen fourth spatial dimension.[1]

Dating initially from the last quarter of the nineteenth century, physicists have been ambitious in attempting to establish *unified field theory*, the so-called "theory of everything": a body of knowledge comprehensive enough to conjoin all physical parameters into a single framework. Three distinctly different types of syntheses must be enabled in order to accomplish this goal. First, unification must be achieved with respect to *modality*: the formulas of physics have diverse, to some degree mutually exclusive, realms of operation according to the scale of interactions contemplated. Second, unification must be achieved with respect to *typology*: four fundamental forces have unique, even contrary, properties divergent in strength of influence within their areas of application. Third, unification must be achieved with respect to *material composition*: the three building blocks originally found to jointly constitute a chemical atom has mushroomed into hundreds of subatomic particles that have now been uncovered, some of which are considered by many physicists to be even more "fundamental."

A principal epistemological tenet necessary for unification of knowledge self-consistently and comprehensively is: *Philosophy must support science, and science must support philosophy*. The data derived from one discipline must not subvert information obtained from the other if Truth is to be gained. This dictum conforms to that uttered by Ludwig Feuerbach in *The Essence of*

Christianity: "What is *naturally* true is *logically* true; what has no basis in Nature has no basis... That which is not a physical law is not a metaphysical law."[2] As suggested in the preface, a prospective theory of everything must be a metatheory; it must derive from a perspective beyond the physical if all of the physical is to be contained therein. Hence, it is appropriate to attempt to rationally connect any physical unified field theory tentatively formulated to the metaphysical speculation delineated in chapter four. A failure to do so would lead to the conclusion that either the physics or the metaphysics, or both, was innately flawed.

Beginning with *modality*, an immediate conceptual difficulty faced by an attempt to formulate a theory of everything is unveiled. Physics divides into three domains according to the scale upon which events in space and time take place. The oldest branch was developed almost exclusively by Isaac Newton in the latter half of the seventeenth century. It accurately depicts the physical properties of objects and their coordinated motions within the spatial and temporal constraints of ordinarily lived experiences.[3]

The second originated with Albert Einstein in two stages: publication of the Special Theory of Relativity in 1905 followed by the General Theory of Relativity in 1915. Special relativity theory applies to the characteristics of objects when their motion approaches the speed of light in a vacuum, which Einstein stressed as the fastest motion to be attained in the universe.[4] General relativity theory refers to his conceptualization of the nature and origins of the specific interactions among object mass, existential space, and the force of gravity.

Coinciding in time with the formulation of Einstein's relativity theories was quantum mechanical theory. Associated principally with physicists Max Planck, Neils Bohr, Erwin Schrödinger, and Werner Heisenberg in addition to Einstein, quantum mechanical theory was established in the first thirty years of the twentieth century. Just as relativity concerns macroscale effects of fast-moving objects over vast astronomical distances, quantum mechanics investigates physical properties on microscale dimensions of space and time: the atomic and subatomic levels of primary matter, chemical elements, which comprise the manifest world.

Any viable creation of a theory of everything must seamlessly unify the physics concepts and formulas utilized in all three spatiotemporal regions. It was explicitly alleged by Newton that space and time are self-existent, things in themselves to use Kant's description, and it is worth observing that this philosophical characterization plays a large role in support of its widespread success. Einstein's special theory remains valid, and the ultimate product of his thought was the $E = mc^2$ formula for interconversion of energy and mass at those almost incomprehensible speeds. Much of special relativity theory's utility derives from its construction atop an important philosophical premise: immaterial energy must somehow be capable of transformation into manifest matter, because the physical demands the metaphysical as its foundation.

An important feature of special relativity theory is that as object speeds approach the speed of light, time slows down; this is known as *time dilation*.[5] An astounding implication is that *time stops* at the speed of light, which, colloquially expressed, means there is an "eternal now." This conclusion is consistent with a property claimed for theological space: as the medium for transfer of information between global consciousness and self-consciousness, or from one self-consciousness to another, it must allow for this transition to occur instantaneously—with no passage of time. It can therefore be inferred that sending and receiving data through theological space occurs at a speed equivalent to that observed for the motion of light.

The General Theory of Relativity, despite its notoriety as an intellectual breakthrough, was inherently flawed. As mentioned in chapter three, Einstein attempted to overturn Newton's theory of gravity by insisting material mass was the primary cause for salient functional aspects of gravitational force. His approach reversed the intrinsic order of creation and therefore *cannot* be epistemologically correct: it proposed an immaterial condition, the eternally present force of gravitational attraction between substantial objects, arose as a result of the influence of the property of mass constituting those objects. Ervin Laszlo arrived at the same conclusion as the present proposition: "It now appears…mass is not a fundamental property of quanta, but a product of its interaction with the zero point field."[6]

Additional, though indirect, support for the notion of gravitational force being dependent for its existence on masses of material particles as incorrect comes from the lack of success in unifying other fundamental force carriers with the gravitational force field equations developed through employment of general relativity theory. This fact was acknowledged by Brian Greene in *The Elegant Universe*: "Calculations that merge the equations of general relativity and those of quantum mechanics typically yield one and the same ridiculous answer: infinity."[7] If all attempts to reconcile Einstein's views on gravity with other principles of physics continue to fail, it can be concluded that development of a theory of everything will be inhibited as long as the main reasons for failure are upheld as necessary postulates.

Quantum mechanics, despite broad acceptance within the community of physical scientists, creates many of the problems it attempts to solve, as will be extensively documented. Like Einstein's ideas about gravitational force, quantum mechanical theory *must* be erroneous in its practical execution and theoretical implications for the nature of reality. Why? Its epistemological ground is logically deficient—a topic initially discussed in chapter two—which is now another temporary assertion that will be justified shortly.

Ironically, given his theory of gravity as a main *deterrent* to unification on the level of modality, Einstein took the initial step *towards* unification of all three domains of physical theory when he proved the existence of a relationship between Newtonian physics and the notions embodied in special relativity theory. He showed his high-speed effects for the movement of light over astronomical distances simplify to Newton's mechanical laws of motion at moderate speeds covering spatial distances typically experienced through a simple ratio of the two velocities, a ratio so frequently encountered it has been given a special symbol: $\gamma = 1/(1\text{-}v^2/c^2)^{1/2}$, where 'v' is the velocity of an object moving much slower than the speed of light and 'c' is the speed of light.

Taking Einstein's triumph as a sign, the physics community in the early decades of the twentieth century was convinced a similar synthesis was in the making between the Newtonian-Einsteinian convergence and quantum mechanical theory. Over the following decades, these hopes were completely dashed. David Bohm explained in *Wholeness and the Implicate Order* why the merging of Newtonian-Einsteinian physics with quantum mechanics was innately problematical. Why, in other words, Brian Greene's above-quoted statement was true and had to be true: "Relativity theory requires continuity, strict causality, and locality.... Quantum theory requires non-continuity, non-causality, and non-locality. So the basic concepts of relativity and quantum theory directly contradict each other. It is therefore hardly surprising...these two theories have never been unified in a consistent way. Rather, it seems most likely that such a unification is not actually possible."[8]

The characteristics of continuity, strict causality, and locality in contrast to those of non-continuity, non-causality, and non-locality must be explained so that sense can be made of this remark. Continuity and non-continuity, or continuity and discreteness, have already been encountered in chapter three. The essential lesson from that discussion was that to connect the manifold of individual events occurring in discrete moments of time or discrete positions in space requires admission of transcendental principles: only the adoption of a meta-perspective allows for cohesion of elements on a particular level of existence.

The transcendental principle for space is Space and that of time is Time; cosmological effects must derive from, and conform to, Causes occurring cosmogonically. "Relativity theory requires continuity" because Newton's Laws, relativity's anchor, posit space and time as entities with real intrinsic being, which means space and time are incorporated into the characteristics of phenomenal existence, and derive their reality from an innate connection to metaphysical concepts of Being, Space, and Time. Central to quantum mechanics is discreteness: the existence of discretely interacting subatomic particles; discretely detectable movements of these particles, as in the motion of electrons in chemical compounds; and, discrete parametric relationships evidently embodied in formulation of Heisenberg's Uncertainty Principle.

The Heisenberg Principle asserts that on the microscopic scales of time and space upon which atomic and subatomic operations unfold, one cannot obtain certainty in accuracy of the measurements necessary to calculate some of the most important quantities related to particles: energy, time, position in space, and momentum, with the latter being the product of mass and speed.[9] Chapter two's discussion of quantum mechanics revealed it to be among the sciences explicitly using the anthropocentric hypothesis, "man is the measure of all things." Since *Homo sapiens* are anatomically constructed to observe discreteness, it would be impossible for human beings using their own assets to perceive continuity even over these short distances or small time periods. Quantum theory's demand for utilization of ideas such as "non-continuity, non-causality, and non-locality" derives from a human inability to rationally connect the manifold through our own efforts.

With respect to causality *versus* non-causality, the former has also been expressed as transcendentally originated and introduced into manifestation commencing with Interpassage. It will be recalled that the logic of causality demands spatial and temporal continuity, a triumph of insight articulated by David Hume. The philosophy of Newtonian-Einsteinian principles of physics incorporates as a subtle premise the existence of a transcendent agent, which both Newton and Einstein explicitly conceived as God, and which has here been called Consciousness. Both scientists regarded the universe as a place in which one event smoothly transitioned into the next and was subject to exact mathematical formulation: God ensures continuity within manifestation both physically and mathematically.

Though Paul Janet was referring to the beliefs held by Gottfried Leibniz, the words applied to the views of Newton and Einstein: "Now, whatever acts is force; therefore, everything is force or a compound of forces. The essence of matter is not, as Descartes thought, inert extension; it is action, effort, energy.... The universe is a vast dynamism, a wise system of individual forces, harmoniously related under the direction of a primordial force, whose absolute activity permits the existence outside of itself of the appropriate activities of created things, which it directs without absorbing them."[10]

Given a Humean-type analysis, both Newton's and Einstein's formulas intrinsically contained the elements for causality to be operational: 'A'+'B' marked spatial boundaries while acknowledging the transcendental property of addition allows for conjunction; 'A' before 'B' implies continuity in time;

'A' always before 'B' implies predictable regularity. Quantum mechanics preaches non-causality expressed through unpredictability as its essence.

The quintessential illustration is the Schrödinger wave equation, whose mathematics is so complex exact solutions are available only for hydrogen, the simplest element. Substitution of some of the parameters for helium, the next simplest element, in place of those for hydrogen in the equation allows for solution only by making several simplifying approximations. Subsequent interpretation of the wave equation stipulates physical properties can only be calculated within the limits of a probability range of possible values. This lack of elegance is due to the equation's failure to include a transcendental component. Parameters used in the formulation are conceived as physical, emphasizing discrete spatial distances in conjunction with discrete intervals of time. Hence, inexactitude and ambiguity are two of the essential traits of quantum mechanical theory. These characteristics are inharmonious with the exactitude and precision expressed both quantitatively and qualitatively by the Newtonian-Einsteinian synthesis.

Lastly, locality was a contingent characteristic of Newtonian-Einsteinian physics. If space is discrete when particular properties of objects in space are measured, within manifestation events can only occur in prescribed regions of adjacent spaces. Without explicit reference to inclusion of Consciousness as a factor in physical phenomena—neither Newton nor Einstein ever sought to include God *quantitatively*—physical effects must be conceived to occur within a limited domain over brief regions of time and short locales in space. Quantum mechanical theory took a different path. It made repeated allusions to non-local phenomena, an example being an electron able to mysteriously "tunnel" through an energy barrier to arrive at a physical location completely discontinuous with respect to its prior location.

It should now be obvious why the Bohm citation professing the innate incompatibility of relativity theory with quantum mechanical theory holds true. What is central to the philosophy of quantum mechanics is completely antithetical to that at the core of Newtonian-Einsteinian postulates. Those most responsible for the formulation of quantum mechanical theory—with the notable exception of Einstein—believed the transcendental agent could be dispensed with, that it was the human observer who determined when and where events began and concluded. Under this premise, it was inevitable the mathematics of quantum mechanics would devolve from precision contained within Newtonian-Einsteinian formulas to approximation and uncertainty, since it is these latter traits that seem to often characterize human experience.

An absence of modifications to the foundations of quantum mechanical theory renders its unification with Newtonian-Einsteinian physics impossible on the level of modality. A radical restructuring of microscale interactions must be developed if a theory of everything is to be attained. If Newtonian-Einsteinian physics and quantum mechanics, as methodologies, cannot be successfully combined, what about unification with respect to typology of the primary forces conceived as operational in the universe? In fact, efforts along the path to a typological union of natural forces predate the historical development of relativity and quantum theories. It was the initial successes obtained in efforts to formulate bridging concepts that originally led to the term *unified field theory*.

It had been known since the mid-nineteenth century that electric currents generate magnetic fields at right angles to the direction of current flow. Thus, physical phenomena involving electricity and magnetism referred to related events. The first successful accomplishment along the path to a unified field theory on the typological level was taken in the 1860s. James Clerk Maxwell united the laws governing electrical and magnetic phenomena through a set of four mathematical equations.

It is significant that an electric current operates according to an *intrinsic duality* consisting of a region of electron accumulation, the negative pole or *cathode*, and a region of electron deficiency, the positive pole or *anode*. The field effects for magnetism are also demonstrably bipolar: there is a specific direction of flow for the magnetic lines of current from the north pole of a bar magnet to the south pole—a nomenclature developed in accordance with the Earth's magnetic field, which has points of origin roughly equivalent to geographic North and South poles.[11] Central to *electromagnetic* interactions is a simple principle: opposite charge types attract each other while charges of similar type repel each other.

It is easy to read a metaphysical justification into these phenomena: the pattern derives from Interpassage of two unique Conscious Spaces, and the omnipresent attraction of opposites is necessary for existential reality to *be* in existence. The mutual repulsion pertaining to similarly charged electrical and magnetic field types would be the cosmological equivalent of cosmogonic Separation, an outcome meaning cessation of physical spacetime. Of further interest, the force of gravity does not display corresponding polarity; it is always an attractive force everywhere. Gravity was also uncovered prior in *time* to electromagnetic attraction and repulsion. Attraction of any Conscious Space for itself has a requisite logical priority before Interpassage, the onset of *supra*physical *Time*, can offer a mechanism for coupling between any two absolute Spaces leading to generation of the options of mutual attraction and repulsion.

Aside from macroscopically important gravitational and electromagnetic forces, there are believed to exist two other universally operational forces, both applicable solely to the microscopic level on which subatomic particles interact. They are the *strong force* and *weak force*. The former is designated as the force needed to maintain the peaceful co-existence of protons inside the nucleus of an atom. Postulation of the strong force is necessitated by the fact that each proton carries a positive electrical charge, and according to the fundamental laws of electromagnetism, particles with like charges repel one another in inverse proportion to the square of their distance of separation. This means, in the small confined spaces of the atomic nucleus, the repulsive force between two or more protons is very great. The strong force serves as a potent brake on this tendency. In the absence of such an overpowering force, the nucleus of every atom would disintegrate causing destruction of every chemical element in the universe with the exception of hydrogen, which has only one proton in its nucleus and therefore exhibits no naturally occurring internal repulsive force. A universe composed of hydrogen could not support living creatures.

The unique property of the strong force is its power to influence protons *increases* as the separation between particles increases—completely opposite to the trends of interaction for the electromagnetic and gravitational forces—thereby ensuring integrity for atomic nuclei almost independent of how many positive charges are confined to that small region of space. In actuality, there does seem to be a limit to how many protons can be "stuffed" inside the nucleus and still have the atom maintain stability. The key constraint is the ratio of the number of neutrons—particles with no electrical charge—to the number of protons. When the stability ratio is violated, which is the case for all elements containing at least eighty-three protons, spontaneous radioactive disintegration of the atom takes place.

The weak force is, in a sense, the opposite of the strong force in that it allows for specific types of radioactive disintegrations of chemical elements. Radioactive disintegration refers to the transformative interconversion of one element into another by means of an alteration in the number of protons, neutrons, and electrons normally possessed by each element in its lowest energy *ground state*. This alteration occurs through a small set of possible pathways denoted as *alpha emission*, *beta emission*, *positron emission*, and *electron capture*.[12]

Distinctive properties of the elements allowing for chemical combination into myriad forms comprising material reality—producing solids, liquids, or gases with varying structures and novel properties—depends upon elemental differences in ground state composition. A chemical basis for transmutation of elements such as the fabled conversion of lead into gold is not medieval myth: the disintegration of plutonium with simultaneous release of enormous amounts of energy is the principle grounding development of the atom bomb.

It is the role of the weak force to allow some transitions between elements to occur under specific unique conditions.[13]

There is a correlation between the four forces and the four teleological options accruing as a consequence of Interpassage: (a) eternal operation of a force of gravitational attraction models a universe permanently existent in spacetime; (b) bipolar attraction-repulsion, such as exists for electromagnetic force, mimics an origin to spacetime, "on state," followed by a termination of spacetime, "off state"; (c) the power of the strong force to maintain stability within limits of proton number and movement depicts an origin to spacetime coupled to a limitation of the temporal aspect due to an absence of motion of existent objects; (d) the weak force promotes disintegration of particles as an analog of an origin to spacetime combined with a limitation of the temporal aspect due to the disappearance of all objects.

These four forces function over distinctly different spatial distances with widely divergent efficacy with respect to their effect on material objects. In *The Elegant Universe*, Brian Greene described the quantitative strengths of the four fundamental forces:

> The electromagnetic repulsion force [between two electrons] is 10^{42} times stronger [than the gravitational attractive force between the electrons]... The only reason the electromagnetic force does not...overwhelm gravity in the world around us is that most things are composed of an equal amount of positive and negative electric charges whose forces cancel each other out. On the other hand, since gravity is always attractive, there are no analogous cancellations... Experiments also have shown that the strong force is about 100 times as strong as the electromagnetic force and about 100,000 times as strong as the weak force.[14]

By the 1970s, three of the four forces had been unified into two separate theories using quantum mechanical principles as the binding technique, *with gravity remaining as the outsider*. Electromagnetic and strong forces were joined through *Quantum Chromodynamics Theory*; the electromagnetic and weak forces achieved unification by virtue of *Quantum Electroweak Theory*. These two theories are "quantum" in nature because, in Greene's words, "all of the probabilistic and uncertainty issues are incorporated from the outset."[15]

Coherence between the strong force and electromagnetism depends upon usage of a particle theory known as the *Standard Model*, about which more will be written when discussion of material composition is broached. The union between electromagnetism and the weak force apparently holds only for the high energies found associated with the temperatures generated in the earliest moments of the Big Bang.[16]

It should not be assumed that coordination between electromagnetic and quantum theories pertaining to strong and weak forces vindicates quantum theory as a whole. Use of approximation, a general feature of all quantum theories, may conceivably smooth over inconsistencies between a theory conducive to explanation of ordinary experiences, electromagnetism, and quantum approaches adapted to atypical situations pertaining to ultra small particles and ultrahigh temperatures. As an analogy, consider two irregularly shaped jigsaw puzzle pieces artificially joined by removing the rough edges defining each piece, whereas retention of the edges would clearly reveal actual incompatibility of the shapes.

The salient reason gravitational force remains resistant to adaptation for unified field theory must be highlighted. After development of the Theory of General Relativity, Einstein spent the rest of his life attempting to unify the force field equations for electromagnetism devised by Maxwell with his own force field equations for gravitational effects.[17] Einstein's disagreement with Newton over the nature of gravity has been highlighted: Newton conceived gravitational influence as operating independently in a continuous spacetime framework; Einstein declared gravity was a dependent construct arising from the influence of an object's mass and the resultant curvature of space. Prior analysis suggested Newton was right: the nature of continuity demanded the independent existence of gravity with a transcendent connection. With two completely different theories concerning the origin and meaning of gravity, if the *wrong* conceptualization for the correct characterization of gravity is the one invoked in unification efforts, it should not be surprising the synthesis of gravitational force with electromagnetic force has been rather elusive.

The theory being argued for suggests invoking global consciousness as the most appropriate means to unify gravitational and electromagnetic forces. This agency invests both forces with their inherently transcendental character of continuity. This agency invests those twin forces with unipolar or bipolar properties because global consciousness represents the immanent analog of a two-in-one merger of Conscious Spaces obtained from commencement of Interpassage. It is this *metaphysical* agency Ervin Laszlo sought *physically*:

> Contemporary physics knows four varieties of universal field...the gravitational, electromagnetic, and the strong and the weak nuclear fields.... There is...evidence of interconnections among phenomena that transcend the known limits of space and time. Gravitational and electromagnetic fields...cannot be responsible... We should entertain the possibility that a 'fifth universal field' may exist in nature.[18]

If the known properties of gravitational and electromagnetic fields can be supplemented through their involvement in *mental* activities incorporating Consciousness and using theological space, they *are* sufficient to explain "interconnections among phenomena that transcend the known limits of space and time."

Current attempts pursue development of quantum theories of gravity, but all efforts have been uniformly unsuccessful due to the inherent discrepancy in fundamental viewpoint: the metaphysical nature of the gravitational force demands recognition of the incorporation of the transcendental into physics, while traditional quantum mechanics rejects that notion. Instead of the failure to unify Einstein's notion of gravity with the remaining three forces leading to efforts at solution through a return to Newton's ideas about the qualities of gravity, the opposite route has been chosen: physicists seek to reconfigure Einstein's notion of gravity according to the dictates of quantum mechanics. According to the characteristics already shown to be germane to the *modal* differences between these domains, it might be anticipated such a procedure could only make the problem worse—which it did. Greene admitted that the physical and mathematical outcome of all attempts support the fundamental incompatibility:

> *Everything* is subject to…quantum fluctuations inherent in the uncertainty principle, even the gravitational field…. *The notion of a smooth spatial geometry, the central principle of general relativity, is destroyed by…violent fluctuations of the quantum world on short distance scales….* The basic principles of general relativity and quantum mechanics allow us to calculate the approximate distance scales below which one would have to shrink in order for the spatial fluctuations to become apparent. The smallness of Planck's constant…and the intrinsic weakness of the gravitational force team up to yield a result called the *Planck length.*[19]

The philosophic difficulty barring inclusion of gravity in field unification efforts is continued conceptualization of it as Einstein did: as a phenomenal product of a wholly localized discrete mass operating on a discrete localized region of space. "Empty" space is not *empty* however much it may be empty of *object mass*. Physical space is alive with the Energy of Consciousness—the isothermal radiant energy of background space, which may be identified with Laszlo's zero-point energy field—and is neither inert nor empty. This Energy manifests as the universal self-subsistent force of gravitation, whose effect is always attractive, meaning always engaged in the process of making space a continuum *just as Newton conceived it to be*. It is this thesis of the philosophical-physical meaning of gravity that can be utilized as grounding construct for its conceptual mating to electromagnetic force, the latter being another form of immanent, immaterial Conscious Energy.

Turning to the final necessary approach to a theory of everything, one constructed on the level of *material composition*, every first-year chemistry student learns the principal subatomic particles are protons, neutrons, and electrons. What is remarkable is that since the discovery of the neutron in 1931 several hundred additional particles have been catalogued as products of a diverse array of high-energy nuclear reactions. After decades of intense effort, a *Standard Model* has been devised that attempts to (a) classify these minutely small particles according to their characteristic mass, energy, and charge; (b) justify their reactivity patterns; and (c) predict future discoveries. The Model is both complex and cumbersome.

Protons and neutrons are no longer considered fundamental particles. It has been proposed they are each composed of a set of three *quarks*, which number six in total and have been given exotic names: *top, bottom, up, down, charm*, and *strange*. In addition, each quark is posited to possess a property called *color*—available in either *red, blue*, or *green*—descriptive of the kinds of interactions the quarks can undergo to form novel particles. A small set of quarks unite to form a diverse array of novel subatomic particles in the same fashion as a relatively small set of unique elements unite to form a diverse array of novel compounds. A further elaboration of quark theory claims each quark has an associated *anti-quark* available in three *anti-colors—magenta* (anti-green), *cyan* (anti-red), and *yellow* (anti-blue)—providing a foundation for the alleged symmetry between matter and anti-matter. Instead of a simple formalism consisting of three elementary particles—protons, neutrons, and electrons—the Standard Model generates, minimally, thirty-six elementary particles: six quarks in three colors plus six anti-quarks in three anti-colors.

It has been mentioned the nuclear strong force is conceived as the agency holding atomic nuclei together in opposition to the repulsive electromagnetic force operating between like-charged protons. The question arises: How does the strong force translate into the Standard Model if it conceives protons and neutrons as composed of quarks? According to the dictates of the scheme, a proton is comprised of two up quarks and one down quark, while neutrons are made up out of a combination of two down quarks and one up quark. The rules for quark interaction state each particle must consist of three quarks and each quark must be of a different color. It is hypothesized that agents called *gluons* are mediators of the strong force, although it must be acknowledged *gluons have never been experimentally observed.* Gluons supposedly consist of a single quark color in conjunction with an anti-color.

If anyone is confused by this *ad hoc*, seemingly arbitrary scheme, clarity will not be attained upon illustration of the kinds of processes contemplated to occur through use of this Model. In *The Physics of Consciousness*, Evan Harris Walker offered an example of the kind of conversion conducted under the hegemony of the Standard Model: "Inside a proton containing...a red up quark, a blue up quark, and a green down quark, a red-magenta gluon forms out of the red up quark. The red color of the up quark passes into the gluon, and magenta (anti-green) forms as the gluon leaves green color behind in this up quark. Now the gluon travels to the green down quark, where the magenta neutralizes its green color, which permits the red to be deposited to form a red down quark."[20]

The net result of this transformation is to convert a red up quark into a green up quark and a green down quark into a red down quark. The strong force serves to continuously transform the identities of quarks inside protons and neutrons allowing for nuclear stability in opposition to proton-proton electrostatic repulsion. Gluons can also interact with each other, and Walker gave an example: "A red-magenta (anti-green) gluon interacting with a blue-cyan (anti-red) gluon turns into a single blue-magenta gluon. The anti-red cyan on one gluon joins with the red on the other gluon to cancel out, leaving just one gluon to carry...blue...and magenta."[21]

The plethora of names and classes of subatomic particles doesn't stop with quarks, anti-quarks, colors, anti-colors, or gluons. Protons and neutrons in an atomic nucleus are *hadrons*, and are different in kind from electrons, a constituent in the *leptons* category, which the Standard Model proclaims are authentic elementary particles. Consistent with expected symmetry between matter and anti-matter, there are also *anti-leptons*. Just as there are six quarks and six anti-quarks combining to form the hadron family, six leptons and six anti-leptons form the lepton family.

This grouping consists of the following particles: *electron*, *muon*, *tau*, *electron neutrino*, *muon neutrino*, and *tau neutrino*. The symmetry employed by the Standard Model is so well defined the six quarks pair up with the six leptons to form three distinct collections of particles with nearly *redundant properties*. The three sets have the composition: (a) up quark, down quark, electron, and electron neutrino; (b) charm quark, strange quark, muon, and muon neutrino; (c) top quark, bottom quark, tau, and tau neutrino. The kinds of redundant properties across the three sets include electrical charge, color, and weak force charge; only particle masses in the three groups are unique to each particle.

Corresponding to gluons serving as mediators of the strong force holding hadrons together, of which there are eight in number, there are three carriers of the weak force mediating interconversion of quarks such as transformation of an up quark into a down quark. Recall that the weak force permits specific radioactive decay modes characteristic of special element forms, *isotopes*, possessing nuclear instability. To account for these transformations using the vocabulary of the Standard Model rather than the previously expressed alpha emission, beta emission, and so on, specialized particle agents are claimed to be needed to assist these processes. The weak force particles are designated W^+, W^-, and Z^0, where the superscripts indicate electrical charge.

In sum, the Standard Model is said to consist of six quarks available in three different colors, six anti-quarks available in three different anti-colors, six leptons, six anti-leptons, eight gluons needed for operation of the strong force, and three otherwise unnamed particles required for operation of the weak force. This leads to a total of fifty-nine *elementary* particles. Inclusion of the electromagnetic force carrier posited by quantum mechanical theory, the *photon*, brings the total to sixty.

Despite the dedication needed to develop the Standard Model, *virtually no physicist believes the Model to be correct as constructed.* The reason most frequently given for its simultaneous acceptance as the best theory currently available and non-acceptance as a final solution is its lack of inclusion of gravitational effects at a subatomic level. The Standard Model requires force carriers to have associated particles mediating activity. This demand would necessitate the positing of a particle corresponding to the gravitational force. No evidence has been found for a gravity particle, a *gravitino*, corresponding to the photon. Although gluons have also never been observed as singular particles, this has not resulted in their exclusion from the Model. While lack of experimental data supporting the existence of gravitinos is a legitimate and damning flaw, there are two other reasons to reject the Model in its current form.

First, there is the Model's lack of articulation concerning the physical basis for the property of color. It is impossible for this variable to relate to electrical charge: first, this property occurs in only *two* forms, positive and negative, while every hadron is composed of *three* quarks; second, not all hadrons are electrically charged—protons are, but neutrons are not. Also excluded as a physical basis is the particle's direction of spin: this feature also exists in *two* versions denominated *clockwise* or *counterclockwise*.

A property is needed capable of appearance in three distinct forms to rationalize the postulation of hadron formation requiring three quarks. With the absence of a physical meaning for the attribute of color comes an absence of a convincing rationale as to why there must be overall color neutrality for each particle.

Is it possible color is related to time, an outgrowth of the Principle of Charge-Parity-Time Conservation referenced last chapter? The three colors could pertain to the threefold nature of experiential time as past, present, and future. Requiring quarks with complementary colors to combine in forming particles with overall color neutrality would then serve to keep particles time symmetric, which the Schrödinger wave equation assumes to be the case. The anti-colors would refer to the passage of time in the parallel universe—more accurately, the anti-universe—postulated to form with our own due to Interpassage. Gluons, composed of a color and an anti-color, would induce symmetry in another sense: the direction of color flow, meaning time flow of past to present to future, in a particle universe would be counterbalanced by the direction of anti-color flow, meaning time flow from future to present to past, in an anti-particle universe. The resultant non-observational status for gluons could then be based on mutual annihilation of matter and anti-matter particles outside the confinement of a quark.

This speculative proposition notwithstanding, the second reason to reject the Standard Model as now constituted is the unexplained triple redundancy of properties possessed by hadron-lepton combinations. If the only physical property differentiating these groups is individual particle mass, it is unclear what the rationale would be for asserting all three as fundamental. This is a powerful criticism given only one hadron-lepton set is needed to explain the composition of a chemical atom—the up quark, the down quark, the electron, and the electron neutrino—as these correspond to the protons, neutrons, and electrons previously considered to be the basic building blocks of materiality according to old-fashioned atomic theory.

Mass *is* a property important to manifestation of gravitational effects. It is possible the redundancy would be broken with respect to some physically meaningful consequence of gravity. Failure to include reference to gravity in the Standard Model would again seem to be a key element in its overall lack of acceptability. It has been repeatedly emphasized that gravity's role in the universe derives from its transcendental cosmogonic significance, and the Standard Model is devoid of allusion to *supra*physical reality.

In review, diligent efforts toward development of a theory of everything on three levels—modality, typology, and material composition—have been uniformly unsuccessful. The principal causes of failure have been: (a) use of Einstein's characterization of gravity, or attempted modifications of the idea by creating a quantum-scale gravitational theory; (b) inclusion of simplifying approximations in quantum mechanical calculations arising from the quest for mathematical analyses of time, space, and energy on increasingly smaller orders of magnitude. These simplifications are accompanied by an inherent uncertainty in producing accurate representative descriptions for existential reality.

It has already been claimed the first difficulty can be corrected by return to Newton's conception of gravity as a thing in itself, a continuous aspect of space having an innate transcendental connection. It is now proposed the second barrier to universal field theory can be amended by a revaluation of the philosophic meaning inhering in small-scale physics. To demonstrate the validity of this contention, a detailed inquiry into the philosophy of quantum mechanical theory must be undertaken rather than focusing on the results obtained from the theory, which has been the case heretofore.

First, what is the Schrödinger wave equation, and what does it indicate? Second, how does one interpret its results? Third, given the types of practical mathematical, and conceptual physical, approximations employed, why does the equation itself as well as its subsidiary calculations yield agreement with experimental observations so successfully?

The wave equation is a mathematical expression correlating physical properties of atoms with the presumed movement of electrons within three-dimensional space over a specified duration in time. Chemistry provides incontrovertible evidence electron movement within and between different elements is ultimately responsible for almost all macroscopically observable features of any naturally available or synthetically made material. Values for many properties can be calculated *if* assumptions are made during solution of the mathematical equations. The Schrödinger equation has been generalized to quantitate the properties of subatomic particles other than electrons.

With respect to the second question, application of quantum mechanical principles and the Schrödinger wave equation have become so central to the physical sciences that four different interpretations of the expression's underlying meaning have been devised. The first—generally found to be the most acceptable of the four—is named after the city hosting the conference at which it was first proposed, the *Copenhagen interpretation.* This version claims the equation expresses a very large number of solutions in its raw form, yet collapses those numerous possibilities into one unique actuality when physical measurements are eventually performed on a system. Peter Coveney and Roger Highfield attempted to explain this remarkable result in *The Arrow of Time*:

> When a particular measurement is made, the electron will be recorded as having arrived at one...position.... The wavefunction...must undergo an instantaneous transformation when...measurement is made.... This transformation from a plethora of potential outcomes to the one observed is known as the "reduction" or the "collapse" of the wavefunction.... It is clear...the collapse cannot be described by the Schrödinger equation itself, for that equation is reversible and deterministic, while the collapse is irreversible and random.[22]

The second interpretive version is the *stochastic* alternative. It rejects the concept of wavefunction "collapse," and alleges all solutions are offered as relative statistical probabilities verifiable in the real world if a sufficiently large number of similar experiments are performed. Formal calculation of, for example, a particular particle mass would, according to the Copenhagen interpretation, be produced from the spectrum of all possible answers only when someone actually goes to measure that mass. In contrast, the stochastic interpretation claims all possible masses would be found experimentally, but the value corresponding to the highest statistical probability would be the one most frequently encountered if enough trials were undertaken to determine the particle mass.

The third suggestion is very different from the prior two. Instead of there deriving from solution of the wavefunction either a single exact answer or a range of probable answers, the *many worlds* thesis asserts every calculated solution occurs in some real, parallel-type universe. Since these universes are distinct from each other, the likelihood of physical enactment of even two, much less all, possible solutions is incapable of proof: we have no access to a parallel universe where the mass of the particle might be different than what we observe it to be in our universe. The argument for existence of parallel-type universes is *not* one similar in form or substance to the invocation of an anti-universe as a consequence of Interpassage.

In the latter proposal, the anti-universe must physically exist *necessarily* because its reality is implicitly derived from consideration of a metaphysical thesis of inherent duality of Conscious Spaces, and evidence for support of the existence of an anti-universe obtains from the known formation of anti-matter and invocation of the Principle of Charge-Parity-Time Conservation. The former proposition contains no necessity for such universes to exist in actuality other than as fulfillment of the principal idea. In other words, the many worlds contention is self-reflexively circular in its construction: the multiplicity of posited universes justifies the existence of multiple solutions to the wave equation, and the abundance of solutions justifies the postulated existence of realms of space in which such outcomes are enacted.

The fourth proposal is denominated the *hidden variable interpretation*. This concept has as a core construct the claim of something internal to the particle that: (a) determines the value of the property in question; (b) enables the value to be accessible to human measurement; (c) invariably produces conformance to one result obtained by solution of the Schrödinger equation. In *The Holographic Universe*, Michael Talbot presented David Bohm's ideas about the underpinnings of this interpretive orientation: "Bohm...believes a quantum wave does not push an electron about... It provides the electron with *information* about its environment...the electron...uses to maneuver on its own. In other words, Bohm believes...an electron is not only mindlike but is a highly complex entity, a far cry from the standard view that an electron is a simple, structureless point."[23]

Essentially, the theory of hidden variables postulates subatomic particles, like everything in manifestation, as being comprised of Consciousness. They might possibly even possess the attribute of self-consciousness. The premise is that, given Consciousness as fundamental, particles have an innate ability to be self-deterministic. By analogy with sentient beings, subatomic particles are viewed as able to communicate, meaning information can pass between them when necessary. Ervin Laszlo appeared to accept this interpretation. In *The Interconnected Universe: Conceptual Foundations of Transdisciplinary*

Unified Theory, he applied his doctrine concerning the existence of a vacuum field—possibly a physically descriptive synonym for theological space—as a needed component in explaining the *Pauli Exclusion Principle*:

> In interpreting the physical events that underlie Pauli's exclusion principle...we can claim that in atomic shells electrons are interconnected by a signal-transmitting field. The wavefunctions of individual quanta within the configuration space of the atom are quasi-instantaneously informed with the wavefunction of the ensemble—an "in-formation" that skews possibilities in the otherwise indeterministic quantum state toward that set of quantum numbers which corresponds to the position of each electron within the ensemble.[24]

The Exclusion Principle, named after its inventor Wolfgang Pauli, states there are a set of four numbers that can be used, like a zip code, to locate any electron positioned around the nucleus of an atom, and no two electrons can have the same four numbers. Thus, every electron has a distinct location in space. Laszlo's statement employed the notion of a "quasi-instantaneous" ability for the electrons to transfer information among themselves. While he did not explicitly claim the "signal-transmitting field" was Consciousness, it is reasonable to interpret it as such. Theological space enables instantaneous communication between source and recipient because information transfer occurs at the speed of light indicative of zero time passage. His remark was wholly consistent with the metaphysical speculation being promulgated.

It is the conflict generated between the first and last of these proposals that has attracted the most attention for physicists. Is it possible to adjudicate between the options to select the one conforming best to available data? One of the earliest repercussions of quantum mechanical theory was an assertion about the preferred way to characterize elementary material objects. Many experiments, it was claimed, were performed indicative of activity associated with discrete particle point-mass forms; other research results seemed more easily explained as the outcome of actions by a continuous energy waveform.

A compromise position was finally adopted: the idea of *wave-particle duality* professed the same microscopic object was intrinsically both, and it was the design of each experiment that determined whether the fundamental agents preferentially "clothe themselves" in one manner or the other. Some physicists began to denote this ambiguity by the term *wavicle*. An illustration of such alleged duality can be found in *The Self-Aware Universe*:

> A beam of light is split into two beams, each of equal intensity—one reflected and one transmitted—by using a half-silvered mirror. These two beams are...reflected by two regular mirrors to a crossing point. To detect the wave aspect of the wavicle, we take advantage of the phenomenon of wave interference and put a second half-silvered mirror near the crossing point. The two waves...are now forced to interfere constructively on one side of the crossing point and destructively on the other... If

> we choose to detect the particle mode...we take away the [second half-silvered] mirror...and put counters past the point of crossing... One or the other counter will tick, defining the localized path of a wavicle, the reflected path or the transmitted path, to show its particle aspect.[25]

Does this procedure demonstrate what it claims? Another straightforward interpretation eliminates any need to posit wave-particle duality. The light beam is always to be formulated as a wave capable of traveling through both normal three-dimensional space and an invisible-to-us transcendental fourth Spatial/spatial domain, theological space. When it encounters an object in physical space, the spatial component appears as a point source *in* physical space.[26] The decision to place or not place the second half-silvered mirror in the beam path had no effect on the continuous wave nature of the photon; it altered what was observed due to a change in Energy/energy conditions for the experiment.

It is improper to consider ourselves the anthropocentric "king-makers" of the universe possessed of powers to determine the real form of light given conditions of our choosing during conduction of an experiment. Rather, we are bystanders to a phenomenon whose description by us varies as we alter our position relative to the event. The situation in a physics experiment is no different in principle from the experience encountered during an automobile accident: observers will describe the collision with noticeable alterations of interpretation based entirely on their individualized perspective as witnesses. The facts don't change; the characterizations of the facts do. Michael Talbot made the point in *The Holographic Universe*: "Perhaps most astonishing of all...there is compelling evidence that *the only time quanta ever manifest as particles is when we are looking at them.* For instance, when an electron isn't being looked at, experimental findings suggest that it is always a wave."[27]

David Bohm developed a similar idea, but employed the neologisms of *implicate order* and *explicate order*:

> Electrons and all...particles...are sustained by a constant influx from the implicate order, and when a particle appears to be destroyed...it has...enfolded back into the deeper order from which it sprang.... The constant and flowing exchange between the two orders explains how...a quantum can manifest as...a particle or a wave.... Both aspects are always enfolded in a quantum's ensemble, but the way an observer interacts with the ensemble determines which aspect unfolds and which remains hidden.[28]

Bohm's contention—"both aspects are always enfolded in a quantum's ensemble"—is, like the notion of a wavicle itself, a violation of the Law of Noncontradiction. In an unmodified form, the proposition is just as innately illogical, against the rules of logic, as a "round square." The "round square" must be either round or square, even if one can be fooled by tricks of lighting or by the angle upon which you gaze at the figure into thinking it fluctuates rapidly between both. Similarly, a wavicle is *always* a wave, always conveys transcendental continuity, although a physicist may be fooled into thinking it is a particle due to the use of our senses with their ingrained discreteness.

Consistent explanations for phenomena stemming from the property of energy as *always* a wavelike field effect—an immanent representation of global consciousness, which is, in turn, a universal analog of transcendent Consciousness fields—underscores a plausible alternative interpretation of the *photoelectric effect.* The experiment demonstrated that the number of electrons removed from the surface of a metal depends upon the intensity of the light focused upon it, while the speed of those electrons is dependent upon the frequency of the light.

Einstein's interpretive conclusion won him a Nobel Prize, and invoked the image of a tiny ball of light colliding like a subatomic billiard ball with an electron, another subatomic billiard ball. This was one of the original justifications for the notion of light as composed of discrete quanta, and helped launch quantum mechanics as a theory of reality. However, the result could just as easily be explained as a series of photon waves transferring energy to the surface of the metal, which then radiates that energy back as a wave *we observe* as particle-like electrons in three-dimensional space, but which in actuality is a four-dimensional wave through theological space.

Recall the purpose for this lengthy discussion on how to best interpret results of processes occurring on minute scales of space and time. Quantum mechanical theory, which has shown itself to be resistant to efforts to merge its strictures into a coherent theory of everything with Newtonian-Einsteinian principles of physics, employs as a core operating premise the Copenhagen interpretation of events whereby human beings are endowed with the ability to determine the numerical values of *quantifiable* properties of subatomic particles through "collapse" of the Schrödinger equation.

Astonishingly, it now appears we are capable of determining the best *qualitative* description of these acts. Anthropocentrism runs rampant through quantum theory, and is completely absent in the postulates of Newtonian mechanics and Einsteinian relativity. Indeed, the latter places emphasis on the subjectivity inhering in interpretation of physical events as a function of one's frame of reference. In other words, relativity theory explicitly claims no single orientation is necessarily preferred or objectively correct. Is it so unreasonable to suggest anthropocentrism prohibits formation of a unified theory?

In responding to the third interrogative—Why does the equation as well as its subsidiary calculations yield agreement with experimental observations so successfully?—the victory of quantum mechanics over Newtonian laws is asserted to rest on two building blocks, one general and one specific. The generalized argument is constructed on the basis of a negative: there has never been a property of a particle, or of a force, or an outcome of a force on a particle, that has not fit with precision the expected values obtained from use of the wave equation. Therefore, the logic goes, if some procedure seems to always give the correct result, the method used to obtain the answer must *certainly* be valid, must be True.

This general argument is substantiated by a specific result referred to as a test of methodological legitimacy: Alain Aspect's experimental rejection of John Bell's statistical theoretical treatment concerning a particular prediction anticipated from quantum mechanical theory. Despite these professions to superiority of a quantum mechanical description of the nature of the universe with respect to a Newtonian one, do the general and specific propositions support the conclusion in a *necessary* fashion? It is better to analyze them in reverse order in order to provide a proper answer.

The clash between perspectives began in the 1920s as soon as quantum mechanical theory emerged. Peter Coveney and Roger Highfield offered some sense of the history in *The Arrow of Time*: "Schrödinger's equation is accepted as the basis for a new, fundamental mechanics... This has profound consequences for determinism and causality. The only alternative is to deny that quantum mechanics is truly fundamental; instead it is...a method of making statistical statements about events of which we have an incomplete understanding—but which at a deeper...level remain strictly deterministic and causal, just like Newton's and Einstein's physics."[29]

It is well known, a longstanding irony in the history of science, Einstein attempted mightily to disavow his connection to the development of quantum mechanical theory, established in part through the photoelectric effect using discrete light quanta as fundamental units of energy. Above all, he desired to repudiate its interpretive conclusions based on the twin pillars of descriptive indeterminacy and quantitative approximation.[30] His most significant effort toward this end was contained in a paper published in 1935 hypothesizing what has come to be called the *EPR Paradox*, with "EPR" referring to the first letter of the last names of the three co-authors:

> Einstein, Podolsky, and Rosen considered a system consisting of two particles ...permitted to interact with each other and then separate. During...interaction, the particles share momentum, angular motion, and position. As they separate, they share...total momentum...angular motion, and a common point of origin.... The question is, "Do these particles really have an exact position and...an exact velocity, which we are unable to measure without disturbing them, or do these quantities not exist as a part of reality at all?"[31]

The physical meaning of momentum and angular motion are unimportant for the ensuing discussion however pertinent they may be within the context of a technical physics journal. Schrödinger's equation contains all solutions as a set in which any particular solution exists merely as a probability until exact measurements are made according to the Copenhagen interpretation. At that point in time, all mathematical solutions are eliminated except the one actually corresponding to measurement. "Collapse of the wavefunction" has already been used to signify the acts of observation and measurement endow a particle and its associated traits with physical reality. It is this view of reality as nonexistent prior to human determination of both its content and extent that Einstein, Podolsky, and Rosen are questioning.

The critical aspect of this paper's proposal was allowing the particles to interact with each other prior to separating. For a variety of technical reasons not necessary for consideration at present, their interaction means that some properties must be physically and mathematically complementary, thereby restricting the possible values each property can take on once measurement commences after separation. This feature also implied measurements made on one particle should automatically allow determination of the properties of the other particle without making any measurements on that second particle due to the requirements of complementarity:

> "If without in any way disturbing a system, we can predict with certainty a value of a physical quantity, then there exists an element of physical reality corresponding to this physical quantity."... We are left with two possibilities.... We can assume that particle B can have a real position or a real momentum because it has such a reality for both position and momentum all the time.... Or...we can say that particle B does not have a reality for position or momentum until we have measured the position or the momentum of particle A.[32]

According to the EPR presentation, either paired particles already existed in possession of specific values for complementary properties before onset of an experiment and associated measurements, or properties were determinable only upon measurement. If the former, the Copenhagen interpretation had to be erroneous. If the latter, the "paradox" kicked in: How could properties maintain complementarity in value if measurement occurred post-separation? How could separated particles know the value of the property to be measured for its partner particle such that each could maintain complementarity with the other? EPR maintained this could only be done if information could be transmitted faster than the speed of light, an event regarded as impossible according to special relativity theory.

In sum, EPR claimed quantum mechanics incorrectly described reality through its insistence on uncertainty and approximation permeating needed observations and measurements on particles and their properties responsible for creating that reality. The Copenhagen interpretation, accepted by most physicists, was being directly challenged by a hypothetical test. Here was a situation where properties of particles could be exactly known even though no direct measurements were performed, or so Einstein, Podolsky, and Rosen thought.

For thirty years after publication of this paper, neither a counter theory nor an actual experiment was invented to test the predictions of the article's authors and their challenge to the Copenhagen interpretation. In 1965, John Bell published a statistical prediction of results an experimenter would obtain if he or she tested the theory proposed by Einstein, Podolsky, and Rosen.[33] As retold in non-technical language by Sylvia Fraser in the oddly titled, *The Quest for the Fourth Monkey: A Thinker's Guide to the Psychic and Spiritual Revolution*: "Electrons spin...either clockwise or counterclockwise around their own axis, and...two electrons can be put into partnership so...their spin in opposite directions always adds up to zero. Though the experimenter can't know in advance the precise spin of either electron, once the spin and axis of one of the pair has been determined by measurement, the axis and direction of spin of the other can accurately be predicted."[34]

In his seminal publication, Bell expected EPR would be vindicated to some degree. Einstein, Podolsky, and Rosen anticipated particle properties would retain complementarity after separation because their relationship was predetermined as a fixed, unalterable aspect of objective reality. If, however, the properties of one of the pair were physically altered after separation they would necessarily lose their characteristic complementarity because the non-measured particle would retain its original property values. Bell's analysis agreed with these conclusions, but offered a different rationale. He was a proponent of David Bohm's hidden variable interpretation, which treated particle properties not as unalterable, but as capable of being determined at any time up to the moment of separation so they *appeared* complementary. In *The Physics of Consciousness*, Evan Harris Walker outlined John Bell's argument for the range of theoretical outcomes:

> Bell believed there had to be...hidden variables that determined what each photon would do when it encountered a filter. He believed the plan made the two photons look...correlated...but...this was nothing but appearance. Bell believed...any such plan had to have been created when the photons were created...have been arranged before the...photons separated at the very latest.... If the photons can send messages back and forth...then any...outcome would be possible... If the photons cannot send any such message...the correlation between what this photon chooses...and what its twin does must be limited to the information it could carry in its hidden variables.[35]

A statistical treatment of anticipated experimental results, regardless of what they predicted, was insufficient to answer the question involved, a fact obvious to all physicists. Although several experimental attempts were made to subject Bell's prediction to a practical test in the first decade of his paper's appearance in the scientific literature, they collectively proved inconclusive because each design was significantly different and insufficiently rigorous to survive criticism by outside experts.[36] Two decades passed before conditions were sufficiently advanced in a technical sense to put Bell's prediction to an acceptable examination. Since the ultimate question seemed to concern the speed of light as the supreme standard for information transmission through three-dimensional space, rather than using pairs of electrons, Alain Aspect chose to study photon pairs. Photons are *the elementary particles of light according to the Standard Model*, and travel at light speed by definition.[37]

Without invoking a lengthy and complex recitation of the facts of the experimental design and its method of data collection and analysis, the short answer is that the outcome was exactly *opposite* to what was anticipated by Bell: complementarity was maintained under *all* variations of experimental conditions.[38] Aspect's result, unlike earlier trials, was accepted by the physics community at large, and was a victory for the Copenhagen interpretation: properties of particles do not exist—indeed, in some instances, the particles do not exist—until an external observer and measurement is included in the process. To some degree at least, reality is what we choose to make it. It is not an exaggeration to suggest Aspect's research results amounted to the scientific coronation of the anthropocentric premise as an epistemological rule of the universe.

It is a fact quantum mechanical predictive calculations consistently agree with quantitative data derived from experiment. Another fact is that Aspect's experimental results seem to demand transmission of information faster than light to maintain the complementarity of properties if the EPR rendition of alternatives is accepted on its face. These conclusions appear antithetical, as EPR intended: the properties are correlated and complementary whenever measurement is made as the wavefunction containing the entire spectrum of possible outcomes is collapsed to only one actual result, but the mechanism of persistent complementarity appears to violate the speed of light as the ultimate determinant of information transfer. Can these oppositional facts be reconciled? Does resolution support the anthropocentric thesis as rigorously as advocates of quantum mechanics claim?

Bell's theoretical treatment is grounded on an implicit misconstruction. It sought for an objective universal reality: hidden variables confirming the eternal existence of physical properties independent of any measurement. On the other hand, it attempted to confirm this premise through employment of a locally subjective measurement device grounded in the requirements for any human sensory apparatus. It doesn't matter whether the experimental sensory apparatus is our eyes, ears, or a measuring device external to ourselves. It is still ultimately sense-based because even a physical stand-alone data recorder is constructed such that it conforms to our sensory tools in determining what data the box records and how it records it so we can decipher its recording. *An inanimate recorder is still humanly calibrated.*

How could we prove the Truth of an objective reality characterized by constancy—as the EPR thought experiment contemplates—given that we are subjective creatures always undergoing flux? The answer is, "we can't." *The only possible way to confirm the Newton-Einstein-Bell hypothesis of eternal, observer-independent reality is through the assistance of other species whose data acquisition methods operate on wholly different principles than do ours. Then, and only then, by observing the same phenomenon, humans and other sentient creatures can jointly confirm Truth.*

Absolutes are reference standards for all variability. It is not permissible to use relativity to determine standards; we must utilize standards to confirm relativity. Just as mariners determined the precise location of an object at sea by triangulation from different vantage points, so we need to coordinate our knowledge with that of other intelligences in order to converge on that which is invariantly True.

Bell's prediction combined with Aspect's experimental rejection of that prediction does not jointly result in confirmation of the quantum mechanical view of reality. What is obtained is a self-consistent self-fulfilling prophecy: (a) we are creatures capable of only detecting discreteness by virtue of our anatomical construction; (b) we employ mathematical tools grounded on the discreteness of integer numbers; (c) to quantify results from experiments we conduct, we build instruments calibrated to measure discreteness in physical properties; (d) we use as the subject for those experiments objects claimed to be essentially discrete—in this instance, a photon. Should we be surprised to learn our discrete calculations conform to our discrete senses as well as to our experimental designs constructed to convey discreteness to us? No. To understand the True nature of reality, more than self-consistency is needed. *It is for this very reason that all scientific laws are provisional only.* We need a "way out of the box," the prison of our own built-in constraints on perception and conception.

Aspect's result *does* give a useful clue as to a potential mechanism for the necessary reorientation. If complementarity can be maintained only by an appeal to faster than light speeds accessible for dissemination of information about particle properties, then: (a) faster than light speed is possible in three-dimensional space, in which case Einstein's theory of relativity is blatantly wrong; (b) faster than light speed is possible though access to another spatial domain undetectable by our sensory apparatus, and Einsteinian relativity is acceptable if restricted to three-dimensional space; or (c) we are not dealing with *particle* properties at all, but rather with wave properties, and theories about information transfer need recalibration in terms of continuous energy motion.[39]

What domain lends itself to characterization from our vantage point as: (a) *transcendental*, able to display continuity as an inherent characteristic; (b) *global*, able to display its phenomenological properties under all conditions of time and space; and (c) *absolute*, able to transmit information at the fastest possible rate, namely light speed? The answer is: global consciousness, the transcendent Dimension also known as theological space. Einstein, Podolsky, and Rosen were correct: the properties of correlated objects *are* predictable through a perpetual complementarity. To comprehend how this could be so, the solution is to regard transcendental theological space as the medium for communication of those parameters. Theological space offers a mechanism for the instantaneous transmission of information between connected objects because that Space/space always contains an Energetic/energetic component of every object, animate or not.

Data concerning physical properties are disseminated through the fourth Spatial/spatial dimension—the fifth overall with the inclusion of time—and these data can *appear to be* transmitted faster than light without violation of Einsteinian relativity establishing light as the ultimate speed limit, which is a statement applicable only to four-dimensional spacetime physical processes. The emphasis on *appearance* is important: data transmission does not occur *faster* than light speed; rather, it occurs instantaneously in theological space, a domain in which *time stops* because all movement is *at* the speed of light. Speed is a function of distance and time; it has no meaning in a Dimension lacking both. The properties can be correlated before and after separation in normal spacetime precisely because they are *never* separated in theological space. In this way, a "non-local" phenomenon is understood to be a local event because Space/space is continuous even as we analytically dissect it into discrete pieces.

The Self-Aware Universe offered a significant portion of this proposal: "Thus, what Bell's theorem and Aspect's experiment imply is not a violation of causality but that simultaneously occurring events in our space-time world can be related meaningfully to a common cause that resides in a non-local realm outside space and time. This...is the act of non-local collapse by consciousness."[40] "Consciousness" can mean the typical quantum mechanical interpretation of the role of *human self-consciousness* serving as the effective agent for wavefunction collapse and determination of particle properties, or be generalized to posit a role for *global consciousness*.[41]

As another illustration of the kinds of conceptual breakthroughs possible if the reality of a fourth, unseen, transcendent Spatial/spatial dimension were universally recognized, consider the quantum mechanical description of the phenomenon known as *tunneling*: one subatomic particle passing "through" another rather than hitting it and causing a rebound or deflection because the particles also have wavelike properties allowing them to diffuse around each other. From a three-dimensional perspective, tunneling is discontinuous: a particle located in position 'X' disappears, then reappears in position 'Y.'

There would seem to be no mechanism by which the change of position can be comprehended except through circuitous assertion of wave-particle duality coupled to assignment of a probability of specific position according to the Schrödinger wave equation. In actuality, the nature of tunneling and its implications have been missed. It loses its mysterious quality if theological space is incorporated as a mechanistic alternative.[42] What we see is a slice of the whole: continuous in four-dimensional space, discontinuous in three-dimensional space. Analogous would be departure of objects from a location in two-dimensional space only to reappear in a different location, such as is said to occur in the novel, *Flatland*.[43]

If the apparent successful adherence of quantum mechanical calculations to measured properties is a function of the self-consistency of discreteness and does not validate its foundational theory; *if* the result in its single most important confirmation can be explained in an alternate manner oriented to a different metaphysical construction for reality; *if* quantum mechanical theory prevents establishment of unified field theory due to its incompatibility with Newtonian mechanics and Einsteinian relativity; *then* an individual should at least *consider* voiding quantum mechanics as the best available description of the means by which the universe is constructed.

Over the last twenty-five years, an entirely new theory, *string theory*, has been developed, which many physicists propose as a more accurate depiction of the intrinsic nature of manifestation. String theory proclaims the universe is composed of microscopic strings that, depending upon their frequency of vibration, generate all known particles, forces, and associated properties.[44] With respect to cosmological ontology, string theory proposes these strings originally occupied ten spatial dimensions, and at the onset of the Big Bang, four became the spacetime continuum of phenomenal existence, whereas the remaining six became "invisible" by "curling up" into infinitesimally small points associated with the normal three-dimensional coordinates of space. As Michio Kaku explained the doctrine in *Hyperspace*:

> Our universe, which originally might have begun as a ten-dimensional universe, was not stable; it...exploded into a four-and a six-dimensional universe.... This process is called *symmetry breaking*.... At the beginning of time, gravity and the weak...

> strong...and...electromagnetic forces were all unified by the superstring.... Because the universe began to split up into a four-and...six-dimensional universe, the universe was no longer symmetrical.[45]

String theory agrees with the proposed metaphysical-physical theory in acknowledging a necessity for a minimum of one non-physically observable dimension. The proposition is weak overall in that there has been offered no justification for instability associated with super-symmetric ten-dimensional universes, and none for postulating its fracture into four- and six-dimensional pieces was the only way to resolve the tension. Further, note the reference to "symmetry breaking." Cosmological theories often presume symmetry from the dawn of creation of the universe subsequently undergoing destruction. They do so because, as already expressed, there is an asymmetric excess of matter over antimatter in the universe, yet the radiant energy background is isotropic and suggestive of symmetry.[46] To include both features, symmetry breaking is considered essential.

Alternatively, asymmetry is easy to find if one begins with asymmetry. A more plausible scenario is the metaphysical speculation found in *TOPP* and chapter four: dual Consciousness fields jointly produce the asymmetric matter-inhabited universe with its known particles and forces, as well as an asymmetric anti-matter parallel universe—the imaginal, or mirror image, counterpart to ours. Interpassage, cosmogonic generator of the cosmological Big Bang, was the source of the instability associated with the onset of the Big Bang. Nothing compelled Interpassage to be symmetrical if viewed from the vantage point of one of the Consciousness fields, but must have been symmetrical if considered from the perspective of both fields simultaneously. Why? Newton's Third Law: for every action there is an equal and opposite reaction.

If space was initially expanding at an enormous rate at the beginning of the Big Bang, as postulated by inclusion of the concept of Inflation, space was not only in motion, but was in accelerated motion. Two facts ensue from this thesis. First, there must have been an energy source fueling accelerated motion. Any assertion of cosmic inflation caused by the Big Bang would be nonsensical: universal expansion was not the *result* of the Big Bang; it *was* the Big Bang. It could not have been self-caused. The Energetic/energetic source fueling Big Bang expansion may be expressed as due to Interpassage. Second, according to the General Theory of Relativity, accelerated motion is due to gravitational attraction of curved space following shortest path lines of motion for objects in that space. Hence, there must be a higher dimensional Space/space possessing enormous strength of gravitational attractive force of sufficient magnitude to cause phenomenal space to expand *uniformly* at an accelerated rate.

Aside from string theory's unsubstantiated cosmology, the mathematics involved in an eleven-dimensional spacetime framework for creation of a unified field theory is simply intractable. In *Hyperspace*, Kaku explained the mathematical difficulty with admirable brevity: "Although the theory is well-defined mathematically, no one has been able to solve the theory.... Solving field theory requires techniques...beyond the skill of any physicist."[47] It is legitimate to envisage this impediment as merely contingent and capable of eventually being overcome. Another plausible implication, though, is that the theory is overly cumbersome. Either way, string theory's direct appeal to dimensionality higher than our normal four-dimensional spacetime in order to generate anything resembling a unified field theory is significant.

In fact, entryway towards a solution to the problem of unification of the fundamental forces has been available for decades. In light of the underlying theme of this chapter, it should not come as a surprise to discover the key element allowing this accomplishment was adoption of a meta-perspective: specifically, positing the existence of another spatial dimension beyond the three we presume to inhabit. *Aside from offering a viable explanation for the results obtained from combining Bell's statistical prediction with Aspect's research data, use of a fourth spatial dimension allows the mathematics of the electromagnetic and gravitational forces to be seamlessly conjoined*: *a result all appeals to quantum mechanics failed to accomplish.*

Kaku was emphatic: "One fundamental theme running through the past decade of physics has been that *the laws of nature become simpler ...when expressed in higher dimensions*.... Laws of light and gravity find a natural expression when expressed in higher-dimensional spacetime."[48] How was unification accomplished? The pertinent notion derived initially from the nineteenth century mathematician Bernhard Riemann: "Riemann concluded that electricity, magnetism, and gravity, are caused by the crumpling of our three-dimensional universe in the unseen fourth dimension.... Riemann found that in four spatial dimensions, one needs a collection of ten numbers at each point to describe its properties.... This collection of numbers is called the Riemann *metric tensor*."[49]

Four years after Einstein's paper on general relativity appeared in the literature—the publication containing his theory of gravity as dependent on object mass and the shape of space—Theodor Kaluza achieved the sought-for synthesis: "Kaluza began...by writing down Einstein's field equations for gravity in five dimensions.... He proceeded to show...these five-dimensional equations contained within them Einstein's earlier four-dimensional theory with an additional piece.... This additional piece was precisely Maxwell's theory of light."[50] This statement is *almost* correct; it omits a detail of critical importance in understanding the meaning of the unification of these forces.

This methodological approach can be illustrated using a simple diagram. Following Riemann's technique for the symmetrical arrangement of numbers describing the different field domains for each force, Kaluza extrapolated it to five dimensions—four spatial and one temporal—yielding a collection of 15 numbers. Using conventional mathematical expressions, this five by five grid, called a *matrix*, can be represented as:

$$\begin{matrix} g_{11} & g_{12} & g_{13} & g_{14} & g_{15} \\ g_{21} & g_{22} & g_{23} & g_{24} & g_{25} \\ g_{31} & g_{32} & g_{33} & g_{34} & g_{35} \\ g_{41} & g_{42} & g_{43} & g_{44} & g_{45} \\ g_{51} & g_{52} & g_{53} & g_{54} & g_{55} \end{matrix}$$

The properties of this matrix are such that a plane of symmetry extends along the diagonal running from g_{11} to g_{55}: the terms above and below this mirror plane are mathematically equivalent; that is, g_{12} is identical to g_{21}, g_{13} to g_{31}, and so on. This means that only 15 of the 25 terms are unique: g_{11}, g_{12}, g_{13}, g_{14}, g_{15}, g_{22}, g_{23}, g_{24}, g_{25}, g_{33}, g_{34}, g_{35}, g_{44}, g_{45}, and g_{55}. Ten of these terms are those contained in Einstein's field equations for gravity, while four others yield Maxwell's field equations for electromagnetism. *The remaining term is usually considered unimportant for the unification of the physics underlying electromagnetic and gravitational forces.*

Where is this extra spatial dimension located? "The higher dimension... not observable by experiment, was different from...other dimensions. It had ...collapsed down to a circle so small...even atoms could not fit inside it.... The energy necessary to probe this tiny distance...is...the *Planck energy*, or 10^{19} billion electron volts.... It is...an energy beyond anything we will be able to produce...the next several centuries."[51] Kaluza's added dimension, and the justification for its unobservable location, was later amended and incorporated into string theory as the hidden *six* extra dimensions.

Consider the underlying logic: an invisible fifth dimension not verifiable by human investigation able to unify into a cohesive unit discrete operations of electromagnetic and gravitational forces in four-dimensional spacetime. In alignment with what has been presented, the conclusion seems obvious: *what was described as a physical spatial dimension is actually a transcendental Spatial Dimension. The remaining term, 'g_{55},' represents a physical aspect of Consciousness identifiable with theological space.*

According to Ervin Laszlo, physicist Thomas Bearden had a similar idea: "Thomas Bearden...considers the...scalar potential...an n-dimensional stress in the quantum vacuum, where 'n' is equal to, or greater than, four.... The vacuum is equal to energy-filled spacetime... The virtual state of the medium determines all that emerges into physical reality."[52] Translating this into more comprehensible English, Bearden conceived Laszlo's vacuum field—the physical analog to theological space—as symbolized by a fourth- or higher-dimensional term in the field equations: the source of "all that emerges into physical reality." This assessment certainly sounds identical to that proposed for theological space, and it is possible Bearden was referring to the scalar term, g_{55}. Thus, Einstein's field equations for gravity were mathematically correct, useful in unification with the electromagnetic field equations of Maxwell, but were tied to an incorrect interpretation of their philosophical meaning.

The electromagnetic force plays a more central role in external observer physics, which explains why "the electromagnetic repulsion force between two electrons is 10^{42} times stronger than the gravitational attractive force between the electrons." Gravity is substantively more significant with respect to subjective physics for which Consciousness, or self-consciousness, plays the principal part, and which is hinted at by inclusion of that remaining field term in theory unification.

Symmetry between transcendent and immanent domains—"as above, so below" as an overarching paradigm for physical reality—suggests that within theological space, gravity is exactly 10^{42} times more powerful than the electromagnetic force. The product of force strengths in normal spacetime would be equivalent to the product of force strengths in theological space. *The vastly larger gravitational field strength emanating from theological space causes isotropic expansion of space during the Big Bang plus Inflation phase of cosmological expansion.*

These pages revive Bohm's hidden variable approach, but reshape it in the form of *hidden constant* theory: the constant presence of Consciousness attributable to every particle determining in the moment, not *a priori* to the interaction, the outcome of each encounter. No superposition of potential states followed by collapse of the results obtained from the Schrödinger wavefunction equation, as used in the Copenhagen interpretation, is needed. Rather, we would measure the one and only existing state representative of the outcome of the experimental interaction, *as chosen by the wave energies involved.* Thus sayeth Evan Harris Walker in *The Physics of Consciousness*: "The direct path to...understanding...forces of nature...is...discovered by understanding what observation and consciousness are...all about and by understanding the measurement problem in quantum mechanics."[53]

Chapter Six: Self-Consciousness

> Consciousness is...the dimension of trans-phenomenal being in the subject.
>
> —Jean-Paul Sartre, *Being and Nothingness*

If, as has been claimed in the prior two chapters, global consciousness is coextensive with all matter and energy in spacetime, is its essence, one can conclude all entities—inert objects, plants, and animals—*are* Consciousness. Whether all, or only some, possess self-consciousness is another question. This theme was originally explored in the second chapter where discussion focused on anthropocentrism and its liabilities. The investigation must now take a radically different course given the tentative plausibility of theological space as a bridge between physical spacetime and transcendental Space, a hypothesis whose physics-based justifying rationale was presented in chapter five. Of principal concern here are the qualities and activities to be associated with self-consciousness.

It is appropriate to address the present inquiry from a human viewpoint, both because it is most relevant to us as human beings and because we are better informed about the details of that orientation compared to points-of-view held by objects, plants, or nonhuman animals. Ultimately, though, we must consider potential perspectives held by other lifeforms; to not do so, to assume such orientation is *prima facie* impossible or extremely unlikely, is to engage in the grossest form of anthropocentrism and beg the question.

Well-documented analytical explorations of the themes of Consciousness and self-consciousness have appeared in prior publications by this author.[1] Some investigators deny the actuality of Consciousness; others merely refuse to acknowledge it as *the* central concept in metaphysics and psychology. A significant portion of the confusion rampant on the topic derives from the failure of researchers to distinguish Consciousness from self-consciousness.

Those with expertise in psychology or physical science have probably never conceived of the necessity for such a differentiation due to a lack of familiarity with the disciplines of philosophy and religion. On the other hand, those with substantial knowledge in the latter subjects may have abstractly considered the difference, but were ignorant of a viable scientific foundation upon which to build a distinction. Those studying Consciousness and self-consciousness having adequate background in *all* the relevant domains of inquiry may have appreciated the need for such a division, but not have noted its full significance for advancement of a comprehensive philosophy.

In illustration of the difficulties encountered in untangling the "problem" of Consciousness as contemplated by numerous researchers, consider Owen Flanagan's mistake in "Conscious Inessentialism and the Epiphenomenalist Suspicion": "Consciousness did not have to evolve. It is conceivable that evolutionary processes could have worked to build creatures as efficient and intelligent as we are...without those creatures being subjects of experience."[2] Flanagan's misstatement was fashioned from the assumption Consciousness evolved and not present at the dawn of creation.

Certainly, creatures did not exist at the onset of the Big Bang; nor did they appear during the early stages of cosmic Inflation. In that sense, sentient creatures came into being at some moment in time during the process of evolution pertaining to universal history, which means *self-consciousness* did as well. That fact of appearance does not imply the concomitant appearance of *Consciousness*.[3] Evolutionary facts support only the conclusion that self-consciousness is a relative latecomer. To fail to differentiate between the two, to presuppose both are dependent upon the existence of lifeforms at a particular developmental stage—according to Flanagan and other researchers of similar mindset, dependent upon specific patterns of neural functioning—is unwarranted.

Leaving aside all metaphysical speculation concerning the relationship between Consciousness and self-consciousness, there is a legitimate reason why the lengthy historical dispute concerning the existence and nature of Consciousness—by which most writers have meant self-consciousness—has been referred to as the *mind-body problem*.[4] On a theoretical level, it has been deemed immensely difficult to explain the magical interaction proposed to exist between metaphysical and physical structures. How does one bridge the apparent mechanistic gap between a thing having extension in space and a no-thing lacking it? How is it possible to compartmentalize operations of mind, an assumed-to-be unquantifiable source of mental and emotional acts, from brain, a physical structure whose qualitative and quantitative functions appear to lie at the root of those conditions?

Flanagan took one common approach to the issue. *Epiphenomenalism*, to which the title of his essay referred, posits self-consciousness as an unneeded byproduct of evolution, as some accidental, possibly unwanted, growth akin to a cancerous tumor, if its reality is to be acknowledged at all. From this orientation, a brain is useful and desirable; self-consciousness is superfluous. Another perspective is John Searle's, who agreed that it was an evolutionary development, but considered it a separate and important adjunct to the basic physical neuroanatomy of the brain. His approach was reflected in another contribution to the compendium containing Flanagan's essay: "The existence of consciousness can be explained by...causal interactions between elements of the brain at the micro level, but consciousness cannot itself be deduced or calculated from the...physical structure of...neurons without some additional account of the causal relations between them."[5]

Searle's argument has as part of its underlying logical grounding support the presupposition that only two modes of deduction are possible, which may be termed *definitional deduction* and *causal deduction*. The former demands something be true by definition; the other derives truth from a causal chain of thought. There is missing from this list another methodology: the possibility of *associational deduction*. Something can be true by virtue of its formal and direct relationship with something that *must* be True. In the present instance, Searle was correct in saying self-consciousness is other than anatomy and physiology, that it could not be definitionally deduced from the enumeration of physical parts. He was incorrect, though, in concluding self-consciousness emanates from the *interactions* obtained from the components involved in anatomy and physiology, is causally deduced from the complex relationships between those physical parts.

A more accurate assessment of the connection between self-conscious mind and physical brain was given by Dermot Moran's characterization of Franz Brentano's position in *Introduction to Phenomenology*: "His view in 1874 was...that mental processes *depend* on the physiological processes, and specifically on occurrences in the brain, but the physical does not *explain* the mental... Brentano dismisses the possibility that laws governing the mental could be deduced from physiological laws and ultimately from physical and chemical laws."[6]

In *The Conscious Mind*, David Chalmers concurred with Brentano, and invoked Gustav Fechner's concept of psychophysical laws: "Psychophysical laws are eternal, having existed since the beginning of time. It may be that in the early stages of the universe there was nothing that satisfied the physical antecedents of the laws, and so no consciousness, although this depends on the nature of the laws.... As the universe developed, it came about that certain physical systems evolved that satisfied the relevant conditions."[7]

The thoughts of Brentano and Chalmers are cited because they mimic the view to be followed here, which is: (a) the connection between mind and brain is associational rather than causal; (b) the essential deduction pertains to the innate relationship between self-consciousness and Consciousness; (c) psychophysical laws pertinent to collective interactions of mind and brain are needed to explain feelings, thoughts, and actions of all kinds.

Prior to the last decades of the twentieth century, discussions on these topics were conducted in a partial vacuum. There was insufficiently detailed information accessible about the anatomy and biophysical operations of the brain, as well as a relative ignorance concerning the laws of physics and their conceivable relationship to metaphysics. Philosophers and theologians from the rise of city-states until today have lacked adequate scientific training in biology, chemistry, and physics; meanwhile, scientists have generally been loathe to ponder on that which they couldn't explore directly with their five senses. The present state of knowledge is at a sufficiently advanced level that the ancient questions can be profitably re-investigated. We are literally at the dawn of a new era in being able to finally make sense out of these issues. This is not to claim we are in a position of omniscience, for there is much likely to remain a mystery for some time. It is to say we can make inroads on an unprecedented level. Hence, to even begin to answer the questions, we must start with our current state of knowledge about the brain.

There are diverse ways to define and characterize brain content and operations. Some basic terminology is needed to provide a context for the ensuing discussion. Many species of creatures with which we are familiar have brains similar in organization and function to our own. Since there are also differences in organ arrangement and degree of development, the forthcoming recitation will refer entirely to the human brain. Justification for this limitation comes from a simple fact: it is that anatomical element of most direct interest to us in discussions about self-consciousness.

It must not be forgotten it has been previously proposed that humans are not uniquely in possession of self-consciousness. Further, it is a gross error to believe *Homo sapiens* are necessarily at the apex of potential development and constitute an appropriate standard for judgment of all other beings with respect to questions concerning the existence of self-consciousness. Nor can it be said we provide the reference point for the precise mechanism of brain-mind interaction in those other species.

Structurally, our brain is composed of two hemispheres, right and left, running from the front of the skull to the back inclusive of both the top and sides of the head. The hemispheres are almost mirror images of each other in terms of their internal components, but are said to specialize in functional utility: the right and left hemispheric-dominated operations are, generally speaking, complementary and not redundant. The hemispheres are connected by the *corpus callosum*, a thick bundle of neurons used to transmit electrical signals. Individuals with a severely constricted or missing corpus callosum do not process information the same way as those having a developmentally normal network. Each hemisphere is subdivided into four lobes: frontal, parietal, occipital, and temporal. The *frontal* lobes cover the forehead and front part of the top of the skull; the *parietal* lobes are on the top of the skull running front to back; the *occipital* lobes are located in the rear of the skull behind the ears; the *temporal* lobes occupy positions on the side of the head from the eyes to the ears.

Within each region of the brain exist numerous organs with specialized function. There are identical structures and redundant functions with respect to the parietal and temporal lobes, and this fact is the source for the judgment of the hemispheres as near-mirror images of each other. Andrew Newberg and Eugene D'Aquili summarized the role played by each lobe in *Why God Won't Go Away*: "The temporal lobe, located along the sides of the head, is associated with language and conceptual thinking; the occipital lobe, at the back of the head, with vision; the parietal lobe, which sits beneath the crown of the skull, is home to sensory perception, visual-spatial tasks, and body orientation; and the frontal lobe, situated...behind the forehead, is associated with attention and initiating muscle activity."[8]

Our ability to think and feel with phenomenological intentionality as a consequence of internal or external stimuli is what we generically include as central to models for the nature of human self-consciousness. As pointed out in chapter two, it is with regard to these two functional domains where we place the criteria most frequently employed in discussions about either the possibility or probability for existence of nonhuman self-consciousness.

Any focus on primary tissue structures and their functional roles must necessarily include the main components of the *limbic system*, which jointly integrate emotional content with cognitive context. The main components of this system are the hypothalamus, amygdala, and hippocampus.

First, the *hypothalamus*: "A bundle of nerve bodies, about the size of a peanut, located just above the roof of the mouth. It contains...centers, which mediate the excitement and inhibition of the hunger, thirst, and sexual drives, as well as emotional arousal.... In conjunction with the reticular activating system, the hypothalamus is...involved in the mediation of sleep and arousal states."[9] About the hypothalamus, Newberg and D'Aquili wrote: "The oldest structure in the human limbic system is the hypothalamus, which sits near the upper end of the brain stem.... One of the major roles of the hypothalamus is to link the operations of the autonomic system to the higher structures of the brain's neocortex."[10] They later noted: "The hypothalamus...also moderates immune function, hunger, thirst, and body temperature."[11]

Second, the *amygdala*: "Located in the middle part of the temporal lobe, the amygdala is...one of the oldest structures in the brain... When a stimulus requiring...attention is presented, the amygdala...directs the mind to pay attention by assigning emotional value to the stimulus.... The amygdala does not exert its influence directly upon the autonomic system.... It activates the hypothalamus, which in turn influences autonomic activity."[12] This organ "is responsible for storing...emotional aspects and personal reactions to events, whereas the hippocampus acts to store the cognitive, visual, and contextual variables.... The amygdala plays a particularly important role in memory and learning when activities are related to reward and emotional arousal."[13] In fact, the amygdala is apparently so efficient in its operation for the human body that it "is capable of processing visual, tactile, gustatory, olfactory, and emotional stimuli simultaneously."[14]

Third, the *hippocampus*: "Located slightly behind the amygdala in the temporal lobe, the hippocampus is greatly influenced by the activity of the amygdala, and the two structures often act in a complementary way to focus the mind's attention on interesting sensory input, to generate emotions, and to link those emotions to images, memory, and learning.... The hippocampus ...has the power to regulate...reactions generated by the autonomic nervous system in order to...maintain emotional equilibrium."[15] The hippocampus is said to be "exceedingly important in memory, acting to place various short-tem memories into long-term storage and retrieving short-term memories."[16]

Andrew Newberg and Jeremy Iversen, authors of "On the 'Neuro' in Neurotheology," summarized the contributions of these cranial areas: "The hippocampus functions in conjunction with the thalamus, hypothalamus, and septal nuclei to prevent extremes of arousal, thereby maintaining a state of quiet alertness. The hippocampus greatly influences the amygdala, such that they complement and interact in the generation of attention, emotion, and certain types of imagery.... The limbic areas, particularly the amygdala, help to identify certain experiences as profound or real."[17]

With just this brief recitation, it is easy to demonstrate the areas of the brain linked most directly to activities of self-consciousness could not have been a late-stage development in human beings. Recall a proposal presented in chapter four: self-consciousness is primarily a state of alertness, and only secondarily affiliated with awareness of internal or external states; prior to active cognition of an object of any sort, one must be prepared to cognize. The oldest parts of the limbic area—amygdala, hippocampus, thalamus, and hypothalamus—are deemed responsible for "maintaining a state of quiet alertness," of exhibiting basal self-consciousness. Even with the supposition that self-consciousness should be associated with internal awareness, it is the amygdala that takes center stage: "When a stimulus requiring...attention is presented, the amygdala...directs the mind to pay attention."

The claim is *not* that self-consciousness is located in the limbic system in some primary or exclusive fashion; the claim is that self-consciousness is associated with operations in all regions of the brain, and to the extent some of those areas were functionally active earlier in human evolutionary history than others correlates with the assertion of self-consciousness as operating through those oldest portions of the brain prior to, in a literal temporal sense, its connection with newer organic structures. In this respect, both Flanagan and Searle were wrong: when the genus *Homo* arrived on the world stage, human self-consciousness did as well. It was not the case that *Homo* arrived on day 'X' and self-consciousness was created on day 'X+Y,' where 'Y' represents the date of development of a new level or type of brain function.

Since other species also possess these limbic structures, this is supportive of another claim made in chapter two: reflexive self-consciousness is likely in other living creatures *to the extent* anatomy and physiology can dictate the operation of any nonphysical process. This last statement can be underscored if the issue is approached from another direction, namely, comparison of our neuroanatomy with that of other living creatures. The *triune structure* of the mammalian brain has been defined and characterized. It consists of three levels arranged in order of phylogenetic development within diverse species over time: the *reptilian*, *mammalian*, and *neo-mammalian* brains.

The reptilian brain still exists within the structure of the modern human skull furthest from the skeletal surface, while the mammalian brain spatially encompasses the older structure, and the neo-mammalian brain provides the outer layer of the current-day cranium containing the other functional areas buried within. In addition, these three layers have been successively credited with excessive specialization of functions sequentially pertaining to actions, emotions, and thoughts when arranged from oldest to youngest with respect to developmental capacities. Concerning classification, Carol Raush Albright wrote in "Religious Experience, Complexification, and the Image of God":

> In mammals...a certain region of the brain greatly resembles the brains of reptiles... This region, which lies above the brain stem, is involved in the need to protect one's own territory or home, to reproduce, to establish basic social structures, and to forage and store food... Surrounding the "reptilian brain" lie structures...called the *mammalian brain*. One of its important functions is to add emotion to the guarding, mating, competing behavior of reptiles, which mammals also share. Emotions label experiences as good or bad, helpful or dangerous, and so they motivate mammals toward behavior that is likely to promote survival.[18]

It is anthropological fact: all three brain structures developed before the first recognizably *human* being ever arrived. Contemporary cranial structural development occurred somewhere between fifty and 280 million years ago, while *Homo habilis*, the human progenitor of *Homo sapiens*, arrived on the terrestrial scene about two million years ago.[19] Antonio Damasio was not subtle in the conclusion he drew: "Although...core consciousness requires ensemble activity that involves regions of every tier and quarter of the brain, consciousness does depend most critically on regions that are evolutionarily older, rather than more recent, and...located in the depth of the brain, rather than on its surface."[20] Since the earliest structural level enabled intentional self-expression through direct behavioral action, then the sense of self could not have been an emergent product of evolution. *Self-consciousness always was* an irrevocable aspect of sentient creatures.

It is anticipated a critic will immediately respond with the argument that self-expression through actions conducive to survival does not mean self-awareness, the latter generally, if erroneously, considered the quintessential characteristic of self-consciousness. An even more convincing argument for existence of self-consciousness than purposeful behavioral action in response to an environmental stimulus would be evidence for the existence of non-externally motivated activities. Evidence of this kind can be discerned in the discovery in 2003 of artwork depicting human faces created by Neanderthals 35,000 years ago. Reported by Jean-Claude Marquet and Michel Lorblanchet in the journal, *Antiquity*:

> The mask, which is dated to be about 35,000 years old, was recovered on the banks of the Loire in France.... The object shows clear evidence, the researchers say, of having been worked—flakes have been chipped off the block to make it more face-like.... The standard view of Neanderthals is that they lacked the thought processes capable of producing art—certainly to any real level of sophistication produced by modern humans.... This object shows that art was not born in the brain of *Homo sapiens* but much earlier in the brains of predecessors like the Neanderthal man and even, no doubt, in *Homo erectus*.... There are claims for even older items, dating back 200,000 years or more, that comprise mainly rock objects apparently sculpted to look like the human form.[21]

Persuasive evidence of artwork—by any criteria, an unnecessary activity for survival, and completely self-motivated—dating at least from the time of Neanderthal Man is indicative of self-expression and self-awareness central to the concept of self-consciousness. The conclusion must be these aspects of human personality are not dependent on brain function activity found only in *Homo sapiens* with their enlarged frontal lobe.

Extending the argument to species very far from the human form: "The last few years have shown many invertebrate groups, including protostomes, to have some capacity for recognizing and reacting against 'non-self.' This effectively disposes of the possibility...discrimination between self and non-self is...accidental or a functionally unimportant peculiarity of vertebrates. Recognition of foreignness and its obverse, self-recognition, may in fact be universal among animals, and even plants, albeit in a crude form."[22] It is possible to argue protostome comprehension of the distinction between self and non-self is a product of reflexive stimulus response, or involves a simple chemical recognition system. The first option is not an explanation unless a detailed account of the structural mechanism for this stimulus response is given; the second choice discounts the fact such systems must themselves be cognized and the incoming data interpreted in order to be effective.

The only reasonable alternative to postulation of some basal level of self-consciousness for all living systems is the assertion that unconsciousness is ontologically more feasible; that the development of unthinking, unfeeling automata is more comprehensible until the arrival of some arbitrary time in the history of the universe when humanity came on the scene. Such a thesis would depict anthropocentrism run amok. If invertebrates, creatures lacking a central nervous system connected to a brain, possess self-recognition, one can conclude, in conjunction with the cited anthropological evidence, human beings must have always had self-recognition ability even if its operational significance were wholly encapsulated within a materialist mechanism. The limbic system organs common to all mammals would seem to be worthy candidates for physiological linkage between brain and self-consciousness.

Under the presupposition of self-consciousness as coincident with brain structure and function, the question of whether or not a definitive center for its operations can be found within a specific cranial area must be answered in the negative. There is no single locus of functional traits salient to that notion co-existing in some proximal spatial location. Hippocampus and amygdala are both involved in functional memory, for example, yet as Michael Talbot reported in *The Holographic Universe*:

> What Lashley had done was to train rats to perform a variety of tasks, such as run a maze.... He surgically removed various portions of their brains and retested them.... He found...no matter what portion of their brains he cut out, he could not eradicate their memories.... If memories possessed specific locations...why didn't Lashley's surgical plunderings have any effect on them? For Pribram the only answer seemed to be that memories were not localized at specific brain sites, but were somehow spread out or *distributed* throughout the brain as a whole.[23]

The passage references are to neuropsychologist Karl Lashley and his student, Karl Pribram. If Pribram was correct, the brain must be integrated to such a high level that widely separated regions covering a diverse array of anatomical elements are conjoined into a set of specific functional behaviors. This suggests any display of self-consciousness associated with particular brain functions must permeate the entire volume of cranial space. Albright called such interconnected domains *convergence zones*, while Newberg and D'Aquili referred to them as *association areas*. Albright noted:

> The most distinctively human part of the brain is the prefrontal cortex... It includes critical *convergence zones* where information from the body...emotions...memory ...thought...holistic insight comes together. This is a key location where judgments are made...scenarios are spun...plans are made, and their execution is monitored....

> Master convergence zones...are critical to thinking, planning...decision-making. Convergence zones for language are...in the left hemisphere, and for understanding of spatial relationships in the right... Other convergence zones are concerned with sensory input and with memory formation.[24]

Theologians, philosophers, and cognitive scientists committed to radical anthropocentrism will often cite the prefrontal cortex as the localized center of self-consciousness most relevant to its operations. They do so *because* the prefrontal cortex is "the most distinctively human part of the brain," and allege it's associated phenomenological abilities typify self-consciousness, thereby "proving" other creatures are incapable of such a display. Albright's next statement is meaningful: "It includes critical convergence zones where information from the body...emotions... memory...thought...holistic insight come together." *The prefrontal cortex is not the point of origin for emotions and thoughts*. It is a region where these activities are conjoined, presumably for more efficient coordination in their product output. *Efficiency has been selected for in development of the cortex in Homo sapiens, not the capacities themselves*.

The principal difference between animal *orders* or *families*—used in the taxonomic sense of the words to include all mammalian or non-mammalian vertebrates and invertebrates—concerns the degree of neural interconnection between disparate anatomic structures—the total number of synapses capable of carrying an electrochemical signal—rather than a yes-or-no conjunction. Animals display approximately the same cognitive and emotional abilities as humans because, in many cases, they possess similar neuroanatomical organs and neural connections. However, efficiency in displays of cognition, affect, and behavior may not be attained to the same degree due to decreased speed in executing outputs, or because diminished complexity of neural networking results in absence of a direct link between a particular thought and a specific emotional response.

Human beings demonstrate a wide spectrum of intelligence and great diversity in emotional range. Nonetheless, we would classify all as having an associated self-consciousness. What we frequently misjudge as an absence of capacity with respect to qualities elicited by self-consciousness may actually produce altered formulations of similar traits. For example, it is unreasonable to expect animals to be able to know '2+2=4'; it is highly likely many have a strong innate concept of number, and know constructs such as "greater than" and "less than." Animals may not define "justice" the same way we would, but it is likely they have a well-developed sense of its consequences. These hypotheses, based on neuroanatomical evidence, were what grounded the arguments for animal self-consciousness presented in chapter two where tool usage or communication skills were discussed.

Newberg and D'Aquili offered more detail in their report of the utility of these convergence zones in *Why God Won't Go Away*. They concentrated on three areas: the *orientation association area*, the *attention association area*, and the *verbal conceptual association area*. About the first, they commented: "The orientation association area...receives...input from the senses of touch, vision and hearing. These give it the ability to create a three-dimensional sense of 'body' and to orient that body in space.... The left orientation area is responsible for creating the mental sensation of a limited, physically defined body, while the right orientation area is associated with generating the sense of spatial coordinates that provides the matrix in which the body can be oriented."[25] Animals and other creatures demonstrating behavior indicative of a precise distinction between self and non-self—Micklen's protostomes, or Gallup's chimpanzees—must have an anatomical-functional equivalent to an orientation association area, since comprehension of spatial extent of one's own body is a prerequisite for any delineation between self and other.

Newberg and D'Aquili's description of the attention association area confirmed Albright's assessment, and included additional relevant data:

> The attention association area...the prefrontal cortex of the brain, plays a major role in governing complex, integrated bodily movements and behaviors associated with attaining goals.... Researchers think of the attention area as the neurological seat of the will.... Damage to this area...results in a loss of the ability to concentrate, plan future behavior... Victims of such damage...frequently exhibit emotional flatness, a lack of will, and a profound indifference to events in the environment.[26]

If a damaged prefrontal cortex induces "profound indifference to events in the environment," creatures lacking a well-developed front cortical area might feasibly display a similar withdrawal from aggressive interaction with surroundings. This is manifestly *not* true: evidentiary citations in the second chapter expressed in excessive detail the alertness attributable to animals in their native habitat. If they were indifferent, they would quickly die: killed by predators, or at the mercy of meteorological disturbances and natural disasters. Only two conclusions are possible: (a) a well-developed attention association area within the prefrontal cortex is not sufficient for normative interaction with the environment; or (b) it is in human beings, and animals have a compensatory system. In either case, that anatomical region alone cannot inherently designate the operational existence of alertness-awareness functions critical to self-consciousness.

The third area mentioned by Newberg and D'Aquili is also marshaled by anthropocentrist proponents as germane to the issue of self-consciousness. This was done by the authors themselves: "The verbal conceptual association area...is primarily responsible for generating abstract concepts and...relating those…concepts to words…. These operations are crucial to…expression of consciousness through language."[27] What the authors failed to say is, "crucial for us humans." Many animals vocalize: domesticated dogs and cats; birds; whales; dolphins; chimpanzees; wild prairie dogs like jackals, coyotes, and wolves. Although their utterances may seem incomprehensible to us, this is hardly adequate justification for discounting these verbalizations. It can be imagined they represent legitimate language, a vehicle for communication within their own community.

Moreover, some animals possess, in common with *Homo sapiens*, brains incorporating neural connectivity patterns necessary for existence of a verbal conceptual association area. Roger Crisp's assessment possessed the attribute of *a priori* credibility: "The general truth that animal brains are anatomically similar to our own must be accepted. And this gives us a reason for believing …the mental events that take place 'within' them are in some degree similar to those which occur in our own brains."[28]

Despite extensively detailed knowledge of our brain and the specialized functions conducted by some of its structural regions, neither singularly nor collectively do those features account for the totality of self-consciousness. Although it is true that: (a) specialists in neuroanatomy and neurophysiology understand some, or many, of the roles collections of brain tissue and neural networks enact in evoking intellectual thoughts and emotional feelings; (b) mechanisms for displays of willfulness and memory have been outlined; (c) sensory modes of perception and the source of some natural compulsions we apparently possess, like survival and reproduction, are comprehensible as attributions of clusters of brain cells plus neurotransmitters working together; it is also true that: (d) *none of these processes lead directly to our inherent capacity for willful behavior expressed through creation of value systems,* or *even* (e) *necessarily induce our deliberate self-assessment of a variety of virtues and vices*, with the former characterizable as character strengths and the latter as character flaws.

A memory system doesn't come with a set of instructions telling us what to remember, or offer a definitive neurological mechanism establishing why we choose to remember certain aspects of experience while allowing other events to recede into forgetfulness. Neuroanatomy and biochemistry alone do not stipulate how to account for the fact person 'A' prefers the company of person 'B' to the company of person 'C.' Nor can it be justified to assert that human beings are superior to other beings through the mere existence of our self-consciousness, or as a direct consequence of the manner in which we appear to demonstrate our unique form of evolutionary development.

The argument can even be legitimately made that our brain hemispheric specialization and greater degree of cortical mass is retrogressive rather than progressive. It is admitted hemispheric specialization enhances efficiency by decreasing possible misinterpretation of signals received by a given region of the brain. In addition, however, human hemispheric specialization suggests other species are much more inherently integrated in utilizing their reception to somatosensory stimuli and the cognitive conclusions to be drawn from those inputs. Other lifeforms are more synesthetic: in intuitive harmony with natural processes in their environment. By contrast, we treat the environment as an adversary.

As was suggested in chapter two, nonhuman creatures may conceivably possess a more highly attuned telepathic ability, which we as a species are just beginning to develop or re-cognize. This could explain the absence of highly sophisticated verbal language skills in other terrestrial intelligences. Human abilities arising from hemispheric specialization serve humans well. It can also be said the same holds for other lifeforms lacking that type of dissection of the brain into segregated arenas. Who is qualified to determine that our talents exceed those of other species? Who is capable of stating with the certainty of evidence that self-consciousness demands creatures whose brains look like ours?

Further testimony favorable to this interpretation of the significance of hemispheric specialization derives from some humans displaying astonishing abilities in music or mathematics or memory capacity appearing to correlate with an enhanced level of integration of cranial regions.[29] The fact that these individuals often have diminished ability with respect to basic living skills, or perform in substandard fashion on intelligence tests, means evolutionary development of a corpus callosum or a thickening of prefrontal cortical tissue is symptomatic of compensatory assets and not preferentially superior ones.

Independent of these specific arguments, the rising mound of evidence clearly points to the propriety of sounding the death knell for anthropocentric bias. It would be highly desirable to cast this epistemological approach upon the dung-heap of outdated theories. The notion that brain suffices to explain mind must likewise be discarded as plausible: sophisticated neuroanatomical and neurophysiological findings contravene the claim that if we can slice and dice organs and neural nets to a sufficient level, we will eventually come to grips with self-consciousness. As has been repeatedly emphasized in these pages, discreteness is not the fundamental law of the universe; continuity is. An increasingly detailed analysis of minute cells and organelles in the brain will only reveal specialized, though valuable, information. The continuity of self-consciousness can be comprehended only in relation to a philosophical and physical meta-perspective.

David Chalmers' call to arms in *The Conscious Mind* for a theory of Consciousness—by which he meant self-consciousness—explaining how it emerged from the physical substructure of the brain is in actuality an innately *inappropriate* demand: "Ultimately one would like a theory of consciousness to do...the following: it should give the conditions under which physical processes give rise to consciousness, and...it should specify just what sort of experience is associated.... We would like the theory to explain *how* it arises, so...the emergence of consciousness seems intelligible rather than magical."[30] Chalmers backtracked on his desire for a theory of emergent Consciousness when he insisted the "account of consciousness" be non-reductive in type, by which he meant not ultimately constructed entirely from physical processes: "A completed theory of mind must provide both an (non-reductive) account of consciousness and an (reductive) account of why we judge that we are conscious, and it is reasonable to expect that these two accounts will *cohere* with each other."[31]

To develop a comprehensive theory of the nature of self-consciousness, some long-held notions permeating the literature that are blatantly fallacious, serving to restrain understanding, should be discarded. First is the doctrine of epiphenomenalism, a topic introduced by the citation from Owen Flanagan. This construct was invented towards the end of the nineteenth century, and proclaimed self-consciousness is an artifact of physical brain processes. This thesis has undergone scrutiny and its weaknesses made clear: (a) it cannot satisfactorily explain the continuous sense of self-presence fundamental to phenomenological self-consciousness; (b) no one has succeeded in detecting a spatially centered or neurologically distributed unique anatomical locus for self-consciousness.

Another proposal concerning the connection of matter to Consciousness, or self-consciousness, has been called *panpsychism* or *dualism*: the view that the source of mental events—Aristotle's final cause—is separate and distinct from the origins of, and the processes underlying, physical phenomena. Max Velmans offered a definition of this idea in *Understanding Consciousness*: "Continuity theorists...do not believe...consciousness suddenly emerged at *any* stage of evolution.... In the...explosion that gave birth to the universe, consciousness co-emerged with matter and co-evolves with it."[32]

The principal difficulty with this proposition, which cannot be rationally circumvented and was a devastating criticism of the Cartesian postulate from which it sprung, was that if matter and consciousness are distinguishable in the ontology of their birth in the universe, there is no plausible mechanism for their subsequent interaction. Velmans attempted to cover this discrepancy by resorting to vague descriptions exemplified by use of "associated sensory *qualia*." Precisely how these associations arise cannot be stated. Why? He slammed the door shut to any solution to the conundrum by writing on the next page, "the physical world is causally closed."[33]

The three contentions cannot be cumulatively correct: either matter and Consciousness *are* separate and distinct, "co-emerging" and "co-evolving," *and* the universe is causally closed for both physical and mental processes, which means the posited interactions can then never come into being, *or* if interrelationship is possible, their separation and distinction coupled to causal closure must be false. A preferred variation to panpsychic dualism is claiming there is a valid distinction between physicochemical processes and mental states exhibited by creatures, but actions, emotions, and thoughts must be directly correlated with material physicochemical events.[34]

This was the position of Franz Brentano cited previously. In other words, there is no causal closure: mental and physical phenomena are different, not separate. This is a reasonable proposal: if continuity is to be fundamental, there cannot be an impassible gap between the variety of possible mental and emotional states and whatever physicochemical support mechanisms exist for expression of those conditions. In *Enlightenment East and West*, Leonard Angel contrasted the doctrines of epiphenomenalism and correlationism:

> On the first view of agency not only is every action the result of prior brain states, but also there is no cognitive state which is not embodied in a brain state. On the second view of agency, there are cognitive states which are not embodied in brain states, and these unembodied cognitive states cause changes in the physical states of the brain, which...cause other alterations in the brain and in the nerves proceeding from the brain, leading to muscle contractions constituting basic action. The second view holds...there is a special kind of causation, which in a...precise sense, is not physicalistic causation.... An unembodied mental state directly causes a physical change at some location of the brain.[35]

There is a problem with this view also. Angel revealed a difficulty in the coherence of this theory even as he gave voice to an error frequently made in arguments offered *against* correlationalism:

> In order for an unembodied mental state to directly cause a physical change in the brain, there must be a violation of the conservation of energy law which (usually) operates in the neurochemical interchanges.... If it takes the expenditure of so much energy to get a neuron to fire normally...as the mind-brain interactionist supposes, there is neuronal firing without...prior expenditure of energy or prior neurochemical triggering process...in violation of the...mass-energy conservation principle. The X amount of energy is not supplied, and yet there is neuronal firing. Thus, mind-brain interactionism requires that there be what I have called "functional preemption" of physical laws.[36]

He and so many others misunderstand the meaning of the law of energy conservation. The law—like all "laws" in science used to describe general statements of fact *as we experience and comprehend them* subject to future revision upon reception of new data—states: in any physicochemical change there is no net loss of energy within the confines of the defined system and surroundings; energy can change forms qualitatively, but neither increase nor decrease in absolute quantity.[37]

The flaw in Angel's passage was the contention, "there is neuronal firing without...prior expenditure of energy." If system and surroundings must be taken into account in determining energetic changes, while he defined the system as brain in general and neurons in particular, he failed to define any composition for the surroundings. Stating the surroundings are constituted by an "unembodied mental state" is not a definition but a characterization, and a nondescript one at that. Second, he presumed the only kind of energy is that perceived under typical physical conditions: light, heat, chemical, electrical, mechanical, *et cetera*. Some have argued for recognition of Consciousness as Energy: "According to Hart, consciousness might itself be a 'form of energy' currently unknown to physics... Conservation of energy would...include the energy of consciousness and transformations from physical to consciousness energies could, in principle, be found."[38]

If one supposes: (a) Consciousness is a sentient transcendental Energy field; (b) its manifest counterpart on the level of individuals resides in self-consciousness; (c) capability exists for exchanges of information, organized Energy/energy, between transcendent and immanent; *then* it can be claimed: (d) the surroundings subject to analysis is theological space acting as bridge between existential and transcendental; (e) the Energetic/energetic exchange between forms of Consciousness occurs there; (f) the information transfer process involves Energy, the psychophysical Energy of Consciousness, being transformed into physical energy inducing neuronal firing in a brain; and, (g) the net change in the sum of Consciousness Energy plus physical energy expended can still be numerically zero. This sequence of thought overcomes the error committed by those who invoke the conservation of energy law as a rationale opposing physical-mental interaction. There is still "conservation of energy" once the Energy of Consciousness is included in the relationship.

Nonetheless, an inherent flaw in a distinct-but-not-separate doctrine has been exposed. *Consciousness infuses continuity between physical and mental operations within the universe*: *they are ultimately non-differentiable.* Lenn Goodman had it right when he asserted: "Regardless of the...complexity of the relationship between consciousness and body, the relationship is not simply...effect to cause.... Their states are not strictly determined by those of the bodies they belong to, even if it proves to be the case that for every mental state there is a corresponding physical state in the body or...brain. Consciousness itself may be a determinant of the states of the body—as, for example, when we discover or develop an idea or come to a decision."[39]

In contrast to David Chalmers' call for a theory of Consciousness giving "the conditions under which physical processes give rise to consciousness," more preferable would be elaboration of a self-consistent thesis proclaiming *the conditions under which Consciousness gives rise to physical processes while providing explicit details for the mechanisms responsible.* William James seemed to understand the connection between the two when he alleged in "Does Consciousness Exist?": "Just so, I maintain, does a given undivided portion of experience, taken in one context of associates, play the part of a knower, of a state of mind, of 'consciousness'; while in a different context the same undivided bit of experience plays the part of a thing known, of an objective 'content.'... Since it can figure in both groups simultaneously, we have every right to speak of it as subjective and objective both at once."[40]

James' assessment conforms to the speculation detailed in chapter four positing Consciousness as the transcendental absolute, global consciousness as its generalized analog within the spacetime universe, and individualized self-consciousness as its representation endowing creatures with vitality and innate intelligence. Justification for the proclamation of self-consciousness as particularized Consciousness comes from consideration of yet another error commonly made precluding recognition of its inherent potency: the failure to acknowledge that, on an experiential level, self-consciousness *is incapable of avoidance*. It is the inherent framework from which all thought, emotion, and impetus to act derives.[41]

Sydney Shoemaker suggested in "The First-Person Perspective": "Where it seems...one can imagine discovering...a putative possibility from the first-person perspective, one should always ask whether the seeming discovery could be confirmed from the third-person point of view."[42] He should have recognized *there is no such thing* as a "third-person point of view." What he undoubtedly meant to say is, "one should always ask whether the seeming discovery could be confirmed from" an *observer's* point of view.

The third-person perspective is *never* third-person from an observer's point of view: when person 'Y' observes the behavior of person 'X,' he or she is making a first-person observation, and cannot trust, or fail to trust, his or her senses—or his or her introspective reflections on the interpretations of that sense data—any more than 'X' can accept his or her own. What Shoemaker is recommending is substitution of one first-person observation for another. There is no reason to believe this will result in the attainment of more reliable information. Consequently, each person must accept their own judgment of the meaning of sensory input based upon their own, presumably, rational thought processes.

Illustrations invoking this logical error can be endlessly multiplied. Max Velmans in *Understanding Consciousness*, as well as David Chalmers in *The*

Conscious Mind, both alluded to this fictitious third-person perspective in their discussions on the nature of Consciousness. Velmans asserted: "If one examines human information processing...from *a third-person* perspective ...consciousness does not seem to be necessary for any form of processing. The operations of minds and brains seem to be explainable...in functional or physical terms that make no reference to...experience.... Viewed from a *first-person* perspective...consciousness appears necessary for the analysis of novel or complex stimuli, choosing what to attend to or do, and most forms of learning and memory."[43] Chalmers utilized terminology resulting in the insertion of "phenomenal properties" for all first-person orientations and "psychological properties" for any third-person perspective.[44]

What Velmans and Chalmers both intended was to contrast objective and subjective orientations. The problem is that while this is a rational mode of analysis for any other topic of inquiry, it is inherently inappropriate for all discussions on self-consciousness. It is possible, for example, to coherently discuss objective and subjective assessments of historical events in the sense that an objective evaluation emanates from someone knowledgeable about the topical issue but not directly involved in the actions or consequences of the event, while a subjective analysis would come from someone who was a participant.

Precisely because self-consciousness is inalienable from the nature of the investigator as much as from all sentient creatures, any statement uttered on the theme derives from an association with the author's self-consciousness. *He or she cannot be objective under any experiential condition.* All human thought necessarily invokes human self-consciousness. It is involved in the moment anyone begins to physically process somatic utterance of the claim, "consciousness does not seem to be necessary for any form of processing." Factual existence *of* the statement contradicts the interpretive meaning given *to* the statement.

Aside from its identification as a manifest, constantly operational version of Consciousness, what can be said about the nature of self-consciousness? Closely affiliated with recognition of self-consciousness as an inalienable attribute of living beings comes the realization that it cannot be a condition dependent upon time indicated by specific temporal associations coincident with diverse physical states of the body. Hence, still another misconception commonly ascribed to the notion of self-consciousness emerges: its constant depiction as synonymous with immediate awareness of self and environment. *Self-consciousness does not have an on-off switch.* Wholly erroneous are all efforts to define it as related to a state of wakefulness in contrast to a person asleep, in a comatose state, or experiencing a hallucination. It does not have levels or degrees; it is a constant, continuous, monolithic whole.

The idea of self-consciousness as somehow dependent upon one's state of arousal has a history dating to John Locke's remark found in *An Essay Concerning Human Understanding* that self-consciousness is "the perception of what passes in a man's own mind."[45] The difficulty with this proposition is it transformed self-consciousness into an object, contents of mind, instead of conceiving it as subjective mind itself. David Chalmers offered a modernized rendition in *The Conscious Mind*: "All...instances of consciousness...I know about are accompanied by awareness."[46] His statement did not reach to the core of the issue: awareness implies the existence of something as an object about which one can come to awareness. Even introspection treats self as an object of awareness, and therefore the *a priori* existence of an object is a prerequisite for awareness.

Johann Fichte long ago made everyone cognizant of the flaw in equating self-consciousness with object awareness in *Foundations Of Transcendental Philosophy*:

> We cannot be conscious of things posited in opposition to us...of external objects, unless we are conscious of ourselves... We become conscious of this consciousness of...consciousness only by...turning it into an object... Our consciousness...would never be explained in this manner. Or else one would have to conclude that there is no consciousness at all—so long...as one continues to treat consciousness either as a state of mind or...as an object... Thus, in this very act of consciousness, I posited myself as both the subject and the object of consciousness.[47]

As long as self-consciousness is conceived as an object, as something to, figuratively, hold at arm's length, one fails to understand its essence. To re-emphasize: *self-consciousness is always first person subjective*. One can be introspective and treat mental and emotional contents as objects for further pondering and analysis; such activity can only be conducted by using self-consciousness as generator of the process. Recall the tautologous Cartesian *cogito*: the "I" must precede the "think," and the "I" *is* self-consciousness.[48]

Worse than associating self-consciousness exclusively with awareness of some objective thing is the proposition that it comes into and out of existence like the illumination provided from a light bulb.[49] Evan Harris Walker erred in *The Physics of Consciousness* when he declared: "Consciousness is an *onset* phenomenon.... If awakened from...sleep and asked what we were just thinking, we may...be able to relate that our brains were active with thought, but this data processing was not associated with any conscious awareness."[50] He acknowledged we exist while asleep: we breathe, possess brain activity, and have a *repertoire* of autonomic system functions. *There is continuity of being*. His physicalist presumption of self-consciousness as dependent upon brain and body is unjustified. Self-consciousness *is* the continuity of being.

The relevant epistemological law is newer developments imitate older ones—in biological terms, phylogeny recapitulates ontogeny—and not the reverse. Self-consciousness emanates from Consciousness, and as the latter is not restricted to adherence with some physical body, neither is the former. The mind-brain relationship is associative not causal, and the mind can and does exist independently of the body. Only when it directly interacts with a body does an individual creature possess life. It is Consciousness in the form of self-consciousness providing the Energy animating atoms, molecules, and cells comprising a biological organism. Absent this Energy, matter doesn't exist, much less the beings composed of it. Since the body still exists even when temporarily dormant in sleep, self-consciousness persists throughout. Thomas Nagel made the point in *The View from Nowhere*: "The subjectivity of consciousness is an irreducible feature of reality...and it must occupy as fundamental a place in any...world view as matter, energy, space, time, and numbers."[51]

If self-consciousness is not a creation coupled to primitive wakefulness or phenomenological intentional awareness, how should it be characterized when interacting with the material atoms, which are also Consciousness, of an existential creature? The answer has been given: prior to awareness is alertness, the capability of *becoming* aware of an object by *being* alert as a subject. "Consciousness is the carrier of conscious thought. It is what is left when thought is removed, when the mind is stilled. When thought is entirely gone...consciousness is still there."[52] Self-consciousness is a fungible entity: it is the same everywhere, indistinguishable from global consciousness and from transcendental pure Consciousness.[53]

What makes self-consciousness unique to each entity is its interaction with the singular brains of living creatures. However, self-consciousness is not associated solely with the brain; it permeates all organs and cells of the body as the transcendental Energy transformable into physical energy. It: (a) alerts the body to take notice of incoming stimuli, and (b) urges the body to transmit those signals to the brain through the peripheral nervous system, where the brain can act as the central processing unit to digest and distribute all information obtained from its outlying sentry posts. Since every human brain is "wired" slightly differently—the total number of neural connections and their pattern of connectivity between organs is different from person to person—an explanation of the potential for displays of diverse thoughts, feelings, and actions derived from sensory input is readily at hand.

Self-consciousness is also responsible for development of the "software program" instantiating the electrical signals with *meaning*; only a mind can contemplate *significance*. Here is the mind-brain interface. Upon completion of signal analysis, the process is reversed. The brain transmits a new set of signals, at the command of self-consciousness, back to outlying regions of the body telling them to respond: through movement of arms, legs, eyes; by smelling, touching, listening; by coughing, sneezing, crying, laughing. All responses are monitored by the self-consciousness associated with every cell of the body. What are sought are external responses to internally generated action, and the cyclic nature of this process continues without surcease every day that person lives. If this view is correct, then *becoming* alive is the initial joining of global consciousness with the body to form self-consciousness. Death, therefore, would be separation of self-consciousness from body and marks its return to global consciousness.

Many theoreticians agree with Walker's judgment: "Self-consciousness is an onset phenomenon." They do so because it is consistently portrayed as demanding awareness as its principal defining trait. It stands to reason that if self-consciousness can be turned "on," it can be turned "off." Consequently, there exists another confusion permeating the literature. It was voiced by Dermot Moran in *Introduction to Phenomenology*: "We have no access to many of the most important acts upon which our conscious life rests, these acts taking place in the brain and being strictly speaking not available to consciousness at all."[54]

This thesis has its origin in Sigmund Freud's hypothesis of the existence of an unconscious along with its attendant repercussions for psychological pathology.[55] Prior to Freud, John Locke's dictum, as recorded in *An Essay Concerning Human Understanding*, was taken to be an obvious truism: "To say a notion is imprinted on the mind, and...at the same time to say...the mind is ignorant of it...is to make this impression nothing. No proposition can be said to be in the mind...which it was never yet conscious of."[56]

If self-consciousness is permanent persistent alertness united with the brain-body of material substance, then Sigmund Freud and all subsequent psychologists have misunderstood brain functioning. Güven Güzeldere's introduction to the volume he co-edited with Ned Block and Owen Flanagan entitled *The Nature of Consciousness: Philosophical Debates* offered an interesting summation of elementary Freudian psychoanalytic theory:

> In Freudian theory, the *unconscious proper* consists of repressed processes, exerting stress on the conscious component of the subject's mind and shaping his or her daily life in substantial ways. This is in contrast to the *preconscious*, which includes those processes that only contingently...lie outside awareness.... Freudian unconscious exists because of past events, explainable by repression mechanisms and the like, and is not *in principle* inaccessible. The cognitive unconscious...exists due to the way our perceptual-cognitive system is constituted and lies *in principle* outside our access.[57]

While self-consciousness is mediated through brain functions, it is not wholly dependent on neurophysiological events. Furthermore, it can at best only derivatively be described as awareness. Explanation for the ubiquity of unconsciousness as a construct lies in the inseparability of self-consciousness from the conscious entity: each self-consciousness is cognizant of all brain functions leading to phenomenal intentionality of an object, only it cannot always provide a report on its cognition through the brain itself because any attempt at articulation would invoke the same brain functions it sought to describe. There would inevitably result some change in the performance of these functions in the same way an experimentalist might interfere with the observation of an experiment by the very means he used *to* observe. There would be an attempt to report a constantly changing situation, which would result in inaccuracies as soon as self-consciousness began the effort to make the report bringing the observations into awareness.[58]

This argument is supported by Güzeldere's admission: Freud's concept of the unconscious implied its contents were ultimately accessible. This fact was obviously acknowledged by Freud and all of his intellectual descendents through their use of the methodology of psychoanalysis to recover the hidden information. This is significant in that it demonstrates the transitory nature of repression and its lack of inherency. Franklin Merrell-Wolff was explicit in "Consciousness and Dualism": "That state...we call 'unconsciousness' is so only with respect to the surface consciousness of the relationship between a knower and a known, but with respect to the Root Consciousness it is an extremely conscious condition."[59] If we fail to retrieve information from memory *in the moment*, and are therefore unable to treat our stored data as an object, this fact in no way impinges on the reality of self-consciousness as a persistent subject.

Güzeldere's argument is false concerning "cognitive unconscious" if he meant we could not comprehend the mechanism for any and all biophysical processes irrespective of whether we do in fact understand them at any given moment in time. There is no *a priori* reason why advances in the physical and biological sciences must have a predefined limit. If he meant we have no control over some functions, that will—an activity of self-conscious mind operating through the brain—does not control ability to perform in certain ways, this fact may be absolutely True or contingently true.

Supposition it is the latter is supported by evidence from specific cases. For example, at one time it was dogmatically held that an individual could not control their breathing, heart rate, and pulse beat because it was believed these were autonomous bodily activities. This belief has been scientifically disproven.[60] Some people are even able to control the production and use of enzymes: "The longer Oscar Estebany held a test tube containing a solution of the...enzyme trypsin, the faster it catalyzed... When Estebany held other enzymes...their reaction rate decreased. Employing other psychic healers... confirmed this... Some enzymes were always accelerated while others were always slowed."[61]

This *in vitro* phenomenon has an *in vivo* analog that has been intensively studied, with results carefully documented, lending further credence to the conclusion Güzeldere's evaluation of what is unlikely or impossible with respect to operations of self-consciousness are premature. There are also numerous reports of influence on the physiological processes of individuals by *other persons* from a non-contact distance, the *Bio-PK effect*, where 'PK' stands for psychokinetic: the ability to move objects over spatial distances by mental effort alone. In *The Conscious Universe*, Dean Radin summarized the results of studies undertaken by William Braud and Marilyn Schlitz:

> In 1991, Braud and Schlitz summarized all their studies...thirty-seven experiments employing seven different physiological response systems, such as blood pressure and muscle tremor.... The thirty-seven experiments combined resulted in odds against chance of more than a hundred trillion to one (10^{14}:1). Fifty-seven percent of the experiments were independently significant (with odds better than twenty to one), where five percent would be expected by chance.[62]

How do we accomplish these exemplary instances of self-control? The results conform to the previously expressed thesis of self-consciousness as an Energetic force capable of transmission through theological space. Certain individuals have apparently discovered how to willfully exercise the ability to direct this Energy towards surrounding objects. Therefore, the alteration of biochemical reactions is probably due to an energetic stabilizing influence, a lowering of *activation energy barriers* facilitating molecular transformations.

If human beings possess an ability to control physiological processes considered beyond the threshold of awareness, it is not unreasonable to assert we can likewise monitor and directly control psychological events thought to be beyond our capacities. Güzeldere's statement about how "the cognitive unconscious...lies *in principle* outside our access" may lack inherency.

These speculations cohere with the idea of self-consciousness monitoring *all* biological processes—those we can and cannot presently command—to ensure their proper functioning. It is also the function of self-consciousness to alert the body when one or more of these processes fail or becomes erratic. With respect to psychological mechanisms for brain function or malfunction, unconscious processes are logically cognitively impossible: if they existed, we could never come to know about them at all and never be capable of affirmatively proclaiming their existence, much less be able to recover their contents. Knowing some psychic event is unconscious is an oxymoron.

To reiterate: *there is a continuum for self-consciousness with respect to our awareness, which is precisely why the two are not the same.* At one end are the just-mentioned "subconscious" or "unconscious" events currently, though temporarily, unavailable from brain storage sites. At the other end are those mental and physical events occurring so rapidly as to arrive before our recognition of their presence. Physically, these are categorized as *reflexes*; mentally, they are organized under the heading of *intuitions*.

Being startled, with its accompanying bodily flinch and chemical release of adrenaline, when someone comes up behind you without awareness of their proximity is an example of the former. Gottfried Leibniz defined the latter in his *Discourse on Metaphysics* as the existence of a condition "when my mind understands at once and distinctly all the primitive ingredients of a conception, then we have intuitive knowledge."[63] The characterization of "at once" is the critical element: the event proceeds in its commencement and in its termination before awareness can register the factual arrival of a thought.

Thus far, there has been a focus on false depictions of the nature of self-consciousness. It is *not*: (a) a separate channel of operation distinct from bodily processes; (b) a recent innovation in the biological development of creatures on Earth; (c) associated with a particular physical condition of the body such as spatiotemporal wakefulness; (d) primarily definable as a causal agent for psychological intentionality towards either an internal or external object; (e) avoidable by some specious claim of accessibility to a uniquely objective third-person cognitive orientation; (f) undercut by an unconscious, subconscious, or preconscious realm in brain and/or body in possession of information and knowledge permanently hidden from retrieval.

Quantification will be the concern of this book's final chapter, hence only a qualitative mechanism of action will be offered here. Self-conscious mind has available two modes for receipt of information it can transform into knowledge and action. The first is the standard biological utilization of the sensory nervous system coupled to the brain to gather physically originating data and develop a mental image or representation of an object.[64] According to Johann Fichte's description in *Foundations of Transcendental Philosophy*: "A thing...presents itself to the I in two different ways: (1) As a being... absolutely present, something present 'as such' or 'in itself,' without any help from me. (2) As a being that can either be present or not be present, and, to this extent, it is a representation."[65]

This procedure has already been referenced in the discussion on the topic of the uniqueness of personal mind-brain interactions. Global consciousness manifesting as an egoistic sense-of-self occupies a state of alertness: it scans the surrounding environment with complete freedom of activity. When the "I" encounters a limitation to that freedom, it becomes *aware* of both itself as a subject and the restriction as a "Not-I," an object. Theological space acts as a conduit for information transfer—a kind of high-access, high-speed tunnel or wormhole—whereby the information content potentially obtainable from the energy comprising the distant object is transmitted in near-instantaneous time. After the exchange of Energies constituting the Conscious interaction between subject and object takes place, self-consciousness transforms that Energy wave used to detect the object into a series of electromagnetic signals capable of being detected and transmitted through the central nervous system to the brain with the intent they be decoded and processed for the purpose of further response.

Processing is comprised of several steps. The signals are: (a) dispersed to the appropriate convergence zones within the brain devoted to singular tasks, (b) coordinated with previously received signals through interface with short-term and long-term memory storage areas also distributed throughout the cranium, and (c) subjected to rules of transformation established by one's personalized self-consciousness. The "rules," in effect, are tables of values constituting the individual's character and personality, the purpose for which is to ensure consistent, reproducible behavior depending upon the situation encountered. Hence, self-consciousness monitors the physical sequence of activity and generates commands the organism performs as the physical or emotional responses engaged in by the various parts of the body.[66]

This proposal conforms to schemes postulated by previous investigators. For example, in the *Critique of Pure Reason* Kant's epistemological chain begins with the *a priori* unity of apperception, which is self-consciousness. Through a *faculty of sensuous intuition*, we make ourselves aware of data received by the perception of our senses. This sense data is collected and transported to another functional entity called the *faculty of imagination*. The role of this latter faculty is to accomplish a complex process: transformation of sensory input obtained over time from the external object into a series of images, representations of this piece of reality, which are collated in such a manner as to allow the *faculty of apperception* to unify them in concluding they belong to the same phenomenon. Another internal organ, the *faculty of understanding*, applies *a priori* categories or rules to further transform this set of representations into formal concepts.

The process of transforming sensory input to concepts, and cognition of these concepts as belonging to a single unified self-consciousness, is what Kant meant by the acquisition of knowledge.[67] Allowing for the antiquated notion of "faculties," this sequence is reasonable as far as it goes. In essence, it claims knowledge is the result of a progression involving data acquisition, data categorization, and data analysis. What Kant omitted was the final step closing the epistemological cycle: data extrapolation to new sensory input to allow the process to begin again, yet at an advanced level from the previous round.[68]

The transition from global consciousness to egoistic self-consciousness allowing for the introduction of images, concepts, and emotions clarifies the phenomenological distinction between the two. The introduction to Jean-Paul Sartre's *The Transcendence of the Ego* explained the meaningful difference: "There is only an ego *for* consciousness. The ego is 'out there,' in the world, an object among objects.... *Consciousness has no contents*.... Consciousness contains neither a transcendental ego nor anything else. It is...a spontaneity, a sheer activity transcending toward objects.... 'Images,' 'representations,' 'ideas,' 'phenomena,' 'sense data,' *et cetera* are objects *for* consciousness, not contents *in* consciousness."[69] Extending this proposal would yield what is here alleged: "'Images,' 'representations,' 'ideas,' 'phenomena,' 'sense data' ...are objects" *for self*-consciousness, and contents *in* the brain.

The second modality utilized by individualized self-consciousness in the generation of knowledge is entirely metaphysical-mental: the mind, coupled to an organic brain, can access theological space to obtain information in the form of abstract ideas. Plato's proposal for an archetypal realm of idea-forms was valid, but not as self-subsistent entities.[70] They exist as Energetic packets transmitted by global consciousness in acts of revelation whose receipt can be depicted as an instantaneous intuitive cognition by each sentient being. The concept of intuitive cognition has been encountered in chapter three, and differs from notions of revelation only with respect to the overtly religious content of accessed information for the latter, whereas the former can pertain to any topic of contemplation.

Jack Cohen recognized the generation of both routes as coming from the same transcendent source. In *Major Philosophers of Jewish Prayer in the Twentieth Century*, he theorized: "The active and passive functioning of the mind...is all part of one system. Immanence and transcendence are two poles of a continuum. Revelation and discovery are the...outcome of thinking that harmonizes immanence and transcendence."[71] Since standard ratiocination is more common than generation of concepts through unanticipated revelatory intuition, it will be addressed first. What will become obvious through this discussion is: (a) immanent self-consciousness generates thought in human brains; (b) self-consciousness is material and formal causal source of reason and revelation; (c) transcendent Consciousness, *as* global consciousness, acts as the efficient and final causal source for each type of cognition.

How do ratiocinative and intuitive events transpire? The electromagnetic field surrounding the human body is a direct effect of the association of self-consciousness with body. *Auras* are scientifically verifiable radiant energy fields, not a theosophical fantasy.[72] In *Infinite Mind: Science of the Human*

Vibrations of Consciousness, Valerie Hunt gave a synopsis drawn from her studies in the laboratory:

> In the physical body there are two primary electrical systems. One is the well-known alternating electric current of the nervous system, the brain, neurons, and the nerves, which causes muscle contraction, nerve transmission, glandular secretion, and sensation. The other is a newly discovered electromagnetic system...emanating from atoms and cells. This energy has been called an aura... We discovered that this energy field is unique. It is continuous, while all the other body electric recordings of heart, brain, and muscle are off-and-on signals... The human field absorbs and throws off energy dynamically. It interacts with and influences matter, whereas fields associated with inert matter react passively.[73]

Hunt's exploration of auras did not constitute some new, extraordinary discovery on her part; the research followed in the footsteps of numerous other investigators. Nearly eighty years ago, Manly Palmer Hall's study of esoteric philosophical systems led him to state: "Man's body is enveloped in an ovoid of bubble-like iridescence...called the Auric Egg.... The supreme consciousness of man is in this aura, which extends in all directions."[74] The auric field of every object projects a gravitational force of attraction into the surrounding space, the theological space identical to global consciousness.

In the Dimension/dimension of theological space, global consciousness is an Energy capable of being bundled in units of information content. In terms of information theory, the entropy—possible combinations of novel messages—contained in the Space is infinite, which means it necessarily incorporates all facts about reality both immanent and transcendent. One can image theological space as an inexhaustible library of data awaiting transfer to some material entity in the form of knowledge.[75] Or, framed as a modified form of Plato's notion of archetypal ideas: "They exist as Energetic packets transmitted by global consciousness in acts of revelation."

The mechanism for Energy transfer within theological space utilizes *both* gravitational *and* electromagnetic forces. It has been proposed the product of these two forces in spacetime is identical to their multiplicative effect within theological space. Hence, gravitational effects are orders of magnitude more consequential transcendentally than within manifestation. General relativity theory invokes the gravitational force as equivalent to acceleration in normal spacetime. This is known as the *equivalence principle*.[76] It implies gravity transmutes observer physics into participant physics: one does not *observe* acceleration objectively; one *feels* acceleration subjectively. *The properties of gravity are central to understanding Consciousness and its psychophysical laws*. Gravity shapes theological space: global consciousness is every-where and every-when simultaneously.

Gravity demonstrates the principle that all matter, all energy, was once originally homogeneous. The nonexistence of gravity particles as gravitinos provides important testimony to the fact the transcendental component of this force is dominant—it is never to be envisioned as discrete, only continuous. Gravity's symbolism contrasts with that of the electromagnetic force, which highlights the options of repulsion and attraction within manifestation.[77] The existence of electrical monopoles—electrons, protons, positrons—coupled to the lack of existence of magnetic monopoles depicts an asymmetry of forces within the universe representative of Interpassage *if* one considers only the perspective from our universe.

If an individual is engaged in deep thought, their electromagnetic auric field binds theological space more tightly around that being. The interaction between the two domains, when properly coordinated, allows information to pass in the form of Energy transfer between global consciousness and self-consciousness. Upon receipt of the incoming Energy, self-consciousness acts in the same manner it did when engaged in sensory perception with respect to information transmission and processing within the human brain. This is the manner in which all ideas and other forms of creative inspiration arise.

Reasoning and intuition occur when the body-associated electromagnetic field that *is* self-consciousness seeks to access theological space and does so successfully. Hannah Arendt's opinion should be taken literally in *The Life of the Mind*: "Thinking annihilates temporal as well as spatial distances. I can anticipate the future, think of it as though it were...present... I can remember the past as though it had not disappeared. Since time and space in ordinary experience cannot...be thought of without a continuum...I could with some justification say...not only distances, but also time and space themselves are abolished in the thinking process."[78]

Crucial to one's self-consciousness seeking entry to the data available in theological space is the functional role to be attributed to one's *will*.[79] Arthur Schopenhauer elevated Will to the status of an *independent* transcendental, thereby conflating the Energy invigorating human will with the Energy of pure Consciousness: "If this thing-in-itself...is the will...considered as such ...apart from its phenomenon, it lies outside space and time, and accordingly knows no plurality, and consequently is one.... It is not one as an individual or a concept...but as something to which the...possibility of plurality...is foreign."[80] Although self-consciousness and Consciousness are intrinsically identical, in practical terms the first is limited in activity whenever it operates in collaborative conjunction with a brain. Hence, the Energy/energy actually commanded by self-consciousness is less than it would potentially command as Consciousness.

Every material object is composed of the Energy of Consciousness. So it would seem as if a "physical" brain would not be an impediment to use of Energy by a self-consciousness that is also this Energy of Consciousness. The difficulty is that manifestation is a composite of dual Conscious Spaces, not descended from a single Conscious Space. *Physical reality is a mixed-Energy reality*. This ontology causes the energy/Energy within the existential universe to be less potent, because less homogeneous, than unadulterated Consciousness. Gnostic philosophy's central tenet can be interpreted to mean the descent into physicality induced impurity on pure Consciousness.[81] The approach has merit in that it describes the Energetic consequences of creation by the Big Bang as an in-the-universe model of Interpassage. Gnosticism was false, however, in its assertion of an intrinsic duality between matter and energy whereby the former is the ontological source of impurity.

Returning to the notion of will, a more exact assessment of its role was given by Hannah Arendt in *The Life of the Mind*: "The Will, by virtue of *attention*, first unites our sense organs with the real world in a meaningful way, and then drags...this outside world into ourselves and prepares it for further mental operations: to be remembered, to be understood, to be asserted or denied."[82] When self-consciousness detects something worthy of attention, it engages the brain. The will focuses that attention and directs the brain to receive and record data about the environment, where "environment" is to be understood as referring to an internal condition in the case of introspection, or externally directed in the case of sensory contact with one's surroundings. It is the transcendental Energy of will that catalyzes the process in either direction.

For classical mystical revelation, a similar process occurs except human willpower is deliberately lessened in its Energetic intensity. This state allows for initiation of contact between self-consciousness and theological space to be directed by global consciousness and the information it seeks to impart to the individual. The gravitational force of attraction exerted by theological space on self-consciousness constitutes the physical mechanism for what in religious terms is often denominated *grace*—the unanticipated reception of esoteric facts pertaining to ultimate reality. In summary, one methodological technique uses a pull of transcendent Energy, global consciousness, towards the immanent Energy of self-consciousness; the other epistemological mode employs the opposite approach: pulling immanent self-consciousness toward transcendent global consciousness.

This model of knowledge attainment fits with prior publications detailing a three-stage theory of conscious judgment.[83] The first stage is designed to orient self-consciousness towards external objects. This occurs by correlating environmental events with representations contained in the brain when an internal sensory nerve center is stimulated. The second level introspects on the external situation in order to determine its meaning to self—taking into consideration its aesthetic value or emotional and cognitive importance—in order to precipitate an internalized reaction to the stimulus, which might or might not result in a formal response to the external event. It is the role of the third stage to accumulate all secondary level experiences so as to be able to discern general patterns, the purpose of which is to establish personal values capable of serving as guideposts for future experiences deliberately sought.

The detailed examination of personal history occupying the attention of the third stage of conscious judgment constitutes the developmental source of an individual's character and personality traits, the aforementioned "rules of transformation." Due to the layered responsibilities adhering to each stage, rational thought and intuitive cognition on routine topics are the province of the second and third stages. Revelation in a mystical, though not necessarily religious, sense of the term is jurisdictionally processed exclusively by the third level.

An important consequence of the enactment of willpower is its ongoing ability to select among three distinct activities: the doing of something, the not-doing of something, and the decision to not choose whether to do or not do something—the latter expressing an attempt at neutrality. For Arendt, the decision not to choose conveyed the essential meaning of human freedom: "The Will can *suspend* itself, and while such suspension can only be the result of another volition...this second volition, in which 'indifference' is directly chosen, is...testimony to human freedom, to the mind's ability to avoid all coercive determination from the outside."[84] For John Duns Scotus, the decision not to choose represented the significance of ethical belief and moral behavior: "The will's...capacity to refrain from choosing according to the dictates of right reasoning places the locus of moral responsibility in the will, and thus allows greater extension to the discussion of the domain of prudence: not...as judgment of means but as directional knowledge of ends of moral living."[85]

Each theorist was correct in his or her evaluation. The inalienability of freedom of the will possessed by all sentient creatures is critical to teleology: participation in resolution of the cosmologic-cosmogonic crisis commencing with the onset of Interpassage demands unimpeded and unbiased access to choice. *GATF* offered detailed justifications, reiterated in the fourth chapter, for the central roles occupied by ethical principles and moral acts in any coherent application of metaphysics. *GATF* proposed, as the fundamental goal to be undertaken in life, the attainment of sufficient knowledge to reach a point where free participation in teleological resolution was theoretically plausible and actually possible. Augustine made the point most articulately in

On the Freedom of the Will: "If human beings...cannot do right unless they so will, then they ought to have a free will, without which they cannot do right."[86]

The qualitative mechanism for human operation herein argued provides clear justification of the necessity for eating and sleeping. The brain, at the will's command, expends enormous amounts of energy in data manipulation. This energy is obtained from chemical conversion of bioorganic molecules: proteins, sugars, and fats consumed as nourishment and metabolized, with release of the needed energy upon demand. Ingestion of food thus becomes a quintessential instance of matter-energy interconversion. Consequently, the only time the body has available to rebuild its reserves is when resting, and more efficiently when sleeping. In such states, the brain and other organs use energy in a minimal manner simply because they process less environmental data.

Self-consciousness, however, has the inexhaustible supply of Energy that is Consciousness, and has no need to rest or sleep—which explains why self-consciousness persists independent of the physical condition of the body—and can remain in that eternally vigilant state called alertness.[87] Alertness can transform into awareness instantaneously, which is how we suddenly awaken from a deep sleep due to a child's cry in the next room, or by the wail of an ambulance siren heard from a distance, or when we have suddenly uncovered the answer to a problem worked on the previous day. It is self-consciousness that receives the information input and awakens the brain.

Any summary of the interlocking relationship between a metaphysical self-consciousness and an all-too-material physical world could do worse than employ the verbiage offered by Dean Radin's *The Conscious Universe*: "Mind and matter...*seem* to be very different creatures. The *mind* thinks; it isn't located in space; its concerned with values; it has free will; it's driven by purpose; it's private; and everything we personally know is through the mind's direct, conscious experience.... As far as we know *matter* doesn't think; it's localized in space...value-free, determined, purposeless, objective; and all knowledge about it is inferred by our mind."[88] The conditional "as far as we know" is a significant admission. Precisely what is being claimed here is that both self-consciousness and matter are, in the final analysis, forms of Consciousness in the existential universe.

Chapter Seven: Mysticism

> Mysticism is defined...as the worldview that seeks...perfection of the individual through union with an agent or force conceived as absolute.
>
> —Livia Kohn, *Early Chinese Mysticism*

If the connection between epistemology and metaphysics is to conform even approximately to the theory outlined in the previous chapters, the tenet of maximal comprehensiveness—presumptively a major indicator of value of a system of thought—must be applicable to unusual as well as commonplace mental and physical accomplishments. This idea will be tested in chapters seven and eight. Here, the focus will be on mysticism in theory and practice. It is a topic experienced but not studied until recently. The definition of the theme presented in the epigraph highlights an intention for the subject to be taken in its broadest sense. Usage of "agent or force" serves this purpose: it allows for incorporation of both Eastern and Western religious sentiments. Any declaration limiting the universality of discussion, such as one provided by Bernard McGinn in *The Flowering of Mysticism*—"Thus, mysticism is characterized primarily by a sense of an immediate relation to God and the transformation this effects in the consciousness of the recipient"—would be unduly narrow and restrictive.[1]

Mysticism takes a path to knowledge and truth, and possibly to Truth, science either denies or ignores. Nonetheless, the two paths are more alike in methodology than some would credence. As it turns out, both require formal and systematic training if the end results are to be efficiently attained. In addition, patient observation to detail is necessary in preparation, execution, and subsequent interpretation of "experiments" of either type. Scientific and mystical practice equally demand an initial abstract impartiality that does not block either perceptual experience or systematic analysis of the import and interpretive meaning after the data has been collected. Finally, both require an underlying faith that the method employed, as well as the goal sought, is intrinsically significant to self and others.

Where they are most at odds, perhaps, is in their histories. Even prior to the modern scientific era, investigators in "natural philosophy" have always attempted to combine theory with experiment, if only in a rudimentary way, by compiling available data—Aristotle's efforts in botanical classification, for example—and despite the absence of rigorous theories allowing designed experiments, such as those in alchemy, to be reproducibly successful.[2]

In contrast, until recently, those with an interest in mysticism have never invested adequate effort in development of its theoretical component, which previously largely consisted of an abundance of anecdotal reports regarding individual experience, or the consequences of experience played out on the world stage, as in the formation of religions. All religions have developed extensive theologies intended to facilitate orientation of interpretive aspects of mystical experience sought and obtained within any particular tradition. A theory of mysticism referring to an intellectual understanding of the precise elements involved needed to explain how these events occur was, however, lacking.

Once upon a time, philosophy and religion were either inseparable, as in the Orient, or susceptible to heavy cross-fertilization, as was the case in the Occident. Virtually all philosophers were also religionists; some were even of a mystical bent. Perhaps because mysticism lacked any semblance of an explanatory theory concerning how and why it works, its publicized activity fell into disrepute in comparison with the steady progression of successes attained by the introduction of scientific methods despite the fact the history of mystical practice in religious settings predated techniques of science by millennia. The seemingly random, mostly unproductive, exercises engaged in by mystics of all backgrounds contrasted greatly with accomplishments generated in the physical sciences.

Much of the difference between the two endeavors has been ascribed to the repeatability and second source verification of scientific experiments in counterpoise to the unpredictable nature of mystical practice. Hence, it could come as no surprise when modern philosophers—many of whom were/are completely divorced from religious influences—became predisposed to favor techniques of science over those available through religion, and as a direct consequence began to undervalue mystical efforts pursued by some deeply religious adherents.

Such a judgment can be discerned from the views of Chauncey Wright according to Edward Madden's articulation of Wright's philosophy in his essay for *Ultimate Reality and Meaning*:

> As an empiricist Wright claimed that to qualify as knowledge a proposition must lend itself to test and verification by the particulars of concrete experience that are public and communicable. Beliefs based on revelation and mystical experience...are private matters, reflecting individual experience or else mere institutional authority; and so they cannot...be tested and verified by concrete experience available to all.[3]

Edgar Brightman not only supported Wright's outlook, he extended the negative view of mystical endeavor. He regarded the fruits of all mystical experiences as lacking truth-value for anyone other than the recipient of the alleged revelation.[4]

The opinions of Wright and Brightman must be reviewed before being accepted. They can be challenged on several levels: (a) Mystical revelation is neither more nor less an individual experience than seeing a tree. Everyone's experience of their interaction with a tree, and their characterization and assessment of it afterwards, will be unique to varying degrees. The singular nature of mystical unfoldings is not phenomenologically exclusive; (b) Many philosophers regard pronouncements concerning Truth disseminated by the scientific establishment as institutionally biased as dogma promulgated by a religion-affiliated church; (c) Mystical events *are* available to all individuals as a consequence of *grace*; (d) Why should scientific experiments be granted recognition as bearing inherent truth-value for nonscientists, while mystical experiences are not accorded the same generic acceptance?

If Chauncey Wright meant "concrete experience available to all" to be synonymous with "shared experience," then it is questionable whether even sexual intimacy constitutes a shared experience in this sense. If, however, he meant the phrase to imply any individual should be able to pursue a specific methodology and thereby induce mystical experiences, he was wrong: *every* religion has maintained an abundance of ritualistic practices promoting mystical enlightenment for those willing to undergo the necessary training.

William James offered a non-religious characterization of the notion of grace in "A Suggestion about Mysticism": "States of mystical intuition may be...sudden and great extensions of the ordinary 'field of consciousness.'... Some persons have...a very wide, others a very narrow field... When by any accident the threshold lowers...the field widens and the relations of its center to matters usually subliminal come into view, the larger panorama perceived fills the mind with exhilaration and...mental power."[5] It should be noted that James' account is consistent with the proposal found in the previous chapter. In writing, "the threshold lowers" and "the field widens" he was unknowingly discussing interactions between global consciousness and self-consciousness leading to mystical events.

To those presenting arguments discounting expression of unimpeachable Truth accessed during mystical experience, mystics have a potent rejoinder: the critics have never had such a transcendental unfolding. Mystics from all cultures and all historical time periods have always vigorously professed their revelations as necessarily True. The *Laṅkāvatāra Sūtra* affirmed: "But the cessation of the discriminating mind cannot take place until there has been a 'turning about' in the deepest seat of consciousness. The mental habit of looking outward by the discriminating mind upon an external objective world must be given up...a new habit of realizing Truth within the intuitive mind by becoming one with Truth itself must be established."[6] In *The Ascent of Mount Carmel*, Saint John of the Cross outlined the stepwise progress to be made by the diligent student in seeking mystical revelation:

> Since the order whereby the soul acquires knowledge is through forms and images ...and the natural way wherein it acquires this knowledge and wisdom is through the senses, it follows that, if God is to raise up the soul to supreme knowledge, He must begin to work from the lowest and extreme end of the senses of the soul, in order that He may gradually lead it, according to its own nature, to the other extreme of His spiritual wisdom, which belongs not to sense.[7]

Both proponents and opponents of mystic Truth raise valid issues. The pronouncements *may be* True, may be judged as True by those to whom such encounters occur, provided no misinterpretation of the information followed after the fact. Under no circumstances should statements made by mystics be *prima facie* acceptable to spectators. Self-consciousness cannot err but brains can; questionable is whether mystics or bystanders are evolved enough to critically and rigorously *know* their assessments of *supra*natural information are True.

There are numerous admonitions in the spiritual literature against undue reliance on revelatory visions, implying mystics are not immune to illusions, fantasies or unwarranted speculations. Thomas Merton offered a warning in *Contemplative Prayer* based upon difficulties of the mystic path he perceived to be innate: "Sometimes prayer, meditation, and contemplation are 'death' —a...descent into...nothingness, a recognition of helplessness, frustration, infidelity, confusion, ignorance.... Imagination and the emotions wander away."[8] In *Jewish Meditation: A Practical Guide*, Aryeh Kaplan emphasized the adverse results potentially accruing from false conceptual interpretations of mystical experience: "In the Kabbalah literature, there are warnings even to advanced practitioners of meditation not to give credence to visions. Even the most impressive visions can be spurious and come from the Other Side. Acting on the basis of images seen while in a meditative state is considered to be extremely dangerous and detrimental to one's spiritual development."[9]

The apprehensions of Merton and Kaplan speak to an extraordinary irony in modern society. Even as the diverse geopolitical nation-states around the globe develop socioculturally through expansion of technological benefits accruing from progress in science, more individuals are engaged in spiritual practices with a mystical focus than at any previous time in history. In part, this result is an anticipated outcome to be attributed to the enormous growth in world population.

In the main, it is due to the increased general availability of literature on mysticism. In digestion of this material, it has become clear that while the founders of religions were incited to a practical end through experiences that could be categorized as mystical in nature, and some devoted disciples of various faith traditions have been reported to undergo mystical experiences of varying frequency and depth, such unfoldings need not and do not occur only to those individuals who are safely within the confines of an established religion. Mystical events occur to ordinary, non-religious people every day, only these instances are rarely catalogued by non-participants. While some of these individuals later associate themselves with a particular religion as a consequence of their experiences, usually to allow them to place the event in some familiar context, allegiance to an established tradition is not, has never been, a prerequisite for mystical experience.

Consequent to an enormous increase in the absolute number of mystical thrill-seekers among a population of otherwise ill-prepared individuals, it has become necessary to find ways to combat the insidious potential of negative effects, including development of extreme forms of neurosis or psychosis. Psychological damage, temporary or permanent in kind, can ensue as a result of a mystical experience due to the inability of the recipient to integrate the data received during an event into their normal personality.

In *The Ascent to Truth*, Merton gave voice to a tenet of mystical practice we should take as a given: "The mere fact that some souls of 'simple faith' have risen, without instruction or analysis, to mystical contemplation does not permit us to infer that discursive knowledge, intelligent study, logical meditation, and other mental disciplines are...useless or positive obstacles to ...growth in the life of prayer.... The exercise of...intelligence in meditation and in the prayerful study of revealed truth is the ordinary way by which we must dispose ourselves for contemplative prayer."[10]

To one motivated toward pursuit of "intelligent study," independent of whether he or she makes the effort through rigorous practice of a particular religious tradition, the cultural climate allowing for attempted development of mystical theories, at least in the Western hemisphere, probably began with William James at the start of the twentieth century. Here was an established philosopher and Harvard University professor who declared mysticism was worth studying as an academic discipline. His remarks are justly famous:

> I will...propose to you four marks which, when an experience has them, may justify ...calling it mystical... (1) *Ineffability*—The subject...immediately says...it defies expression...no adequate report of...contents can be given in words... (2) *Noetic Quality*—Mystical states seem to those who experience them to be also states of knowledge... Illuminations, revelations, full of significance and importance...and as a rule they carry with them a...sense of authority for after-time... (3) *Transiency*—

> Mystical states cannot be sustained for long... (4) *Passivity*—Although...mystical states may be facilitated by...voluntary operations...yet when the characteristic sort of consciousness...has set in, the mystic feels as if his own will were in abeyance.[11]

More recently, William Braud has expanded on this set of characteristics in "Brains, Science, Nonordinary and Transcendent Experiences." Included beyond the four enumerated by James were, "feelings of unity, numinousity, loss of ego, an altered appreciation of space and time, perceptual experiences and alterations, changes in affect, transformative changes, paradoxicality." Braud acknowledged both short-term and long-term aftereffects of mystical events: "Changes can range from fresh perceptions, through major changes in one's worldview, to callings and changes in vocation, to experiences of profound transformation, rebirth, and 'conversion'."[12]

Ineffability is probably the most frequently asserted characteristic of such experiences by students of the subject as well as those who have undergone these occurrences. *Every experience can be claimed to be ineffable.* If one tries to describe eating a vanilla ice cream cone on a hot summer day, one can use lots of words and none may completely convey that exact experience to the listener. Identical twins partaking of identical vanilla ice cream cones side-by-side simultaneously will undergo some differences of experience one is unable to articulate with adequate precision. People and their experiences are unique. Language can at best approximate what actually transpires during an event, and while some descriptions may be more picturesque in providing a bystander with a clearer picture, words can never be an exact substitute for experience. By extension, a written account of the contents of an experience will omit aspects possibly central to the event because the writer lacks the requisite vocabulary. Mysticism is not unique in the ineffable quality found at the root of the experience.

Noetic quality also cannot be alleged to be peculiar to mystic experience. *All experiences convey information convincing the participant of something previously unknown.* Some data is comprehensible in the moment and some only upon subsequent contemplation. This fact has special validity for mystic expression in light of frequent claims by reporters of such experiences that they were not in a cognitive frame of mind during the experience itself. Thus, some data must have been stored in some manner for later retrieval and analysis. In light of the claim in *GATF*, and repeated in chapter four, that the major life purpose of every sentient creature, whether they are aware of it or not, is to gather information so as to learn what is true, and eventually True, about physical and nonphysical reality, it cannot be a surprise to read about claims of *noesis* associated with mystical events.

Nor does *transiency* convey something special about mystical experience serving to differentiate it from other events. *With very few exceptions, all experiences are transient*. They come and go in time. Some happenings are *essentially* temporal in that they involve time as part of the event. Others are not *concerned with* time subjectively however much they may *involve* time objectively. Mystical unfoldings are neither longer nor shorter in duration than other kinds of episodic events. Some spiritual adventures have been reported lasting seconds; some have been claimed to last for days, months, or even years.[13]

Of all the characteristics mentioned in the passage by James, *passivity* has the best claim to being unique to mystical experience. Some autonomic physiological responses—cell division, metabolism, protein synthesis, and the like—occur to the individual without their active participation in bringing about the event. People are "innocent bystanders" when unexpected, perhaps undesired, events occur in their presence without direct personal involvement in bringing on those events. Yet these situations are not really what is meant by "passivity" when the term is used in the mystical literature. The opposite of passivity is *not* activity. Seekers of mystical experience have a variety of techniques at their disposal to facilitate putting themselves in a physical and mental state conducive to such happenings. *By passivity is meant that no one can force a mystical event to occur at a time and place of their choosing by creating an environment frequently allowing for such experiences* according to the anecdotal evidence.[14]

In some vague analogical sense, this definition can also be said to apply to other experiences—for example, no matter how much one may want to experience pleasure from eating that ice cream cone, he or she cannot force him or herself to enjoy it. While one cannot force oneself to derive *pleasure* from that ice cream cone, he or she can guarantee they will experience the *ingestion* of that food. Mystic passivity means conditions for experience can be premeditatively arranged to the ultimate degree, and no revelatory event will necessarily transpire as a causal result.

What is really meant by passivity is that, unlike *most* other experiences, mystical revelation only occurs as a condition of grace, now employed as a theological term.[15] Precisely because one is presumably in contact with some transcendental agent, the source of the information received, the individual simply has no control over whether the experience will occur. As mentioned, it is the phenomenon of grace permitting otherwise untrained individuals to have mystical experiences in the first place. The emphasis on "most" is used to signify it will later be shown how mysticism is actually not unique in its reliance on grace for information acquisition.

Turning to Braud's supplemental traits, *unity* suggests an experience of non-differentiation: sometimes between subject and surroundings, but more generically between the immanent experiencer and the transcendent agent or force, to invoke Livia Kohn's terminology.[16] This element is indeed unique to mystical experience, since typical relationships in the world are ordinarily of the subject-object variety. Even the notion of empathy, usually used with respect to interpersonal interactions—some individuals, like Saint Francis of Assisi, claim to be empathic towards nonhuman living creatures—maintains a duality of personality during the experience.

In Islam, the most potent experience of union with a transcendent agency is termed *fanā al-fanā*, which literally means "annihilation of annihilation." This rare event, and the stages leading to it, was delineated in *The Mystics of Islam* by Reynold Nicholson:

> 1. A moral transformation of the soul through the extinction of all its passions and desires. 2. A mental abstraction...from all objects of perception, thoughts, actions, and feelings, through its concentration upon the thought of God. Here the thought of God signifies contemplation of...divine attributes. 3. The cessation of all conscious thought. The highest stage of fanā is reached when even the consciousness of having attained fanā disappears (fanā al-fanā). The mystic is now rapt in contemplation of the divine essence.[17]

All esoteric religious doctrines regard union with whatever is conceived as transcendental, whether it is a personal deity or an impersonal potency, as the acme of mystic experience. The traditions differ on this issue only insofar as the likelihood of attainment of this level.[18]

Both *numinousity* and *loss of ego* follow from the notion of unity with the transcendental agency and add nothing to the characterization of possible experience.[19] Similar sentiments can be expressed for inclusion of *perceptual experiences and alterations*, *changes in affect*, and *transformative changes* among distinguishing traits. Unless one is an automaton, activities ranging from the most trivially mundane to the most extraordinarily spiritual will be accompanied, during or after, by one or more overt sensory, emotional, and intellectual responses. Moreover, many of these occasions could have long-lasting significance throughout a person's lifetime. Almost any kind of event can be designated "a life-changing experience." Consequently, nothing more need be said about these conditions as they apply to mysticism.

The remaining two characteristics on Braud's list are more pertinent to a mystical experience, though here too an exclusive attribution of singularity *vis à vis* other types of events overstates the case. *Paradoxicality*, or at least the appearance of paradox, is fairly common in philosophy and physics. *An altered appreciation of space and time* also occurs during a psychedelic drug experience, having a dream while asleep, or if suffering from acute paranoia.

With respect to paradoxicality, the noteworthy point is the "appearance" of paradox. Almost invariably, what this phrase suggests when invoked is that insufficient information is available to resolve the abstract theoretical or concrete practical discrepancies presently discernable on any given issue to the complete satisfaction of all specialists in the field. Assertions of paradox in mysticism have the same basis. Peter Moore had a sound rationale for his conviction that the probability of real paradox in mystic practice should be rejected:

> There is...no evidence...mystics are forced by their experiences to break the law of noncontradiction in their writings.... Where the real meaning of a paradox remains obscure, it would seem reasonable to assume...the paradox is not literally intended unless the mystic asserts as much independently of his use of the paradox itself. If mystics really did have literally contradictory experiences of the kind suggested, one would expect to hear...more about them in addition to the...presence of paradox in their writings.[20]

Steven Katz, in the same collection of essays containing Moore's article, was also critical of all claims made for ineffability and paradoxicality within mystic endeavors: "If the terms 'paradoxical' and 'ineffable' mean anything, do they not cancel out all other descriptive claims, thus undermining any and all attempts at a phenomenological typology of mystical experience based on post-experiential reports?"[21] If spiritual experiences were *truly* engulfed in paradox and inherently ineffable, all of the descriptive constructs presented by William James or William Braud would be nonsensical: they could not be offered as valid qualifiers of experience without, at the same time, invoking contrary assessments.

Asserting an altered appreciation of space and time as a representative characteristic of mystical experience, although not specific to that condition, is significant in the context of the epistemological theory promoted in this volume. Theological space has been posited as the conduit for information transfer between Consciousness and self-consciousness, the connective link providing the mechanistic basis for the processes of thought, intuition, and revelation. The spatiotemporal characteristics of theological space *have been claimed to be different* from those of normal spacetime: it is an everywhere and everywhen, where distance is meaningless and time ceases to pass with a specific directional flow.

The words of Evan Harris Walker in *The Physics of Consciousness* are significant: "To connect consciousness with space, it should span space... To connect with time, it should span all time."[22] Samuel Alexander said much the same thing in *Space, Time, and Deity*: "Since God's body is the whole of Space-Time, God in respect of his body is all inclusive... In their continuous connection as pieces of Space-Time and linked by spatiotemporal continuity they are fragments of God's body, though their individuality is not lost in it."[23] Taken in combination, the resultant expression is the equation of global consciousness with God with theological space, a position argued for in the chapter on metaphysics. Ann and Barry Ulanov confirmed this representation of the relationship in *Primary Speech: A Psychology of Prayer*: "In one way of speaking, we can say...all time and all space are accessible to us in prayer, and in another way, we can say...there is neither time nor space in prayer."[24]

Upon completion of data transfer, the mystical revelation terminates and the process is reversed. Self-consciousness commands its associated brain to regain *awareness* of the distinction between itself and transcendent Other; synchronization of energy with Energy is lost; "descent" to normal spacetime conditions follow. Secondary aspects of experience commence immediately: (a) astonishment at being granted a revelation; (b) acknowledgement to self and others of the essential communicative ineffability of the nature of the unfolding event; and, (c) the insistent attempt to place the experience in some philosophical and/or religious context sensible to oneself and outsiders. Each phenomenological stage of a typical mystical account occurs in its proper chronological sequence.

In "Contemporary Epistemology and the Study of Mysticism," Donald Rothberg offered an example of a spiritual unfolding containing many of the suggested features of mystic experience that may serve as a useful summary of the qualities enumerated by William James and William Braud:

> There is...in concentrative training, a suspension of processes of categorization of phenomena. At a later stage...a decentering of experience as...the self collapses.... The meditator no longer experiences phenomena through the categories of time and space.... There remains awareness and cognition.... Knowledge is...essentially...a mode of experience in which categorization is drastically reduced...and in which the knower might, in...rare moments of speech, make statements about the inadequacy of language... Processes of abstraction and objectification...which make possible naturalistic knowledge, are greatly reduced.[25]

What initiated the recitation of characteristics of mystical experience by William James in *The Varieties of Religious Experience* was a compilation of reports from the mystical literature. What transformed the esoteric mystical experience into a discipline capable of rigorous analytical study analogous to any science was, as has just been implied, the adaptation of phenomenology to its exploration. Although Edmund Husserl, founder of the method, was focused on using the technique to explore more common experiences, many researchers found it an appropriate and effective tool in its application for mystical events, an attitude Husserl himself later adopted.[26]

The first element of phenomenological methodology turned out to be of utmost importance when applied to research on mysticism: the acceptance of first person reports as legitimate sources of data about something real. The contrast with the strictures of supposedly objective science was stark. Prior to that conditional agreement, mysticism was often thought of, when it was thought of at all, as the raving of delusional religious fanatics, as occurring rarely and only to a few privileged individuals. Such events were believed to hold little meaning for those non-mystically inclined. In fact, a declaration of value adhering to personal experience encouraged many to pursue spiritual practices for the purpose of attaining the benefits allegedly accruing from the mystical path.

The second element of phenomenological technique, the *epoché*, was mentioned in chapter one: the willingness for researchers to remain neutral during investigations, to weigh available evidence gleaned from the reports of individuals who have written about personal mystical experiences without prejudgment of degree of credibility or likelihood of verifiability. Adoption of the *epoché* is what, at the same time, made the study of mysticism both identical to scientific method and completely different from it. The similarity was the neutral, dispassionate willingness to explore truth, potentially Truth, wherever found. The difference was acceptance of mystical experiences *as* unrepeatable and unverifiable, yet not discountable for these reasons alone.

As a rule, scientists are reluctant to accept unrepeatable and unverifiable data as useful data. There are, however, numerous exceptions. Under certain circumstances, important principles, theories, or conjectures in, for example, evolutionary biology, astrophysics, and particle physics are dependent upon conditions unrepeatable by definition. For instance, no one can recreate the birth of a species believed to be extinct. Occasionally, results of experiments have been accepted without additional verification by a second investigator, such as occurs in postulated existence of infrequently encountered subatomic particles. Yet, the investigation of mysticism is different from research in the sciences because the dogma of unrepeatable-unverifiable-unacceptable is the general guideline in the latter and of no importance for the former.

The last two steps in phenomenological method are usually coupled and wholly constitutive of orthodox scientific procedures: *bracketing*, or isolating the most meaningful conditions leading to an experience by discarding what is considered nonessential; and *reduction*, or finding as critical factors in any experience the lowest common denominators that seem to have had the most direct bearing on the fact of the event and its interpretive outcome.[27]

These four stages of phenomenological method have leant themselves to the study of mysticism in a number of ways. One clear usage has been its instrumental role classifying transcendent experiences according to the type of trigger inducing the event. A significant division is between internal and external triggers. Mystics are always conscious of the central role occupied by grace: there are reports of spontaneous revelatory experiences occurring while gazing at oceans or sunsets; while sitting upon mountaintops or under trees; or like Saint Paul, traveling on a dusty road to Damascus. All events of this kind are classifiable as being precipitated by external triggers.

On the other hand, the established religions have collectively developed an abundance of techniques designed to deliberately effect some mystical unfolding, and the methodologies would be categorized as internal triggers if they led to an actual experience. Some of the approaches employed make use of sensory deprivation or sensory entrainment through repetition of special words, or require one to continuously stare at religious artifacts or images. Some involve physical control: rhythmic breathing exercises or positioning the body in pre-specified postures. Some involve combinations of sensory and physical discipline, such as is in prescribed forms of intricately stylized dancing or praying. Others are intellectual exercises, as observed in abstract contemplation on religious doctrinal issues, attempts at solution of written or verbal logic puzzles, or the permutation of letter and word combinations to uncover philosophically or theologically significant variations.

Left out of this classification is the generic use of chemical stimulants to induce mystical experience. It is not obvious the results obtained through this process are comparable to other modalities mentioned. Ingestion of various drugs may constitute proper modes of access to transcendental knowledge on *a priori* grounds. The problem is the use of chemicals has consequences for physiological operation that introduces new variables into the analysis. When experiential conditions change as dramatically as they do with drugs, the data obtained cannot be compared to other procedural techniques for inducing the mystic event without prior rigorous investigation of the effect of the change on the type of data collected in isolation from all other factors. This step has not been taken for pharmaceuticals.

In fact, the relative success rates proclaimed in precipitating successful events are discordant. Transcendental revelation is rarely the outcome of the noninvasive methods outlined. Millions throughout history have employed them—not once, but over years or an entire lifetime—without being able to claim acquisition of substantive information even though each experience may be regarded as extraordinary and of profound personal significance. By contrast, relatively numerous are proclamations of success through chemical enhancement. It is the entire basis for the position of the shaman in various religious traditions, where narcotic-type agents are viewed as near-unfailing sources for guaranteeing not only successful access to mystical realms, but of generating reception of information not otherwise available.

The likely source of the discrepancy in reported outcomes is clear: when an individual's brain chemistry is altered through the introduction of non-physiologically normal chemicals, it is difficult to assess whether their after-the-fact expression of both the nature of the experience and its interpretive meaning represents a valid recollection of events and a proper reconstruction of value. In fact, this possibility is a major reason why some individuals still consider mysticism *as* psychosis. Hallucinations accruing from the result of a "bad trip" should be, theoretically, epistemologically distinct from mystical revelation, yet in practice they share similar characteristics: ineffable to some degree; noetic to some degree; certainly transient in duration; and ultimately passive for the participant since he or she is, literally, out of control during the event. The need for reemphasis is transparent: ingestion of various drugs *may* constitute proper modes of access to transcendental knowledge.

Phenomenology has also been employed as a tool in event classification according to conceptualization of common characteristics. It is apparent from comparison of different constructions that the categories chosen are arbitrary, as there is much crossover between researchers. A trait associated with one type of event can be equally and accurately reliably linked with another kind of event under another researcher's classification scheme. The typologies of William James and William Braud have already been extensively discussed, although the former, of course, did not use phenomenology as a guide in compiling his list.

Other than James' early effort, the scheme best known and accepted by other researchers used the terms *extrovertive* and *introvertive*. The concepts were invented by Rudolf Otto, but most widely disseminated by Walter Stace in *Mysticism and Philosophy*, and explained clearly by Bernard McGinn in *The Foundations of Mysticism*:

> The common characteristics of extrovertive mystical experiences are unifying vision (all things are one), concrete apprehension of the One as life in all things, a sense of objectivity and reality, blessedness and peace, feeling of the holy, paradoxicality… alleged ineffability. The common characteristics of the higher form of introvertive mystical experience…are…unitary consciousness, nonspatiality and nontemporality …objectivity and reality, blessedness and peace, feeling of the holy, paradoxicality, and alleged ineffability.[28]

One can see at a glance the overlap between the Otto-Stace breakdown and the James-Braud analysis. The major distinction between the two sets is that the former augments the recitation of state properties with a subjective evaluation of one class as a "higher form" than another. Still another scheme, expressed in *Mysticism Examined: Philosophical Inquiries into Mysticism*, segregated mystical unfoldings into categories of *nature*-mystical experience and *depth*-mystical experience. Richard Jones explained:

> The first type occurs when sensory or other awareness remain present but the filters that our minds…provide for ordering the content of awareness are significantly weakened or totally absent…. A sense of unity overcomes our normal…duality of

> subject and object.... The second type of mystical experience occurs when the mind is completely stilled.... There is no awareness of any object or of any content of any kind and yet there is still a realization of an ultimate reality.... The "nature-mystical experience" has a sense of unity, but this experience is free of all content. This type of mystical experience will be referred to as the "depth-mystical experience."[29]

Once again, the slight shift in focus by the investigator causes a subtle rearrangement of the division of concepts. Comparing the Otto-Stace classes with that of Jones, it is clear that the "sense of unity" indicative of the nature-mystical experience is identical to the "unifying vision" of the extrovertive mystical experience. From an epistemological vantage point, being "free of all content" would seem to indicate no operational characteristics significant to creation of a salient distinction from "nonpatiality and nontemporality."

Jones asserted depth-mystical experiences are "free of all content." This proposition leads to the next role phenomenology can play in unraveling the intricacies of mysticism. Of more value than schematization, phenomenology can aid resolution of issues of import to a rigorous study of mysticism, which can lead to an overarching theory of explanation. One central topic concerns the possibility of mediation *during* an experience. Highly controversial is the reality of a *pure Consciousness event*, a state in which, for the duration of the unfolding, the usual thought processes are held in abeyance. What is said to occur in this state has been alluded to: a transfer of information content of infallible certainty with regard to its Truth without the recipient displaying intentionality toward its acquisition.[30]

Stace defined this potential state in *Mysticism and Philosophy*: "Suppose that, after having got rid of all sensations, one should go on to exclude from consciousness all sensuous images, and then all abstract thoughts, reasoning processes, volitions, and other particular mental contents; what would...be left of consciousness? There would be no mental content whatever but rather a complete emptiness... What emerges is a state of pure consciousness—'pure' in the sense that it is not the consciousness of any empirical content. It has no content except itself."[31] In opposition to a claim of feasibility for pure Consciousness events is the doctrine instigated by the publication of Steven Katz' highly influential essay, "Language, Epistemology, and Mysticism," in which he wrote:

> Neither mystical...nor more ordinary forms of experience give any indication, or any grounds for believing, that they are unmediated. That is to say, all experience is processed through, organized by, and makes itself available to us in extremely complex epistemological ways.... This "mediated" aspect of...our experience seems an inescapable feature of any epistemological inquiry, including the inquiry into mysticism, which has to be properly acknowledged if...investigation of experience, including mystical experience, is to get very far.[32]

Katz did not say the experience *itself* cannot be unmediated; he made a much more limited claim by asserting the entirely reasonable thesis that our conceptual *understanding* of mystical experience, first-hand or second-hand, must include interpretive features. These are, Katz believed, a product not of the experience itself, but of the sociocultural and religio-philosophical ideas a person brings to the event, or comes to consider after the fact.[33] This is the meaning of "experience is processed through, organized by, and makes itself available to us in extremely complex epistemological ways."

Stace acknowledged legitimacy in this view with regard to interpretation of mystical unfoldings; he was, however, adamant in his belief concerning the event itself: "It can hardly be doubted...sensation is one thing and its conceptual interpretation...another thing."[34] Stace's judgment recognized that experiences of all types, including those most appropriately characterized as mystical, are phenomenologically distinct from any report of the events, and that the retelling must be further differentiated from any subsequent analysis.

Katz' essay touched off a firestorm of debate among philosophers that has not abated—or been satisfactorily resolved.[35] This dispute has never been of concern to mystics. They seem nearly unanimous in their agreement that unmediated experience is not merely possible, but may be the norm. The sole categorical rejection of this perspective emanates from Jewish mystics. On doctrinal grounds, they have been unwilling to accept the actuality of pure Consciousness events.[36]

Those not mystically inclined appear reluctant to accept the reports of those directly involved at face value. Many analysts are convinced the gap between experience and interpretation is so subtle as to elude any mystic's ability to detect, especially when a common language is used to express both. Donald Rothberg, in agreement with Katz' idea, unrestrictedly proclaimed in "Contemporary Epistemology and the Study of Mysticism": "All experience (and hence all knowing) is mediated by cultural factors such that experience and knowing always show contextual traces and never...escape from such influences.... All experience...occurs within the context of a tradition, a form of life, a cultural situation, a set of social and historical relations."[37]

Phenomenology, as Husserl conceived and developed it, prejudices the outcome of its inquiries in a direction *opposite to* that favored by Katz and Rothberg with respect to the mediation issue. Husserl's approach to normal operations of self-consciousness must first be explored to understand why this is the case. In *Introduction to Phenomenology*, Dermot Moran conveyed Edmund Husserl's grounding presuppositions: "Husserl rejects all traditional representationalist accounts of perception which require the mediation of an idea, an image, or a sensuous appearance.... Husserl...emphasizes that, even though each perception is only of an aspect...of the thing, nevertheless we clearly perceive the thing itself and not merely one part of it."[38]

For typical mundane experiences, Husserl's dismissal of representation as a legitimate avenue for apprehension was misguided. It is reasonable to assert *both*: (a) sensory perception of a material object leads to perceiving the object directly, and (b) we do not usually concentrate on concept formation accompanying perception of the object. What must be clearly understood is our brain is a digital tool: synaptic connections in the brain transmit electrical signals possessing two possible states, open/on or closed/off. Their primary contribution to comprehension is transformed by one's mind *via* conversion of these digital signals into an analogical representation: self-consciousness manipulates symbols when creating an image literally intended to depict an object, which is subsequently given, and identified by, a unique linguistic label. Thus, the epistemological debate between *direct* versus *representative realism* should be resolved in favor of *both*: mind in conjunction with brain represents; self-consciousness alone realizes directly. This is the justification for a previous statement: "Self-consciousness cannot err, but brains can."[39]

If the object cannot be perceived in its entirety—only one side is visible, or we only hear a distant train whistle, but do not see the train—we *perceive only a part*, but immediately and simultaneously *conceive the whole* just as Husserl suggested. Our brains are naturally programmed to expect continuity even when we have legitimate reason to accept discreteness because our self-consciousness is affiliated with continuity inherently attributable to global consciousness. As an epistemological philosophy, *idealism*, the "third leg" in the triangle, *is* that continuity contemplated in isolation from the phenomenal world.[40] Discreteness and continuity are seamlessly interwoven in mind-brain interactionism, which was proposed to be a comprehensible psychophysical process for the cognitive modes described as thought, intuitive cognition, and mystical revelation.

How does Husserlian phenomenological method bias development of a theory of mysticism and the interpretation of mystical experiences in favor of pure Consciousness events? After we *presuppose nothing* about the nature of an experience for self-consciousness and take its report as an unblemished record of description (step one); after we *suspend judgment* of its possible relation to any real-world entity in terms of truth value (step two); after we *bracket* out of consideration all naturalistic interpretations of the experience that would make it explainable by common phenomena (step three); then, we are *reduced to* consideration of experience itself: its structure and intrinsic meaning (step four).

The problem is, if one brackets out of consideration: (a) whether it is a real or imagined event; (b) everything that gives it a spatiotemporal context;

(c) everything which undercuts the uniqueness of one phenomenon compared to another; how would it ever be possible to arrive at an *essential* difference between one mystical experience and another? Such a reduction *necessarily* leads to essential sameness, an unmediated Consciousness event where the thing in itself, transcendent force or agent, is intentionally comprehended—using the phenomenological definition of intentionality as a direct reference to an object of thought—without benefit of images. Husserlian bracketing prejudices interpretations of mystical experiences concerning the issue of mediation because phenomenology's standard model for analysis demands determinations of commonality and uniqueness through its subject-object orientation.[41] It is precisely the *absence* of subject-object duality that is the claim of advocates of unmediated mystical experiences.

Misunderstanding the focus of Katz' assertion caused some theorists to defend the authenticity of pure Consciousness events even though that was not the issue. In "Are Pure Consciousness Events Unmediated?", Stephen Bernhardt declared:

> It is hard to see how one could say that the pure consciousness event is mediated, if ...it is meant that during the event the mystic is: employing concepts; differentiating his awareness according to religious patterns and symbols; drawing upon memory, apprehension, expectation, language or the accumulation of prior experience; or discriminating and integrating. Without the encounter with any object, intention, or thing, it just does not seem that there is sufficient complexity during the pure consciousness event to say that any such conceptually constructive elements are involved.[42]

Bernhardt's statement represented a quintessential straw-man argument. He prejudiced his contention by describing the event as one involving "pure consciousness." If this is in fact the nature of the event, it cannot possibly be mediated by definition. It is entirely possible, however, to have a revelatory experience that is also mediated: these notions are not mutually exclusive. Consequently, there would appear to be at least three cognitively distinct components of a mystical experience: (a) the event, which may or may not be mediated in the sense of being accompanied by conscious awareness of normal spacetime, maintenance of a sense of ego-self and surroundings, and perception of metaphysical Consciousness; (b) a straightforward recounting of the unfolding without an intentional, or even unwitting, contribution of pre-experience notions of the nature and meaning of what transpired; and (c) a deliberate reframing of the mystical event to ensure its conformity with some preconceived religious doctrine or philosophically developed spiritual belief system.[43]

When perusing any first-person account of mystical experience, evidence for any or all of these elements must be separately sought. With respect to this classification, Katz would then be questioning the realistic possibility of (b), not (a). *If* a pure consciousness unfolding is possible, and *if* step (b) is a legitimate procedural option—both contentions supported through the use of typical character trait descriptions such as "non-cognitive," or "lacking in awareness of time and space"—*then* such an event must be unmediated. The grounds for doubting the accuracy of any report when such language is used lacks justification, and can only be the product of a prejudicial predisposition to view such experiences as impossible without empirical proof in defense of the allegation, *unless* one has independent reason to question the veracity and integrity of the reporter.

The mechanistic theory postulated in chapter six proclaimed the meeting of self-consciousness with global consciousness during any event, mystical or otherwise, involving information transfer is *always* unmediated because the transcendental Dimension/dimension of the theological space does not admit entry by a material brain. When (a) is claimed to be mediated, it is because the individual involved is engaged in an act of *thinking*: the zero time lapse needed for data transfer during the *revelatory* portion of the event makes it appear as a function of awareness as if the brain—the monitor for time passage—was always involved. An act of *intuitive cognition* has been described, where information comes immediately and clearly.

The pure Consciousness event controversy leads to the next major topic pertaining to an analysis of mystical experience with the intent to construct an all-encompassing theory. If, as contended, unmediated experience is both possible and reportable, and if the facts contained within the mystic's report are distinguishable from the mystic's incidental or deliberate incorporation of step (c), then all such events *must* then be cross-culturally identical. At stake is validity of historical assertions pertaining to a *perennial philosophy* said to be the foundation for all religious expression.[44]

Richard Jones clearly favored the idea of a perennial philosophy when he wrote in *Mysticism Examined: Philosophical Inquiries into Mysticism*: "A methodological assumption to be made here is...depth-mystical experiences are of one type regardless of the understanding employed by individual mystics."[45] If one reads a written description, it is impossible to tell whether the actual event was mediated or not unless the report specifically provides the phenomenological facts. Any assertion of mediation or non-mediation by the mystic must be accepted at face value, since there is no possible means to independently confirm or deny the statement. This is why proper utilization of phenomenological method in studying mystical experience demands the *epoché*, the neutral position by the researcher. It is also why all conclusions derived from any investigative study should be accepted hesitantly and re-examined critically.

Many fail to take the responsible approach and prejudice their analysis with preconceptions, perhaps without awareness of doing so.[46] Such was the case with Joseph Dan's statement in the introduction to *The Heart and the Fountain: Anthology of Jewish Mystical Experiences*: "Comparative study of mystical texts is even more misleading because there is no way to verify that the various sources represent the same or similar experience. If two sources use similar terminology, this does not denote that they have 'seen' the same thing; the similarity of words and images is the result of shared cultural background rather than proof of the identity of experience."[47] Note the last phrase: a positive assertion of "is the result of shared cultural background" undercuts the previous sentence's expression of uncertainty regarding the theme of "similar experience."

Does acceptance of a mystic's report about the nature of the experience demand that a bystander accept the mystic's interpretation of the experience? The added step of interpretation does *not* confer on the mystic a privileged position. His or her thoughts about meaning and significance need have no greater adherence to Truth, or be more useful in extrapolation to a general contextual truth, than another's opinion on possible implications obtainable from an unfolding. A mystic interested in understanding what has transpired should not even give excessive weight to their own judgment of meaning.

To this extent, William James was half-right and half-wrong when he suggested in *The Varieties of Religious Experience*: "Mystical states, when well developed, usually are, and have the right to be, absolutely authoritative over the individuals to whom they come. No authority emanates from them which should make it a duty for those who stand outside of them to accept their revelations uncritically."[48] Conviction in the meaning of a revelation should not, *ipso facto*, be endowed with unimpeachable certitude even by the person to whom the event occurred. When approached phenomenologically, a mystical unfolding may lead to truth, but under no circumstance should the experiential product be considered True. Collaborative evidence—use of a different mode of cognition through the efforts of a single being or group of beings of the same species, or preferably, coordination with the knowledge possessed by other species—is needed to endow *any* piece of information with the label of Truth.

With this analysis of the philosophy of mysticism in hand, consider what mystics profess to be true. An overarching theory of mystical practice must commence in its development from these original source materials. Jan van Ruysbroeck expressed his position on ultimate union with the transcendent in *The Sparkling Stone*:

> We feel ourselves to be one *with* God; for through the transformation in God, we feel ourselves to be swallowed up in the fathomless abyss of our eternal blessedness,

> wherein we can nevermore find any distinction between ourselves and God.... And as long as we stand idle, with an inclined spirit, and with open eyes, but without reflection, so long can we contemplate and have fruition. But, at the very moment in which we seek to prove and to comprehend what it is that we feel, we fall back into reason, and there we find a distinction and an otherness between ourselves and God, and find God outside ourselves in incomprehensibility.[49]

Catholic mystics in the Middle Ages were unanimous in their agreement: unmediated experience was not merely possible, but actual. Based upon her experiences, Angela of Foligno constructed a sequence of transitory stages in mystical ascent culminating with divine union. Bernard McGinn referred to "the three transformations" in a citation from *The Book of Blessed Angela*. The first transformation depicted the traditional concept of *imitatio dei*, so prominent in Catholic theology, which illustrated the *purgation* stage "when the soul attempts to imitate the works of the suffering God-man." From there one proceeded to the second transformational level, designated *illumination*, "when the soul is united to God and has great...delights from God which still can be conceived or expressed in words." Finally, one arrived at the highest possible stage of mystical ascent, the third transformation known as *union* "when the soul is transformed within God and God within the soul by a most perfect union. It feels and tastes the highest matters of God...they cannot be conceived or expressed in words."[50] Note her distinction between the second and third transformations: an ability to distinguish these interactions *after* the event might indicate an ability to perceive them differently *during* the event. If so, it would mark a clear transition point between cognitive intuition and revelation.

Whether the testimony of both mystics speaks to the factual reality of the situation, or whether, as thinkers like Katz and Rothberg attest, the unanimity derives from unintentional encroachment of "preconceived religious doctrine or philosophically developed spiritual belief system" is unclear to those who wish to maintain an unbiased view. It must be observed that another Catholic mystic from this time period, Meister Eckhart, spoke specifically about the need to *forego* doctrine during mystical experiences: "He must be so lacking in all knowledge...he neither knows nor recognizes nor feels that God lives in him...he must be free of all the understanding that lives in him."[51]

Is the statement an exaggeration or an exact description of an inviolable requirement uttered by someone who *knows* because he has experienced? How can the validity of either judgment be incontrovertibly proven? One approach to attempt to resolve both controversies—mediation *versus* union; universal perennial philosophy *versus* distinct, mutually exclusive theologies among religions—is to uncover supporting testimony from mystics who do not outwardly conform to the same philosophy or theology.

Chapter one contended Truth is attained from truth by development of a convergent meta-system, a doctrine harmoniously combining less inclusive truths. Unless skeptics are prepared to claim "conspiracy" or "plagiarism"—assertions impossible to substantiate and justify—authentic documentation from diverse cultures and time periods would lend support to a commonality of position upheld by mystics, if any legitimately existed.

Parallel to the recitation of Angela of Foligno is one given by al-Ghazālī in *The Niche of Lights*, where he presented a hierarchy of cognitive modes within each person:

> The first of the luminous spirits is the sensible spirit. It is the one that receives what the five senses bring.... The second is the imaginal spirit. This spirit seeks to fix what the senses bring in, remembering it as something stored within itself, in order to present it to the rational spirit above it when there is need for it.... The third is the rational spirit through which you perceive meanings outside of the senses and imagination.... The objects of its perception are universal self-evident knowledge... The fourth level is the reflective spirit. It takes pure rational knowledge and brings about combinations and pairings, deducing therefrom noble knowledge.... The fifth is the holy prophetic spirit...singled out for the prophets and some of the friends of God. Within it are disclosed flashes of the unseen, the properties of the next world, and some of the knowledge of the dominion of the heavens and the earth, or, rather, some of the lordly knowledge that the rational and reflective spirits cannot reach.[52]

Al-Ghazālī opined a distinction between the four lower "spirits" and the "holy prophetic spirit" in the same manner as Angela differentiated between her second and third transformation within a mystical unfolding. Although he declared the latter to be a restrictive endowment to "some of the friends of God," it is clear from the passage it was operational for the select few during a mystical revelatory experience. Precisely because al-Ghazālī alleged that mode of apprehension is not generally available lends credence to the notion it is other than cognition through the senses and other than intuition. An unmediated process would satisfy this criterion.

Any abstract, intellectual exploration of comparative religious mystical doctrine could be extended indefinitely without reaching answers satisfactory to all. Literature citations could be extracted to substantiate proponents and opponents of mediation or the existence of a perennial philosophy. However, Livia Kohn's epigraph heading this chapter is a concise expression of what most mystics aim to accomplish: they seek perfection in their being, which they contend can only accrue by merger with a transcendental force or agent. This mission statement implies: (a) they regard the transcendental as real; (b) it is perfect, which means whole without flaws; (c) the physical person, or some part of them, is able to attain union with the nonphysical absolute; (d) the individual is worthy of accomplishing this endpoint.

In part, the inclination to believe without reservation in the reality of the transcendental as innately perfect, and humanity's capability in participating in that perfection, probably derives from influences of the particular religious tradition in which the mystic was reared in youth or chose in maturity.[53] To this extent, Katz' argument about the relevance of prior notions on the post-experiential shaping of a mystical encounter with the transcendent is likely valid. It is plausible another part of this universally held faith, this perennial philosophy, derives from philosophic contemplations on the nature of being and its relation to the universe. Such ponderings frequently lead to an innate belief that life and materiality were deliberately created for some teleological purpose, not randomly emergent.[54] Though they draw a different conclusion from the data, even cosmologists admit the conditions for life on Earth, and possibly elsewhere, are highly unique in that there is a very limited range of allowable physical-environmental parameters capable of supporting sentient creatures.[55]

These religio-philosophical factors are not sufficient. The largest portion of the credibility for this construct as a certainty derives from understanding, even if never articulated to oneself or others: *each person is Consciousness, is this transcendent reality in manifest form, and can be nothing else*. This perspective precedes all intellectual comprehension of its truth-content when formulated as a rational statement. It is an inalienable intuition instilled at birth only later made intelligible as a cogent thought.

In *Sermon 6*, Meister Eckhart proclaimed, as if it were the most to-be-expected, incontrovertible assertion possible: "What is life? God's being is my life. If my life is God's being, then God's existence must be my existence and God's isness is my isness, neither less nor more."[56] Remove the allusion to deity and you have Hegel's philosophical *credo* in the *Phenomenology of Spirit*: "Raised above perception, consciousness exhibits itself closed in a unity with the supersensible world through the mediating term of appearance, through which it gazes into this background lying behind appearance.... This curtain of appearance hanging before the inner world is...drawn away, and we have the inner being gazing into the inner world...self-consciousness."[57]

If it is True that self-consciousness is Consciousness, why is it manifest at all, and why does it seek to re-establish complete union through mystical experiences? Why has manifestation introduced an imperfection such that the self seeks some sort of re-connection it does not currently have? The answer has already been given in the chapter on metaphysics: Interpassage instigated an intersection between one Conscious Space and another; initiated Time in relationship to Space; informed, literally as analogs, physical space and time with its actual contents as discrete objects.

The techniques developed for, and utilized by, aspiring or accomplished mystics can now be comprehended. As a single class, all procedures have as their immediate goal the sharpening of mental concentration on one thing—breathing, posture, a sound, an image, an idea, a dance movement, and so on—so as to enhance individual focus and eliminate the diversity of potential distractions. Increases in concentration over a smaller area of activity allows for accumulations of mental power. Pierre Teilhard de Chardin expressed the underlying intention with precision in *The Phenomenon of Man*: "The degree of concentration of a consciousness varies in inverse ratio to the simplicity of the material compound lined by it.... A consciousness is...more perfected according as it lines a richer and better organized material edifice."[58]

When applied to mystical practice, this formula holds the same as it does when applied to chemical reactions. Increasing the concentration of an agent enhances probability of successful reaction through interaction with another agent *if* the first agent in question is part of the rate-determining step for the overall reaction. For mystical experiences, what is the reaction of interest? What is the other agent? The reaction and agent of pertinence have been mentioned in initial expression of the mystical goal: coalescence of self-consciousness with Consciousness in act as much as it already is in fact. For what purpose? Metaphysical speculation suggested human participation in resolution of the cosmologic and cosmogonic crisis initiated by Interpassage of dual Conscious Spaces is the ultimate teleological quest.

It is possible to gain insight into all the subsidiary issues surrounding mysticism and mystical practice from a completely different approach. Over the last thirty years, research inquiries into spirituality has gone beyond the phenomenological—scientific method conceived in the abstract as a product of an intellectual process—to inclusion of physical science experimentally. Specifically, imaging instruments created to facilitate medical diagnosis and treatment of brain injuries have been adapted—in conjunction with the long-standing medical techniques of monitoring blood pressure and circulatory system electrical conductivity—to study individuals engaged, in real time, in an array of techniques employed as a means to gain revelatory experience.

It is to be expected that this emerging field, which some have dubbed *neurotheology*, is as yet highly contingent in the relationships it has tried to establish. These methods are highly dependent upon the subjects of these studies, the would-be mystical practitioners, for assessment of their internal spiritual state while linked to scientific instruments.[59] Consequently, physical scientific investigations return to phenomenological first-person reports as an important component of analysis. Another limitation pertains to the operating paradigm typically *assumed* by scientists involved in this kind of research: a presumption of direct correlations between mystical states and physiological parameters. Scientists devise relationships from their data without some prior justification for the propriety of what they are attempting to accomplish. To operate in this way is to presuppose mystical states are innately reducible to brain states and conjunctive biochemical events.

Despite the problems associated with these studies, such explorations are anticipated to enable scientists to compare these physiological effects with results obtained from ordinary activities. A finding of significant similarities or differences would enhance likelihood of development of a comprehensive theory on unique changes occurring during specific spiritual exercises and their interpretive meaning in the context of mystical practice. It is thought the resultant compilation of data could be employed to refine the classification and characterization of various mystical methodologies and facilitate cross-cultural, inter-religious comparisons of diverse techniques.

One comparison, obtained through the efforts of Andrew Newberg and Eugene D'Aquili, was offered by John Horgan in *Rational Mysticism*. It is an illustration of the potential power latent within the formal mating of science and mysticism. It also implicitly contains a suggestive warning not to let the significance of the uncovered relationships overwhelm or overshadow use of objective scientific methodology. In the present circumstance, the concern is possible premature overgeneralization without a sufficient body of evidence accumulated to support the propositions:

> Newberg and D'Aquili divided...methods of achieving unitive experiences into two categories: top-down and bottom-up. Top-down methods...include meditation and prayer, achieve transcendence through relaxation...focusing and calming the mind. Bottom-up techniques...include dancing, hyperventilation, chanting, and vigorous yoga, approach the same goal through excitation. Each method taps into a different component of the...autonomic nervous system, which regulates heartbeat, blood pressure, respiration, metabolism, and other physiological functions. Top-down methods exploit the...quiescent component of the autonomic system, which limits the body's expenditure of energy and maintains its equilibrium. Bottom-up methods exploit the arousal component of the autonomic system, which triggers the body's fight-or-flight response, causing adrenaline to be pumped into the bloodstream, boosting heart rate and respiration.[60]

The Newberg-D'Aquili proposal for classification conjoins neatly with a typological categorization of states invented by Roland Fischer before their studies were undertaken. Fischer formulated a continuum of physiological energetic states according to the responses recorded for parameters such as heart rate, brain wave patterns, and bioelectric flow through epidermal tissue on the hands. Divergent techniques like meditation, prayer, and the ingestion of hallucinogens were included on his scale. On a gross level, he found there were two extremes called *ergotropic*, "high energy," and *trophotropic*, "low energy."

The former characteristically displayed high amplitudes and frequencies for brain wave patterns and rapid heart rates suggestive of intense efforts being sustained to achieve a mystical state. The latter showed the opposite traits—reduced brain wave activity, an extremely low heart rate, infrequent breaths, *et cetera*—indicative of extreme relaxation with little directed effort necessary to maintain the condition.[61] Even a superficial comparison between the two systems will reveal equivalencies: the "top-down" methods correlate with trophotropism, and "bottom-up" modes with ergogtropism.

These two extremes mirror the two modes through which a mystical state can be reached by attaining connection to theological space: one can, with maximal effort, generate enough focused personal energy to "push into" the barrier typically existent, or conversely, achieve a level of minimal effort and allow theological space to draw the seeker into its Dimension. Hence, there are two general approaches: "me push" or "you pull." Ergotropic techniques specialize in using individual will to "bring down" the fourth Spatial/spatial Dimension/dimension in order to establish a link. Trophotropic methods let grace take over in unifying global consciousness with self-consciousness.

Though this proposal seems to suggest existence of a barrier between three-dimensional space and theological space, such an implication should not be taken literally in its physical sense. A more appropriate sense of the nominal barrier existing between the two can be understood with reference to *resonance*: there has to be a harmonization of the Energy/energy between self-consciousness and Consciousness such that there is attained a uniform homogeneity across the two domains. The lack of normal attunement is due to the coupling of self-consciousness to material brain and body. The former must disengage from the latter before it can link with global consciousness.

The resonance energy gap can be likened to the activation energy barrier theorized to exist in chemical reactions. In chemistry, before reactants can be converted into products, the starting material must gain sufficient energy to reach a level where the reaction proceeds spontaneously. The *activated state complex* is a transitory state where reactants are oriented properly to each other allowing for direct conversion to final products. In mystical endeavors, self-consciousness and global consciousness—that is, theological space—need to attain satisfactory Energetic alignment to establish a link whereby information can be transmitted in either direction, but usually from global consciousness to self-consciousness.

The purpose subserving the practice of diverse techniques of meditation and contemplation is to enable self-consciousness to decouple from the body. Alternatively put, to attain the activated state complex. An analogy would be striking a tuning fork to generate a frequency of vibration that would then induce another tuning fork to begin to vibrate in rhythm with the first. Once personal space and theological space have been Energetically synchronized *via* ergotropic or trophotropic procedures, there can ensue a transmission of information, which can in turn be redistributed to the individual brain for transformation into thoughts and associated actions after-the-fact; that is, after disengagement of self from global consciousness and realignment of self-consciousness with material brain accompanied by a renewed awareness of standard spatiotemporal reality.

Paul Weiss, who first conceived of theological space, offered support to this interpretation for the process of interaction. He envisioned the mystical encounter as reciprocally motivated by both transcendental agent and mystic: "The mystic submits himself to God across the vasts of theological space.... He tries, and to some degree succeeds, in emptying his finite natural self into the longing which stretches towards and terminates in God.... God thereby knows theological space, not...by having reached out beyond Himself, but by having the mystic carry the meaning of it into Him.... The mystic must ...know the nature of that space as well, contemplating it as he contemplates the rest of God."[62]

Another avenue of investigation made feasible by adaptation of scientific instrumentation and research methods to spiritual practice is exemplified by a study performed by Michael Persinger. He applied magnetic fields directly to a person's brain with the intention of, as the title of one of his summaries brazenly put it, "Experimental Simulation of the God Experience":

> Most people report a "sensed presence" of a Sentient Being. The interpretation of the "presence" was a function of: (1) the person's temporal lobe sensitivity; (2) the available cultural attributions or beliefs for the experience; and (3) the shape of the magnetic fields.... Our working hypothesis is...the sensed presence is the prototype of the God Experience and is actually...transient awareness of the right hemispheric equivalent of the left hemispheric sense of self. Intrinsic complex electromagnetic patterns within the left hemisphere that generate the "sense of self" are experienced as "the other" when their right hemispheric equivalents intrude into awareness.[63]

Experimental details can be found in the Notes.[64] Several points worthy of mention pertaining to scientific research on spirituality are elicited by this passage. Persinger believed the origin of "sensed presence" or "experience of God" is as a survival mechanism developed to avoid or overcome anxiety produced by contemplation of one's death. He claimed this response, when coupled to the depiction of God as infinite and eternal, relieved anxiety by causing the person to become convinced some part of him or her was also eternal and immortal. Persinger thereby positioned God as solely a functional output of the brain.

It is important to recognize *none of this background theory derives from the accumulated data. It was the consequence neither of measurement nor of quoted statements attributed to the subjects of the investigations, but a view Persinger brought with him into the study.* His subjects may have given voice to their experiences, in written or verbal form, as invocations of God's presence. Extrapolation from this subjective interpretation to the contention God resides in the brain is unwarranted and speaks to a predisposition on Persinger's part to believe this. How else is one to explain why he was so willing to accept expression of a feeling of divinity by his subjects as literal rather than metaphoric, poetic license symptomatic of hyperbole precipitated by induction of an unanticipated experience?

Scientists must scrupulously examine predispositions before, during, and after completion of research and conscientiously filter out personal biases in interpretive assessment of results and meaning. The crucial role occupied by *epoché*, consistent neutrality, is underscored, just as it was during discussion of phenomenological method as applied to philosophical or psychological evaluations of mystical first-person reports.

Persinger employed a broad range of magnetic field strengths, from ten nanotesla, 10^{-8} tesla, to one microtesla, 10^{-6} tesla, in stimulating the cerebral hemispheres.[65] On prior occasions, he used electromagnetic fields indirectly applied—naturally occurring geomagnetic fields greater than fifty nanotesla; naturally occurring electromagnetic radiation greater than two hundred and fifty nanotesla—as well as direct stimulation of the right hemisphere using magnetic field strengths ranging from one to five microtesla.

It was often the case that *stronger* electromagnetic fields were applied the *closer* the field was located to the brain in order to induce the experience of "presence." In addition, long treatment times with the fields, on the order of a quarter of an hour, were needed before *any* response was noted by the subjects of the studies. If the brain were the *origin* of numinous experiences, one would hypothesize applied stimulation of the relevant area in a direct fashion would require *lower field strength* and be noticed by the subject *sooner*. Therefore, it is unclear whether his data support his conclusion.

On the other hand, Persinger's thesis—stimulation of a particular area of the brain induces a sense of "otherness" one could potentially associate with deity—is consistent with the conclusion rendered by Andrew Newberg and Eugene D'Aquili on the functional role of the *orientation association area* mentioned in *Why God Won't Go Away*: "The primary job of the orientation association area is to orient the individual in physical space—it keeps track of which end is up, helps us judge angles and distances, and allows us to negotiate safely the dangerous physical landscape around us. To perform this crucial function, it must first generate a clear, consistent cognition of the physical limits of the self. In simple terms, it must draw a sharp distinction between the individual and everything else."[66]

Research of the type conducted by Persinger is problematic because the impetus to pursue such studies tacitly assumes spiritual/mystical experiences elicited in a laboratory will follow the same course of unfoldment as it would in a natural environment. Extensive investigations into changes in levels of significant biochemicals—primarily hormones or neurotransmitters—during religious practices are not ultimately insightful unless these alterations can be shown to be unique to the specific practice studied.[67] If novelty is neither sought nor discovered, devising a correlation is not in and of itself especially informative. To understand mystical revelation scientifically, one must seek *unusual* physical, emotional, and mental adjustments, whether temporary or permanent, an individual undergoes.

Presuppositions concerning the science involved—a theme distinct from the first comment pertaining to personal prejudices—must be revealed and rigorously validated before neurotheological research can be accepted as a legitimate method for inferring conclusions about somatic physiological and neurological correlates of mystical experience. Do measurable increases or decreases in electrical or biochemical activity from some regions of the brain during meditation in a laboratory derive *entirely* from the act of meditation itself? Could the effects be coupled to variations in general attitude stemming from awareness that one is being intensely and minutely observed? Could the metabolic alterations be a function of individual response to atypical and unfamiliar environmental settings compared to locales usually employed for meditative or contemplative pursuits?

Control experiments—monitoring the same subject for extensive periods of time when they are not meditating, but are engaged in some other activity such as reading or listening to a symphony or watching television—need to be run. Though Newberg and D'Aquili proclaimed, "studies have shown that participating in...prayer, religious services, meditation...can lower blood pressure, decrease heart rate, lower rates of respiration...and create positive changes in immune system function," the result loses significance if it can be shown that taking a hot bath, getting a therapeutic massage, or watching a romantic movie induces similar or identical effects.[68]

As a final point to make on the subject of scientific studies into spiritual practice, each investigator must be careful *not* to presume all activities of a certain type are identical across religious traditions. Prayer by Buddhists may or may not have identical physical, mental, and emotional effects as when Catholics pray. It is *a priori* possible Tibetan Buddhist prayer has different physiological consequences than Zen Buddhist prayer, and more plausible still that Hindu mantra meditation techniques will induce distinctly different functional responses when compared with Sufi whirling dances. Can these distinctions, if they exist, be studied and confirmed or denied? Absolutely. In fact, in a positive sense, ecumenical correlations demonstrating presence or absence of specific, allegedly universal, influences on physiologic functions may shed light on the issue of whether there exists a cross-cultural perennial philosophy inclusive of all religions.

Neurotheology is seen to be at best an adjunctive method complementary to phenomenology, and cannot be envisioned as a more effective substitute. Compilation of quantitative and qualitative data by scientists is insufficient. First-person evidence, subjective empirical data, must be incorporated into any scientific analysis of mysticism. Hence, purely descriptive statements are illuminating for mystic and non-mystic alike. A useful example is provided by this selection from the *Enneads* of Plotinus:

> Many times it has happened: lifted out of the body into myself; becoming external to all other things and self-encentered; beholding a marvelous beauty; then, more than ever, assured of community with the loftiest order; enacting the noblest life, acquiring identity with the divine; stationing within It...poised above whatsoever within the Intellectual is less than the Supreme: yet there comes the moment of descent from intellection to reasoning, and after that sojourn into the divine, I ask myself how it happens...I can now be descending, and how did the Soul ever enter into my body, the Soul which, even within the body, is the high thing it has shown itself to be.[69]

It is difficult to judge this episode as a mediated event: (a) "Becoming external to all other things" is a definite suggestion of lack of concurrent awareness of normal spacetime and its contents; (b) "Acquiring identity with the divine" means unambiguous mystical union; (c) "The moment of descent from intellection to reasoning" clearly implies a phenomenologically non-cognitive framework different from the state of mind found during a merely intensive period of contemplation lacking an overt transcendental revelation.

Compare these words of the Gnostic Plotinus—Gnosticism was viewed as heresy by the Roman Catholic Church of his historical period—with those from that paragon of later Catholicism, Saint John of the Cross. Saint John reported this experience in *Dark Night of the Soul*: "For God now begins to communicate Himself to it, no longer through sense, as He did aforetime, by means of reflections which joined and sundered its knowledge, but by pure spirit, in which consecutive reflections enter not... He communicates Himself to it by an act of simple contemplation, to which neither the exterior nor the interior senses of the lower part of the soul can attain."[70] Do not Plotinus and Saint John agree on non-mediation of the event during information transfer in some non-ratiocinative mode?[71]

Another meaningful implication drawn from both passages would be that the participants seem overwhelmed and in complete awe of what transpired. They seem as astonished by the fact of the event as by its nature, however indescribable that nature might be afterwards. In the context of the essential nature of grace—being granted permission to experience the transcendent, especially as a unitive event—this response is highly rational. Consider what anyone's initial reaction would be to the unexpected winning of fifty million dollars in a lottery determined by randomly picking a string of numbers: an initial shock of surprise so complete as to defy one's comprehension prior to "coming down to Earth" such that the winner could subsequently vocalize a coherent reaction. Other emotional reactions of varying intensity are possible as the outcome of mystic experiences.[72]

Beyond an understandable emotional response of this kind, can anything more analytically informative be gleaned from a mystic's report? Franklin Merrill-Wolff's remark in his major work, *The Philosophy of Consciousness Without an Object*, is provocative:

> Throughout the whole period of this supreme state of consciousness I was self-consciously awake in the physical body and quite aware of my environment. The thought activity was not depressed, but...alert and acute. I was...conscious of my self-identity in two distinct senses. In one sense, I was, and am, the primordial Self and coterminous with an unlimited and abstract Space, while at the same time the subject-object and self-analyzing consciousness was a sort of point-presence within that Space.[73]

If one accepts the thesis of a pure Consciousness event as one in which there is no distinction between subject and object, the complete mergence of mystic with transcendent agent, Merrill-Wolff did not have that experience in this instance. His experience was mediated: he was cognitively aware of self and surroundings, able to simultaneously perceive and conceive.

The difference in language used to convey the experience of Plotinus or Saint John compared to Merrell-Wolff's lends added credibility to the notion there is a distinctive contrast between the intrinsic nature of these events, and supports division into mediated and unmediated unfoldings. Otherwise, one would have to rationalize the difference in reportorial style as due to personal predilection, a function of observational prowess during the event, or perhaps to recall capacity. As none of these suggestions are *a priori* out of bounds, it is clear one comparison is insufficient for decisive conclusion, and further examples must be culled from the now-abundant, readily available literature. It seems plausible to posit that exhaustive inquiry will eventually accomplish the goal of providing definitive answers to central questions in mysticism.

Merrill-Wolff's commentary about his relationship to "an unlimited and abstract Space" is of keen interest: it is consonant with remarks concerning the nature of Being as Conscious Space. It is unfortunate Merrill-Wolff made no mention of Time. It is unclear whether his use of "at the same time" was meant to indicate an awareness of the passage of time during this experience, or whether the phrase was merely a linguistic effect employed when forced to explain events in linear chronological sequence.

It was stated at the beginning of the chapter that mystical experiences could no longer be considered the private domain of a few religious zealots studying esoteric literature, practicing arcane techniques designed to initiate an ecstatic experience. Over the last century, mysticism has moved into the public arena, where anyone can have a transcendental unfolding at any time. The pure Consciousness event inferred from statements made by Plotinus and Saint John, and absent in Merrill-Wolff, can also be discovered in some nameless person's remarkable chronicle, cited at length, contained in Robert Forman's *The Problem of Pure Consciousness: Mysticism and Philosophy*:

> After about two years, my experience of the transcendent started to become clearer. I would settle down, it would be very quiet...and then I would transcend, and there would just be a sort of complete silence void of content. The whole awareness would turn in, and there would be no thought, no activity, and no perception, yet it was somehow comforting.... I did not yet identify myself with this silent, content-free inner space. It was a self-contained entity that I transcended to and experienced. Then, with increased familiarity and contingent on the amount of rest I had, the process of transcending became more and more natural. The whole physiology was now accustomed to just slipping within, and at some point...the breath would almost cease, the spine would become straight, and the lungs would cease to move.... At this point I began to appreciate that this inner space was not an emptiness but simply silent consciousness without content or activity, and I began to recognize in it the essence of my own self as pure consciousness. Eventually...the thin boundary that had previously separated individuality from unbounded pure consciousness began to dissolve. The "I" as a separate entity...started to have no meaning. The boundaries that I put on myself became like a mesh, a net; it became porous and then dissolved; only unbroken pure consciousness...remains.... In this state the experiencer is not experiencing as it normally does. It is there ready to experience, but the function has ceased. There is no thought, there is no activity, there is no experiencer, but the physiology after that state is incredible. It is like a power surge of complete purity.[74]

The language in this recitation alternates between in-the-moment report of events—"There would be no thought, no activity, and no perception"—and after-the-fact interpretation—"The process of transcending became more and more natural." Since the entire passage was recorded afterwards, it is uncertain what degree of interpretation filtered into the phenomenological description.

Lack of clarity in assignment of a precise boundary between experience and interpretation notwithstanding, note three elements of this confession all in support of the theory being promoted in this volume: (a) characterization of global consciousness as "content-free inner space"; (b) the subject is "ready to experience, but the function has ceased." This remark sounds like an advocacy of the state of alertness considered the primary hallmark of pure Consciousness; (c) physiological changes do not *cause* a pure Consciousness event; instead, they are *associated* with it. The physical elements appear to be indicators of mystical experience, not a prerequisite. This is what would be anticipated if Consciousness, in both global and individualized forms, did not originate with the physical organism, but interacted with it in manifest spacetime.

There is no mental impression generated within the person suggesting, "If I sit like so, and breathe like so, and focus my thoughts like so, then I will have a revelatory experience." A more accurate depiction would be, "I hope to have a revelatory experience, an event I am not able to guarantee of my own volition, so I will minimize as much as possible the energy taken up by my physiological and mental processes so as to leave the greater balance of Energy available for a spiritual ascent should one be granted me." This line of thought implies the foundational construct guiding disciplined efforts at mystical experience is an Energetic one. Moreover, in an intentional sense, it is the application of physical energy in the service of metaphysical Energy.[75] The physical energy conserved by minimizing physiological and biochemical activity can be redeployed to pierce theological space, the conduit linking the transcendental with the manifest.

If the mechanistic theory outlined is to display comprehensive validity, it must be argued that the *type* of technique employed to hopefully induce a mystical experience is irrelevant to the actual occurrence of the unfolding. Consider prayer as a type of contemplation occasionally engaged in for the purpose of receiving transcendental information. Prayer might be regarded as both a prelude to a contemplative process leading to mystical revelation and as a specific version of a contemplative process.

This kind of prayer is denoted *petitionary* prayer: praying for receipt of some beneficence as means to comprehend aspects of the transcendent and the manner of its unfoldment within physical reality. Such a prayer contrasts in format to other modes, in particular: (a) prayers of *thanksgiving*, a form expressing gratitude and appreciation for the world's bounty or beauty; (b) prayers of *glorification*, a means to express devotion to the divine sanctity of the transcendental agent however it may be conceptualized within a given religious tradition.[76]

Petitionary prayer implied two distinct propositions according to Jerome Gellman: "The 'impetrative view of petitionary prayer'...maintains that: (a) when praying a petitionary prayer one is making a request of God, believing ...sometimes God grants what has been asked for...and (b) sometimes God grants what has been asked for."[77] Gellman later expressed an extraordinary assertion with respect to his second contention: "The closer the relationship between God and the person praying, the stronger the obligation upon God to meet the request."[78]

As a measure of the disagreement found among theorists—theologians, philosophers, psychologists—concerning the purpose, efficacy and meaning of this type of prayer, Gellman's insistence on mutuality of obligation on the part of petitioner and the transcendent agent intrinsic to any act of petitionary prayer is a position in diametric opposition to that of Ann and Barry Ulanov presented in *Primary Speech: A Psychology of Prayer*: "Unless we lose our images of an endlessly yielding God who will perform all our miracles for us, we will lose instead the whole venture of praying and what it means."[79]

Despite the contrast between these two visions of the circumstances for petitionary prayer, the Ulanovs quoted Simone Weil in her advocacy of Gellman's thesis: "Desire directed toward God is the only power capable of raising the soul. Or rather, it is God alone who comes down and possesses the soul, but desire alone draws God down. He only comes to those who ask him to come, and he cannot refuse to come to those who implore him long, often, and ardently."[80]

How does successful petitionary prayer operate? The process begins with active intent by petitioner to establish union with transcendent commencing with a willful attempt to penetrate the Energy barrier separating the mundane world from theological space.[81] What must be pierced is the asynchronization of the energy of self-consciousness, *when associated with a physical being*, with respect to the Energy that is global consciousness. This is where the hint provided by Gellman's assertion, supported by Weil, that connection through development of Energetic harmonization is only likely for the most diligent, sincere individuals "who implore him long, often, and ardently" establishes its meaning.

Will *power* is not a figurative description, but precisely literal: an effort to focus one's Energy of self-consciousness leading to its realignment into a more coherent form of electromagnetic energy. In this instance, coherence is meant in the physical sense usually applied to laser light as a more intensely powerful form of normal electromagnetic energy. The amplitude, frequency, and wavelength of the Energy/energy waves are aligned, reinforcing each other, rather than interfering with, and thereby diminishing, the strength of the total.

Initially, following the lead of Angela of Foligno and Robert Forman, the individualized self-consciousness retains awareness of self, time, place, and presence of Other because it is still conjoined with the physical body. If the effort of the individual petitioner towards coherence is sufficient to the task, then when the attempt at merger with the transcendental Energy of global consciousness is granted as an indicator of grace, the two fields unite and the quintessential mystical revelatory experience begins to unfold. The physical part of the being is no longer part of the equation, and information transfer is now possible from petitioner's self-consciousness to global consciousness, or the reverse.

An exact description of this process was conveyed by Thomas Merton in *The Ascent to Truth*: "When the will begins to be absorbed in the mysterious and delightful love...God produces passively in it...this faculty becomes momentarily independent of reason, at least in the sense that it now begins to be guided directly by God in the intimacy of an experience of values...not made known to it through the intelligence.... It (i.e., the intelligence) is now playing a role secondary to...the will.... The reason would have nothing to do but acquiesce in whatever information might come to it through the... will."[82]

A series of questions posed by Leonard Angel in *Enlightenment East and West* can act as an outline to summarize the general conclusions pertaining to mystics and mystical reports argued for in these pages: (a) Do descriptions of mystical states necessarily lead to logical contradictions? (b) Do descriptions of mystical states necessarily invoke metaphysical claims? (c) If descriptions and claims are distinct, and if mystics enunciate metaphysical claims while describing their experience, are the descriptions and sets of claims obtained from different mystics externally consistent with each other? (d) Do mystical experiences provide *prima facie* support for the validity of any metaphysical claim?[83]

With regard to the first question, it seems clear the answer must be "no," unless one automatically believes connection to a transcendent—whatever it may be with respect to issues of personality or impersonality and associated characteristics—is impossible by definition. Materialism rests on a dogmatic presupposition having no solid basis in abstract logic, and which cannot be substantiated empirically because one cannot prove the negative allegation that a transcendental does not exist in reality.

If one claims connection between transcendent and immanent is *a priori* impossible, one would be hard-pressed to provide explanation as to how the manifest came to be. If one denies the reality of any form of transcendence and insists manifestation did not come to be, but always existed—the classic Aristotelian solution—that proposition is contrary to cosmological evidence supporting an ontology for spacetime. Some cosmologists *do* continue to advocate a thesis proposed over 2500 years ago by the Buddhist tradition: an *oscillating universe* theory where all existent astronomical and evolutionary records suggesting the age of the universe to be about fifteen billion years represents one cycle of an infinitely continuous process of manifestation.[84]

The principal difficulty with that thesis is there is lacking in the proposal any reasonable justification for an origin of time as we experience it however much the proposition may avoid the need to define a point of origination for space. As pointed out in chapter four, space and time cannot be treated as innately conjoined constructs. They are distinct in their operational function and must be the product of different ontological interpretations. Hence, there are no rational reasons to claim mystical states are *innately* self-contradictory or externally incoherent notions.

The preferred answer to the second question is, "sometimes." Some descriptions of mystical unfoldings are purely phenomenological and are not simultaneously meant to offer utterances of metaphysical import. Some post-experiential reports offered by mystics do, however, offer interpretations that go beyond the phenomenology of experience. *Interpretations, by mystics or by bystanders, are always metaphysical claims*, for here the ground has shifted from what the experience was, which only the mystic involved can remark upon, to what the experience means, which suggests perspectives on possible facts not directly apprehended, but were extrapolated or interpolated afterwards.

Such after-the-fact judgments become academic exercises just like those in any other field of inquiry. Scientists can report exactly what was observed or done during a particular experiment, and only those who were part of the group doing the work can argue with statements of operating conditions. On the other hand, the responsible scientists can interpret the data collected and propose a scientific theory to explain why the data acquired was what it was, but all theories make claims open to agreement or disagreement by anyone other than the scientists invoking the original interpretation.

The issue of consistency between mystical reports, phenomenological or interpretive in kind, was too vaguely stated to be answered definitively. In that all reports suggest a relationship between a person and a transcendental agent, then obviously they are all consistent. Are all experiences alike to the degree that all phenomenological accounts of experience yield the same information? No. Does lack of consensus render them inconsistent? No. It may be the case that one mystical experience was predominantly visual while another was entirely auditory. Comparison in this instance would be "apples and oranges."

Would two reports of different visual events lead to the conclusion they *must* be inconsistent? No. One person may report a vision involving a figure reminding them of Jesus Christ, while another might say he or she saw the Prophet Mohammed. There need not be any necessary inconsistency in such reports: as pure phenomenological statements of experience, interaction with Jesus Christ *vis à vis* Mohammed are clearly incompatible; if reports are intended to be predominantly or partially interpretive, differing assessments may ascribe the same properties to transcendent figures assigned different nomenclatures due to background beliefs *extrinsic* to the experience. Thus, for example, both Jesus and Muhammad may be *perceived* as radiant figures of light exuding peace and good will—phenomenological facts—to be later *conceived* as one or the other religious figure—interpretive attributions. Are these inconsistent mystical reports?

To draw a parallel from another field: assume two psychiatrists examine a patient at the same time so there is no doubt they are hearing exactly the same information. Based on these findings, one analyst concludes the patient exhibits paranoid delusions, while the other says the patient is suffering from alternating cycles of mania and depression. Are these evaluations innately inconsistent and contradictory? No. The relationship of these descriptors to any organic brain disease could depict different, but correlatable, functional consequences. Such may be the case with mystical reports that are essentially interpretive in format.

Alternatively, *could* two reports be necessarily and innately inconsistent? Yes. One report might claim it was revealed to a mystic that the Earth will explode in the year 2050; another report from a different source might assert a revelation professing the Earth will be destroyed by explosion in the year 2100. The propositions must be alleging definitive information on the same topic in the same manner for them to be set up in opposition to each other. This is the *only* situation covered by the Law of Noncontradiction.

Unlike the third query, the fourth has an easy answer. Do metaphysical claims provide *prima facie* evidence for proof? Of course they do. In what other manner could *prima facie* evidence be gained on any topic unless it was derived empirically? Does that imply mystical reports contain *sufficient* evidence for metaphysical claims? No. They are pieces of evidence, and may or may not be corroborated by data sought from other experiences. Mystical reports cannot be automatically granted as authoritative *except with respect to their precise phenomenological descriptions*.

However, they are not to be unremittingly rejected as incongruities and irrelevancies. Rather, mystical reports, *if interpretations of experience*, are to be considered as data on par with all other information. Why are they not to be invested with extra weight? Interpretations are fallible, and mystics enjoy no privileged status making their interpretations superior to another's. A personal declaration of conviction they are reporting "God's word" is not incontrovertible irrespective of how diligent he or she might be in professing to accurately report those words.

Chapter Eight: Parapsychology

> All matter is connected... In a...filled space every movement has an effect upon bodies in proportion to their distance, so... not only is every body affected by those...in contact with it, and responds in some way to whatever happens to them, but also by means of them the body responds to...bodies adjoining them, and their intercommunication can be continued to any distance at will.
>
> —Gottfried Wilhelm Leibniz, *Monadology*

All atypical modes for retrieving useful information can be categorically denoted as *parapsychological.* A particular subset of the cumulative class of all methodologies could then be regarded as *mystical* in type—the subject of investigation in chapter seven. Of concern now is the utilization of methods not associated with themes of interest to traditional religious doctrine, but to information about non-specialized subjects. Thus, focus will be on modalities typically classified as parapsychological, such as telepathy, clairvoyance, and precognition. *Telepathy* is the ability to communicate non-verbally by direct mind-to-mind contact; *clairvoyance* is the ability to describe people, objects, and events occupying places distant from the clairvoyant; *precognition* is the accurate forecasting of future temporal events. Within the mystical literature, reference has often been made to occurrence of these abilities as relatively unimportant consequences of advanced spiritual practices.[1]

Telekinesis—the ability to move solid objects without the use of direct instrumental mechanical contact—will not be investigated further: it is not usually employed as an information gathering technique. Other events of no interest are incorporated under the heading of "things parapsychological"—often thought of as synonymous with "occult happenings." These include the topics: (a) ghosts and haunted houses; (b) reincarnation with the consequent remembrance of past lives; (c) all derivatives of precognition traditionally considered *fortune telling* using techniques such as astrology, numerology, tarot deck readings, palmistry, *et cetera.*

What is the relationship between the three principal parapsychological modalities and Consciousness? Under the assumption they are real methods and not parlor tricks, how do they occur? What factors make their operation more or less feasible? Can relevant parameters and mechanisms of operation be formalized into rules elaborated mathematically?

Systematic parapsychological research began in the second half of the nineteenth century in England under the auspices of the London Society for Psychical Research. Credit should more properly probably belong to Gustav Theodor Fechner, one of the scientists responsible for separating psychology from philosophy as an academic enterprise in the first half of that century, predating the SPR by decades. Fechner, it will be recalled, invented the term *psychophysics* to describe all phenomena whose operation is subject to both psychological, meaning subjectively mental, and objectively physical laws.

Any event involving a living creature is to be considered psychophysical in contrast to the functioning of *seemingly inanimate* things, like rocks or rubber balls, which appear to depend solely upon the rules of physics for their energetic alterations in time and space. In illustration of the distinction: if study of a thrown baseball is to focus on the mental and physical factors involved in the movements of the pitcher and the laws upon which those movements are based, the outcomes learned will be applicable to refinement of psychophysical law; in contrast, if the study of a thrown baseball is to focus on movements of the ball, both before and after it has been released by the pitcher, the outcomes learned will be pertinent to refinement of physical law alone.

In whatever way history will eventually view contributions by Fechner to parapsychological research, there is little doubt the efforts of the SPR were primarily responsible for stimulating the growth of this field. In the United States, the main inspiration for exploration of these topics was J. B. Rhine at Duke University. J. B. and Louisa Rhine studied one of the simplest forms of psychic functioning: the ability to predict the next card in a deck of cards. They began with simple playing card decks familiar to everyone, but quickly realized prediction using these decks demanded foresight into two variables simultaneously: number and suit.

Reasoning it would be best to start with one variable, they instituted use of a collection of twenty-five cards now known as *Zener cards*: a set of five simple symbols—circle, square, cross, five-pointed star, and three horizontal wavy lines—each imprinted on five cards. The Rhines conducted millions of trials over several decades, among hundreds of participants, in accumulating a massive amount of data. They tracked: (a) overall accuracy of prediction; (b) the relationship between accuracy and time spent during a single sitting to determine a fatigue factor; (c) the effect of continuous result feedback on individual performance by interpreting the successful rate of prediction, the cumulative accuracy during a trial run, as a function of the confirmation or denial of specific predictions.

It is impossible to overstate the influence of the Rhine studies on future investigators. From that time until the present, the number of researchers has mushroomed, as has the creativity expressed in experimental design and the employment of rigorous controls to prevent trickery either on the part of the subject acting alone or in consort with one or more scientists conducting the test. Despite the still-persistent claims of some critics of parapsychology as a discipline of legitimate pursuit, it is now generally accepted and respected by traditional scientific organizations. For example, the American Psychological Association not only extends membership eligibility to parapsychologists, but also acts as sponsor for parapsychological conferences.

Most criticisms still levied against the research and its conclusions center around the issues of replicability and/or fraud, with skeptics professing that falsity in experimental design, data collection, data reporting, or all of them together, explain the successes claimed by some researchers.[2] The broadside attacks on credibility *did* have substance in the early days of the discipline: some SPR enthusiasts were dishonest, promoting the field through deliberate deception. C. E. M. Hansel gave details for several examples in *The Search for Psychic Power*, one of which will be offered to illustrate the point: "The investigation described in that report concerned the five young daughters of an English clergyman, Reverend A. M. Creery. During sitting after sitting, the girls and a young servant...convinced the investigators of their telepathic abilities. Six years later, in 1888, they were caught using a code and admitted to having deceived the researchers."[3]

As a direct consequence of this and/or similar incidents, contemporary investigators are so aware of the potential seriousness of any hint of fraud they take inordinate precautions against its possibility through the imposition of very stringent and very sophisticated experiment design controls. William Braud, one of the most respected scientists in the field, directly addressed the possibility of fraud in an essay entitled, "Brains, Science, Nonordinary and Transcendent Experiences":

> Ostensible psychical experiences such as telepathy, clairvoyance, precognition, psychokinesis...may be explained on the basis of: fabrication or hoaxing of the experiences; misperceptions and distortions of observation or of memory; subtle sensory cues, whether consciously attended to or not, which might betray the nature of the event...to be known or perceived; rational inference through which the nature of the...event might be determined...chance coincidence; or, quirks of the brain that simulate the psychical experience.[4]

A healthy skepticism seeks convincing proof of claims and is objectively scientifically proper. Doubt is inappropriately extended when every positive experimental result is presumed to be attributable to cheating in one form or another, whether deception or fraud are convincingly demonstrated to be the immediate cause or not. Hansel's *The Search for Psychic Power* expressed unhealthy radical skepticism. It selectively reviewed experiments performed to test for telepathy, clairvoyance, and precognition. While some criticisms about experimental rigor and repeatability were legitimate, he did not in any case *prove* intentional cheating and deception; he simply alleged corruptions were possible and felt justified in discounting results obtained from a diverse array of experiments as a consequence.[5]

It is of doctrinaire skeptics such as Hansel that a witticism made by Paul Weiss in *Being and Other Realities* holds validity: "Nothing is to be placed beyond the reach of critical examination, but nothing is to be supposed to have vanished just because it had been dismissed by those who could find no place for it in their accounts."[6] Dean Radin very effectively rebutted critiques arguing for wholesale condemnation of parapsychological methodology in *The Conscious Universe*: "The real issue…is whether claimed flaws *in fact* caused the observed results. If they did not, then claims for dirt in the test tube are inconsequential, whether or not the dirt…existed in the first place."[7]

Aside from claims of intentional or unwitting false experimental design, or a lack of adequate control safeguards to prevent fakery, the single most common argument against either the generalized class of parapsychological studies, or specific instances reported in the literature, is an allegation of data non-replicability. The simple fact is that traditionally accepted scientific data, with respect to experimental design or interpretation of the data collected, are also not precisely replicable. *Replicability presumes a perfect knowledge of all possible influences on an experiment.*

Seymour Epstein offered a cogent observation in *American Psychologist*: "The traditional solution of attempting to obtain a high degree of control in the laboratory is often ineffective because…human behavior is so sensitive to incidental sources of stimulation…adequate control cannot be achieved." In direct consequence, he wrote: "Not only are experimental findings often difficult to replicate when there are the slightest alterations in conditions, but even attempts at exact replication frequently fail."[8] A study, also published in *American Psychologist*, of *mainstream* psychological research literature, the respected branch of the American Psychological Association, discovered less than one percent of the reports ever attempted to replicate previous results.[9]

Infrequency of replication is not confined to "soft human" sciences like psychology or parapsychology. It can occasionally become a major issue for "hard" sciences such as physics. Hence, *a priori* claims of non-replicability cannot be properly argued as legitimate reasons for discounting conclusions obtained from results difficult to duplicate. Dean Radin mentioned a specific example: "The omega minus particle, Ω^-, was considered to be 'found' on the basis of only *two events* out of…nearly 200,000 experimental trials…. An extremely poor replication rate…was…considered sufficient to convince most physicists that the particle was real."[10]

Radin offered a rationale for the inconsistency with which the comment regarding replication is often selectively raised: "The degree of replication required to establish an effect is closely related to how strange that effect is supposed to be.... 'Strangeness' of an event is directly related to how much the observed phenomenon deviates from theory." In other words, based on theoretical predictions, the Ω^- particle was *expected to be real*; therefore, a paucity of detection was not deemed detrimental to acceptance of the result. Parapsychology suffers from an absence of reliable and compelling theories enabling explanation of singular research results to fit under their umbrella. This deficiency has the unfortunate consequence of allowing critics to claim reports of results suggesting the reality and nature of psychic functioning are inadequately substantiated.

It should be universally recognized some results are innately impossible to replicate. For example, the recovery of pottery shards from archaeological excavations is unique by definition; identical artifacts are not encountered *en masse*. Cataloging the find with respect to historical era and presumptive cultural authorship is not a repeatable result. Why? Dating techniques are not sufficiently accurate to denote historical time with precision; hence, multiple attempts by different researchers will generate only approximate, not exact, agreement. Determinations of cultural authorship presume knowledge of all potential candidates is complete. Even if other relics are found at the same site, speculative identification neither unambiguously supports nor denies all conclusions drawn from the first discovery. As independent pieces of data, they are not reproductions of prior results; *corroborative findings are not replicated findings*.

Even when research results are *in principle* capable of replication, this does not mean duplication is realistically attainable. As Seymour Epstein implied, any unknown or supposedly insignificant factor surrounding the first collection of data may prove to confound all attempts at *exact* repetition of the result. A meaningful pragmatic distinction is to be drawn between replication of a result and duplication of all elements leading to the result.

Nevertheless, because the criticism of non-replicability has been claimed so often as a justification to reject the feasibility of parapsychological events, more effort has been put forth in replicating reported results than in virtually any other field of inquiry. Dean Radin devoted a major portion of his book to a recitation of the results obtained from countless thousands upon thousands of experiments performed in the hope valid parapsychological events would be confirmed and accepted as proof of the ability to gain information *via* atypical means. The table summarizes data obtained from some studies found in *The Conscious Universe*. The "odds" column indicates the probability of attaining the number of successful results relative to random chance guesses. Where necessary, general experimental procedures are cited in the Notes.

Category	Test Type	Dates	Replications	Odds
general ESP	ESP card tests	1882–1939	> 4,000,000 trials	$> 10^{21}:1$
telepathy	dream-telepathy	1966–1973	450 sessions	$< 10^{8}:1$
clairvoyance	remote-viewing tests	1973–1988	> 1000 trials	$> 10^{18}:1$
precognition	'forced-choice' experiments	1935–1987	2,000,000 trials	$\sim 10^{25}:1$
telepathy	*ganzfeld* or total field effect	1974–1997	2549 sessions	$> 10^{15}:1$
precognition	random number generation	1959–1987	152 references	$> 10^{12}:1$

In explanation: (a) *Extrasensory perception* or ESP was a neologism invented during the early years of Rhine's research. *General* ESP was an all-inclusive category label intended to include skills in telepathy, clairvoyance, and precognition. The indicated studies involved the Zener cards previously mentioned.[11] (b) The telepathy studies were conducted by Montague Ullman and Stanley Krippner in the Maimonides Hospital Dream Laboratory in New York City. The sessions consisted of sleeping subjects attempting to receive images of pictures intentionally transmitted by an agent located in a separate room. Upon awakening, test subjects were asked to draw memorable dream content.[12] (c) Studies indicative of clairvoyance were conducted at Stanford Research Institute in California. The remote viewing tests were held under the direction of Harold Putoff and Russell Targ. These experiments could be considered the "awake state" analog to the Ullman-Krippner investigations. Subjects were requested to draw pictures of remote geographic locations, distinguishable by some large and unusual nearby structure, at which an agent either was present at that moment in time, or would be present at a specified time in the future.[13] (d) "Forced-choice" precognition experiments required determination of a future-time condition for an object among a limited number of available options. Test techniques might be as simple as determining heads or tails for a coin flip, or which lamp among a series of random lights would be the next one to stay on for more than five seconds.[14] (e) *Ganzfeld* or total field effect studies are refined explorations of telepathy where all external sensory input is blocked during a wakened state through the use of coverings over eyes and ears. In addition, the test subject is made to sit in an enclosed room sound proofed against exterior noise.[15] (f) Random number generation tests for precognition are a technological refinement of forced choice studies. Process outcomes, to be predicted in advance, were generated by machine-dependent variables: an electrical signal voltage; the radioactive disintegration decay rate of a particular atomic isotope.[16]

Given: (a) the broad breadth and large number of experiments reporting positive results at odds astronomically greater than could be obtained from random guesswork; (b) a complete absence of empirical evidence supporting contentions of universal conspiracy conducted by fraudulent operators acting in league with cooperative subjects to confound the public; then it can be concluded *some* individuals in *some* circumstances possess *some* measurable ability to collect and interpret information about people, places, and objects by *some* means not obviously reducible to ordinary methods of perception and conception.

The preceding emphasis was intentionally placed indicating, despite the conclusion, mysteries surrounding the specifics still remain as outstanding problems to be solved. Needing most to be elucidated for development of a plausible foundational theory underlying parapsychological talent are: (a) the personal characteristics of individuals likely to produce more positive results in trial experiments; (b) some delineation of environmental factors relevant to maximizing success; (c) the mechanism of accomplishment.

Moreover, it must be considered *a priori* possible that different theories are necessary to separately explain the distinctive manifestations of ability in telepathy, precognition, and clairvoyance. To assume one mechanism and one type of individual must necessarily fit all permutations is a prejudice not supported by current levels of knowledge, and could serve to prematurely constrict the creativity needed to explain the data adequately. On the other hand, if a unified mechanism and personality type could be proven to explain all expressions of parapsychological talent, such a finding would not only be a phenomenal accomplishment on its own, but would lend itself to a potential future conflation with normally accessed cognitive modes leading to a meta-epistemological framework.

On the negative side, one point of significance about the accumulated data has been overlooked. Cheating aside, the experiments usually cited as offering positive indications of psychic ability are all predicated on outcomes of subjectively correct *guesses* or intuitive *hunches* beyond that expected by chance probability. It is more rational to hold as a conviction that *legitimate talent in telepathy, clairvoyance, or precognition is the product of contingent factual knowledge or certainty, not guesses coupled to hopes one is correct.* This must be the case if the mechanism for reception of information about outcomes is to be regarded as authentic.

Using normal visual perception, when an individual claims to see an automobile, he or she is certain about the nature of the object and does not experience any degree of hesitancy as to whether it is in fact an automobile instead of a train, a goat, or some other randomly observable thing. Although they may express hesitancy, or simply be wrong, when professing knowledge as to definitive make, model, year, color, body type, or any of a number of other relevant data, they do not doubt it is an automobile. Given a standard vocabulary allowing for general differentiation among cars, trains, and goats, one cannot be wrong in the fact of their perception however much they can be wrong about its precise correspondence to a particular real-world object.

Even if there is no real-world object as the basis for comparison—if I feel angry, I cannot deny I feel angry—the certainty is given as part of the experience. The same degree of certitude arises when we obtain information from our other sensory modes or when evaluating our conceptual knowledge. All school tests, for example, are given to students under the supposition that each one *knows* when they know the correct answer, *knows* when he or she doesn't know with certainty, and *knows* when they are guessing rather than responding out of conviction. Collectively, these experiential certainties are the foundational basis for the claim that one cannot lie to oneself however much he or she can lie to another.

Anecdotal reports and statistically correlated evaluations of experimental results in parapsychology point to a particular hypothesis: success rates are *not* arising from a conviction the respondent possesses certain knowledge.[17] Radin offered two comments regarding parameters that might influence the outcomes of these researches. The first: "One of the discoveries made with ESP cards was...psi performance invariably declined with repeated testing. This is not surprising because the ESP card test is tedious to begin with, and being asked to guess the same card symbols over and over...is...boring."[18]

Note the acknowledgment of "guess" as a characterization. Moreover, the remark demands reciprocal queries: Would a child's ability to correctly recite the multiplication table be likely to decline due to repetition? Does repeated practice of the same piano concerto cause one to make more errors at the end than at the beginning? Is boredom a rational justification for errors of fact? There are experiences where lack of mental stimulation may, with justification, lead to mistakes—an assembly line worker engaged in the same bodily movements over prolonged periods of time may make more errors later in the day than earlier simply due to bodily fatigue—but such situations usually involve physical tasks as opposed to mental ones.

The second relevant declaration was: "Trial by trial feedback tended to result in better performance... In forced-choice tests, if feedback is held off for more than a few minutes, the motivational effects of the feedback are lost."[19] If precognitive ability is an authentic talent and not the consequence of better than average guesswork, it is unclear why feedback as to current successes should affect, positively or negatively, the predictive reliability of future trials. Each experimental data point is independent of another, and the mathematical probability of being correct for attempt 'n+10' is exactly the same as for attempt 'n.' A student knowing he or she has the correct answer to question number one on a twenty question test in history does not give the test taker a greater likelihood of answering the last question successfully.

The fact tested subjects show signs of mental fatigue leading to declining performance over time, and are not certain of the correctness of responses—indeed perform better when given feedback during the experiments—may be evidentiary support for the contention they *lack* authentic psychic gifts. More successful subjects may simply display greater than average gifts at *guessing* in a profitable manner. As good guesswork is likely to be a transitory, non-controllable event, it might be expected that, over time, an individual would become mentally fatigued resulting in a loss of concentration. This would explain the time trend in declining performance as well as the motivational boost obtained from receiving positive feedback on previous guesses.

A viable alternative to this sobering interpretation is to posit psychic skill as an evolutionarily acquired capacity at the infantile stage with respect to development in humans. Therefore, while successes are in fact the outcome of legitimate parapsychological skill, they are not yet fully developed even in the most accomplished subjects. If such were to be the case, confidence born of knowledge obtained by these channels should not be anticipated. In this scenario, people are "feeling their way" much like toddlers taking their first walking steps; hence, confirmation of successful outcomes encourages the individual subject he or she is on "the right track." *If this thesis is to become demonstratively provable, any proposed mechanistic theory for enhancement of parapsychological talent must explain the evolutionary element.*

One piece of evidence supporting the potential establishment of a newly emergent talent for human beings obtains from consideration of genetics. In a contribution to *Neurotheology*, Rhawn Joseph mentioned that much of the human genome is currently unexpressed; that genes lie unutilized within our chromosomes: "The simplest cells...single celled organisms, may contain as many as six thousand genes (which may be located along and within a single chromosome).... The more complex human nuclear genome consists of anywhere from 30,000 to 40,000 genes—97% of which are 'silent' and have not been expressed, and/or which come to be expressed only in reaction to changing environmental conditions."[20] *Introns* are viable, although currently unexpressed, gene sequences; expressed genes are *exons*. Since ninety–seven percent of the human genome consists of introns, enhancement in a person's exercise of consistently reproducible and accurate psychic powers might accrue in some future time period.

Interestingly, in *The Quest for the Fourth Monkey: A Thinker's Guide to the Psychic and Spiritual Revolution*, Sylvia Fraser proposed an argument in favor of the *opposite* conclusion, namely that parapsychological capabilities, at least with respect to telepathy, are a relic from our evolutionary past rather than our future:

> Scientists who accept telepathy usually see it as a regressive trait, more relevant to our species' evolutionary past than to our present. This view is based on the fact that the human brain is triune—actually three brains fitted together with the outer two encasing the third... All...remain active in our daily lives, with functions roughly corresponding to instinct, emotion, and reason.... Telepathy is hypothesized to be a function of the lower two brains... As evolution favored language over telepathy and individuality over tribal consciousness, our nervous system learned to block psychic messages from other tribal members so...each of us could concentrate on the task at hand.[21]

Reference to the triune brain model was made in chapter six during the discussion of the reasonability of the proposition of self-consciousness as an evolutionarily emergent phenomenon dependent on brain development rather than an independent essence necessarily associated with all living organisms. Fraser's statement, if proven valid, would be evidence supporting a thesis voiced in chapter two concerning the unjustifiable nature of homocentric superiority, where it was argued many animals have language capability just as we do, though not as sophisticated because they have compensated by having superior skill in nonverbal communication—that is, in telepathy.

In summary, our current level of parapsychological skill may be at a comparatively infantile level either because we are learning how to employ these talents for the first time, or because we are re-learning how to employ them for the second. Either proposal accommodates the experimental results on utility of success-rate feedback and inutility of mental fatigue. Both also justify the conclusion that the data attained to date refer more to successful guesswork than to certainty/absolute Truth or knowledge/contingent truth.

As a bonus, each option explains the essential difficulty with replication. To return to a previous query: "Would a child's ability to correctly recite the multiplication table be likely to decline due to repetition?" Yes, *if* it were the case the child was just beginning to learn the multiplication table and their correctness of recitation was, at best, erratic because the necessary skill for perfect reproducibility was not yet firmly implanted in the interactive mind-brain relationship.

Before attempting to develop a plausible mechanistic theory, one more consideration should be included as pertinent. The opening paragraphs of this chapter made allusion to religious writings, mystical and mainstream, hinting at their inclusion of numerous references to the exhibition of psychic talent. An obvious example would be Old Testament prophets who displayed gifts of clairvoyance and precognition in a public arena. As there would seem to be some connection between mystical revelation and the manifestation of parapsychological ability, it is natural to hypothesize a common mechanism may be involved. Stanley Krippner noted common characteristics uncovered in the work of his peer, Lawrence LeShan:

> LeShan also discovered...[Eileen] Garrett and other psychic sensitives perceived "clairvoyant reality" in much the same way...mystics wrote about their experiences and in a similar fashion to the descriptions of reality produced by certain theoretical physicists. From all three vantage points, there is a unity of all things, relationships are more important than single events, observers cannot be separated from what they observe, cause and effect relationships lose...meaning, time and space limitations are minimized, and one's ordinary perceptions are seen as not producing a valid picture of reality.[22]

The passage is informative in several ways. The consensus description of "unity," and "relationships more important than single events," jointly speak to the concept of the transcendental, since it is an axiom of physical reality that its manifest discreteness can only be conjoined into a continuum through incorporation of a higher perspective, a literally meta-physical orientation. "Time and space limitations are minimized" can only be observable features of experience if one accesses the unbounded expanse of cosmogonic Space and the infinite duration of cosmogonic Time found in theological space.

The citation also displayed an undesirable characteristic: the tendency to generate conclusions in either/or form. "One's ordinary perceptions are seen as not producing a valid picture of reality" is an illogical depiction: someone perceiving the universe differently if a "psychic sensitive" or mystic does not properly lead to the determination that the more commonly held view of the universe is *invalid* however much it might be *incomplete*.

If I am extremely competent at doing algebra, but am not familiar with calculus, my knowledge of algebra is not thereby rendered invalid; it merely means my knowledge of mathematics is incomplete. If during experience 'B' an individual functions differently, exhibits atypical behavior, than he or she would in comparison to their manner of action under conditions found during experience 'A,' the two experiences could not then be related as either/or absolutes. Too often declarations are made in violation of the Law of Noncontradiction. Only for those items viewed in the same way at the same time is it the case that 'X' and 'not X' cannot both be true simultaneously.

Krippner followed his mention of Leshan's work with a recounting of a compilation of mental and physical tendencies displayed by those exhibiting psychic ability, a features catalog created by William Braud. Krippner rather dramatically declared, "William Braud has described a 'syndrome' which appears to be conducive to psi," and continued:

> It consists of muscle relaxation, a slowing down of brain waves, and a reduction of heart rate, blood pressure, oxygen consumption, breathing, blood lactate level, and electrical conductivity of the skin.... Other accompanying psychological processes include an increased ability to think in terms of images and wholes, to have access to intuition and the unconscious, and to turn one's attention inward...a decrease in logical, abstract thinking, in analyzing the environment, and in making judgments. One's time sense changes and the boundaries between past, present, and future fade away.[23]

Of interest is the confirmation of parapsychological abilities as similar to mystical experience with respect to the effects it has on the human body and on one's mental states. Compare the physiology of a person described in this citation with Newberg and D'Aquili's recitation of responses obtained from their research on energetic techniques used to induce ecstatic mystical events presented in chapter seven. In addition, Braud's list of mental states may be favorably compared to qualities of personality exhibited by introvertive and extrovertive mystics, as expressed by citations from Stace or Jones.

Such states can be construed as preparative for a successful experience, the conditions pertaining to one's body during an experience, or after-effects resulting as a consequence of experience. The relationship between mystic or psychic sensitivity and physiology may be causal or associative, and the data obtained to date do not allow for a clear determination. It was hypothesized in the preceding chapter that association was likely to be the more accurate description of events because the joining of self-consciousness with global consciousness in theological space occurs outside the domain of the physical body.

After over one hundred years of research in parapsychology, admittedly of varying quality and reliability, there is a sense of profound disappointment in the lack of construction of a successful theory explaining the accumulated data that would satisfy the most responsible investigators. R. B. Roberts tried to outline what he construed to be the necessary elements in an acceptable theory in "A Theory for Psi": "To be useful a theory...needs to: (a) arrange the facts in a pattern that is logical and...easy to remember, (b) explain numerous specific effects in terms of one unexplained phenomenon, and (c) make predictions subject to disproof and thereby to suggest experiments to support or disprove it."[24] While this proposal is epistemologically sound with respect to many theories generally found to be agreeable to the scientific establishment, it is naïve to suppose it would be sufficient for a discipline materialists view, as they do mysticism, as inherently absurd and fanciful. To attain an aura of scientific respectability both mysticism and parapsychology will require theories incorporating a viable mechanism of action.

A major constraint to creation of a theory for parapsychological action, which needs to be overcome, is the frequent assertion—especially popular among critics of topics invoking transcendentalism in some form—presented by Max Velmans in *Understanding Consciousness*: "If mental events are to influence physical ones, physical energy must be created...and the total physical energy of the universe thereby increased.... For physical events to influence mental ones, energy must be drawn from the physical universe. However, according to the conservation of energy principle, energy can neither be created nor destroyed."[25]

As first articulated in chapter three, the conservation principle applies *only* if the physical world is innately construed as a causally closed system. If it is an open system, the total energy of system plus surroundings is to be conserved. It is begging the question to *assume* the universe cannot include Energetic/energetic exchanges with a nonmaterial Dimension. One cannot allege the principle precludes elaboration of parapsychological or mystical explanations for hypothesized phenomena without first proving the physical universe is an isolated system. Perhaps the transcendental realm is real and not a mere intellectual construct; perhaps exchanges of Energy in the form of information are possible.

It cannot be overemphasized: *materialism is a theory to exactly the same degree as the proposition of the supraphysical as a real existent is a theory.* Each is a plausible *a priori* grounding for philosophy, religion, and science. Each is a speculative scenario until proof is ascertainable that only one is True. Evidence has been offered claiming available data are more consistent and comprehensibly incorporated under the thesis of metaphysical reality. It is acknowledged the available evidence does not rise to a level sufficient for proof, and may not do so until humanity develops the capability to correlate its findings with other sentient beings.

Aside from objections stemming from misconception and misapplication of a conservation of three-dimensional energy principle, attempts to devise coherent theories for operation of psychic skill must overcome an added rationale underlying the standard reluctance, expressed by natural scientists, to acknowledge parapsychological ability as authentically real. Undeniably, the feature making these kinds of phenomena unique is their alleged facility in inducing feats of information gathering at distances remote to the source of the data being obtained.

In this respect, one can propose the existence of a continuum relationship concerning type of cognitive experience needed as a function of the degree of spatiotemporal separation between a data recipient and an informational source. At one extreme are parapsychological events, where huge gaps in spatial, and possibly temporal, proximity between transmitting object and the receiving subject neither prohibit nor inhibit communication. In the middle spacetime region exist sensory experiences used in accumulating evidentiary information at varying degrees of separation between points of origin and termination. These range from senses operating at an intermediate distance in exploration of the environment—visual, auditory, and olfactory organs—to those—taste and touch—whose receipt of tactile data require direct contact with objects. At the opposite extreme are mystical experiences, where zero distance, complete unification, between transcendent agency and immanent mystic is a prerequisite for gathering unimpeachable revelatory information. It is also tempting to speculate there is a direct mathematical relationship between the degree of certainty attained from transfer of information and the distance between sender and receiver: less spatial distance and time lapse produces greater certainty. While the latter would appear to be *prima facie* true, there is an unanswered question as to whether this relationship need be inherent.

Doubts pertaining to the legitimacy of parapsychological phenomena are grounded on application of the inverse square laws for both gravitational and electromagnetic forces. The former has been known and accepted since its inception by Isaac Newton. Centuries later, it was discovered that the form of the law also applied to the electromagnetic force. In words, the law states the gravitational or electromagnetic force generated between two objects varies in inverse proportion to the square of the distance between them; that is, increases or decreases in relation to the distance between them multiplied by that same distance. As a subject ripe for metaphysical speculation, *the fact the two known universal forces abide by the same form in their mathematical treatment is not coincidental.*

Mathematically, these laws differ only in the involvement of object mass for gravitational force in contrast to inclusion of object electrical charge in determination of the strength of the electromagnetic force. Expressed using now standard symbolism: $F = Gm_1m_2/d^2$ for gravitation, and $F = kq_1q_2/d^2$ for electromagnetism, where 'F' is force; 'm' is mass; 'q' is electric charge; 'd' is distance; and, 'G' or 'k' are empirically determined constant values.[26]

It is a measure of the universal acceptance of the inverse square laws and difficulties intrinsic to formulation of a plausible parapsychological theory that Russell Targ jumped to the *wrong* conclusion in *Miracles of Mind*: "The data for...psi research show...no significant decline in the accuracy of any kind of extrasensory perception with increasing distance between the viewer and the object viewed. We also know it is no more difficult to look a short distance into the future... We can conclude from this...it is very unlikely... any kind of electromagnetic field is involved in carrying psi signals."[27] Is the apparent violation of inverse square laws sufficient to overturn or reject the results announced in reports of parapsychologists?

If the *only* means psychic phenomena can be postulated to occur pertains to three-dimensional spacetime involving three-dimensional physics, then the answer must be "yes." Neither gravitational nor electromagnetic events have been found to operate under conditions inducing consideration of violations to the inverse square law. Parapsychological research is not now sufficiently grounded in its theory of operative mechanism to give sufficient warrant to accept either of two theoretically plausible alternative views challenging the preeminence of accepted physical law: (a) there exists another force, beyond electromagnetism and gravity, psychic events employ as a medium for their expression; (b) there exists no other applicable force operational on universal scales beyond these two, yet there are occasions, such as in the production of parapsychological events, when inverse square laws fail to hold, suggesting they are incomplete as typically stated and must be amended to incorporate violations. Either of these two hypotheses may be True; yet, with regard to contingent truth, neither alternative is acceptable if reliance is to be based only upon presupposition of the reality of parapsychological events.

Unlike Targ, William Braud focused on the nature of a proposed carrier of the alleged force rather than on the force itself. Such an approach would be analogous to having some physicists ponder specifics involving electrons as the carrier of electromagnetic force instead of concentrating their efforts on explanation of the role of that force in normal physical operations. It is at best only of partial utility to consider the issue from this perspective because gravitational force lacks a force carrier, a gravitino, so its involvement in psychic operations, if any, cannot be understood in this manner.

Braud came to the same conclusion Targ had reached: "Many...carriers have been proposed for...psychic transmissions, including electromagnetic waves, neutrinos, tachyons... Transmission models...face serious difficulties in explaining the operating characteristics of psychical effects. These effects occur over great distances, and they do not seem to decline appreciably with distance, as do conventional forces. No method has...been found to...shield or prevent psychical effects or to amplify them."[28]

Recognition of the difficulties involved in explaining effects not overtly conforming to the physics germane to all other known event sequences in spacetime led David Kahn to an extreme position: he alleged experimental designs to investigate psychic phenomena were uniformly flawed:

> To be taken at all seriously a putative spontaneous ESP occurrence must carry with it precise and discrete information that could not be obtained by ordinary means. In designing methods to capture such spontaneous events in the laboratory, tests have almost universally used the model of the perceptual process by requiring discrete specific information, defined as the "target," to be perceived. This is well and good

> if psi is in fact primarily a perceptual process. If it is not, however, the very design of the experiments may serve to factor out the phenomenon which we are seeking.[29]

Kahn's assessment, "discrete information that could not be obtained by ordinary means," reveals again the logical flaw hidden in most analyses of parapsychological events. They rely upon an initial premise—the inviolable presupposition that psychic phenomena must pertain to three-dimensional space *exclusively*—that is, in reality, contingent and not known with the certainty of Truth. What would be the outcome if this premise were rejected? Are there existent grounds for its dismissal as an unnecessary constraint? Kahn sought return to a proposal first made by Myers in the nineteenth century: the notion of a group mind coordinating the thoughts of all Earth lifeforms.[30]

It is not an outlandish notion. When mental events are centrally involved, consideration must be given to Consciousness and its derivative in the form of self-consciousness as playing direct participatory roles. Both refer to the transcendental, and transcendental always implies a fourth spatial dimension unifying discrete three-dimensional spatiotemporal elements. In particular, addition of another spatial dimension—resulting in five dimensions if time is counted separately—was found to produce mathematical harmonization of Einstein's field equations with Maxwell's leading to unification of the force of gravitation with electromagnetism.

Therefore, any appeal to violation of the inverse square law as providing sufficient grounds to reject parapsychological data is tantamount to alleging that the inverse square law applies not merely to four-dimensional spacetime physics, but to five-dimensional SpaceTime psychophysics. This proposition has no more evidence to support it than the alternative hypotheses mentioned previously. *If parapsychological phenomena use a fourth spatial dimension through which data can be sent without concern for distances separating sender and receiver, the inverse square law is not violated because it need not apply. It would also justify the notion of a continuum of action for parapsychological, sensory, and mystical cognition through proposal of a common mechanism.*

This idea effectively circumvents Targ's assessment: "We can conclude from this...it is very unlikely...any kind of electromagnetic field is involved in carrying psi signals." Furthermore, the inclusive domain of this proposal is not limited to *Homo sapiens*. All sentient creatures utilize the fourth spatial dimension uniting individual self-consciousness with global consciousness. In effect, *theological space offers a metaphysical grounding to Myers' group mind*. Behaviors of other creatures become interpretable under the construct of global consciousness serving as group mind. A statement made in chapter six, further addressed in these pages, gains added relevance and credibility:

"Nonhuman creatures may...possess a more highly attuned telepathic ability, which we as a species are just beginning to either develop or re-cognize. This could explain the absence of highly sophisticated verbal language skills in other terrestrial intelligences."

Sylvia Fraser provided an example of possible animal-based telepathy in *The Quest for the Fourth Monkey: A Thinker's Guide to the Psychic and Spiritual Revolution*: "Evidence of...group mind operating among insects is even more striking. For example, colonies of harvester ants build radiating roads to food sites half a mile away; when a road becomes blocked, police ants arrive in a phalanx to construct a detour. Experiments...have verified... these ants appear...before an alarm could be relayed by any known sensory or chemical means."[31] "Group mind" could be conceived as a euphemism for access to global consciousness to obtain relevant information. Fraser made a salient comment about the issue of distance with reference to mental events:

> The fact...two brains are a thousand miles apart does not mean...two minds are, and the question of how thought travels physically from consciousness to consciousness becomes a specious one. Thoughts are not things but processes.... Inside the brain you can't meaningfully describe one thought as being an inch away from another... Telepathy—if it exists—operates independently of space so..."here" and "there" become irrelevant... It is a property of consciousness, unbound by Newtonian science based on sensory experience.[32]

According to the hypothesized mechanistic theory, Fraser was correct in *almost all* respects and the advocated model provides the context to take her comments seriously. She was wrong in her conviction, "how thought travels ...becomes a specious one." If confined solely to a three-dimensional space, the mechanism of thought transference evokes mystery. If one includes a unifying continuum binding all discrete entities—thoughts, emotions, acts, and objects—a process for thought transmission becomes comprehensible.

The proposed mechanism for psychic functioning *rejects* Russell Targ's disavowal of electromagnetic radiation as the medium for transmission of thoughts at a distance. Several separately developed pieces of evidence offer strong support for this conclusion:

(a) Valerie Hunt's detection of an aura surrounding living organisms using modern scientific instrumentation: a *permanent dynamic energy field* distinct from the better-known electromagnetic pathway associated with the neurological underpinnings of brain function.

(b) Sylvia Fraser's assessment of the auric field *as* electromagnetic: "A string of researchers have found evidence...the body does radiate an energy that pulsates, flares, and changes color. Strongest around the head and fingertips, it is said to intensify with mental activity...and to be affected by electricity and magnetism."[33]

(c) In *Stalking the Wild Pendulum*, Itzhak Bentov remarked on the energetic coherence of auras with the Earth's magnetic field: "The resonant frequency of the earth's ionspheric cavity is approximately 7.5 Hertz, and the micromotion of the body is approximately 6.8–7.5 Hertz. This suggests a tuned resonant system.... It is the ideal medium for conveying a telepathic signal."[34]

(d) Hunt cited research by Bob Beck, who measured the dominant brain wave frequency for *mystics* and found it to be *exactly the same* as this resonant frequency:

> An experimental physicist, Bob Beck...developed sensitive instruments...he used to record the brain waves of proven mystics. When he processed these data he found ...their brain waves during their mystical work were [operating at] 7.8 cycles per second.... Schumann...discovered a magnetic field...thrust upward from the earth rather than downward, as does the gravitational field.... Schumann recorded this resonance as between seven and eight cycles per second. Beck found it to be 7.8 cycles...exactly the frequency he recorded from the brain waves of mystics.[35]

The existence of an energetic frequency radiating from Earth in a natural fashion, which is identical in value to an energetic frequency associated with human brain and body possibly serving as mediator for *both* telepathic and mystical experience, offers a plausible explanation for the common linkage of mystic revelation with psychic powers found in both exoteric and esoteric religious literature. It also serves as preliminary evidence for the continuum hypothesis for all cognitive modes of knowledge acquisition.

(e) Confirming Hunt's reference, the research of neurophysiologists Gellhorn and Kiely published in the *Journal of Nervous and Mental Disease* revealed the most pronounced brain wave pattern exhibited by proficient Zen Buddhist meditators lies in the theta band range, which is in the range of four to eight Hertz.[36]

(f) A different research group studied practitioners of Transcendental Meditation and supported this finding. According to the paper published in *Psychosomatic Medicine*, they concluded: "High intra- and inter-hemispheric electroencephalograph coherence in alpha and theta frequencies, especially in frontal areas of the brain, accompanied by periods of low metabolic rate, respiratory suspension, and stable autonomic activity have been reported and have been found to be highly correlated with the subjective experience of transcendental or pure consciousness."[37]

Similar neurological data from two meditation techniques can be viewed as preliminary evidence lending credence to the thesis of a common cross-cultural pattern grounding all mystical approaches to revelation. Additional research across a greater spectrum of religious practices would be needed to upgrade the hypothesis from credible to likely.

(g) Young children are more proficient than adults when tested for telepathic ability, and these results have been correlated with enhancement of theta band brain wave activity:

> The dramatic decline of psychic ability in very young children as speech develops has been verified by a number of researchers. In a five-year British study involving one thousand subjects and using simple guessing games, parapsychologist Ernesto Spinelli found...children aged three...had telepathic scores seventeen percent above chance; by four and a half to five years, this had declined to fifteen percent above chance; by five to seven years, to four percent; from eight years onwards, to a level expected by chance.[38]

Sylvia Fraser later noted, "By electroencephalographic experimentation, it was also discovered...children aged three to eight had a predominance of theta and alpha waves, corresponding to...high scores in tests of telepathy."[39] The ever-present nature of the body-associated aura as a resonant system compatible with the Earth's magnetic field throughout life is confirmed by extending to young children the significance of theta band usage during expression of telepathic ability. This finding eliminates any supposition the frequency or coherent nature of this energy is dependent upon training, such as is endured by those undertaking a meditative discipline.

Greater telepathic ability in younger children is also consistent with the contention this skill can be correlated with *less* hemispheric specialization. It supports the thesis that other animal species having a less pronounced frontal lobe development may offset a structural-functional advantage *Homo sapiens* have in this region with enhanced telepathic capacities enabling nonverbal forms of communication.

(h) Despite Targ's overt denial of a role for electromagnetism in the performance of parapsychological skill, he reported in *Miracles of Mind* that extraneous electromagnetic radiation *does* impede the demonstration of these abilities: "Recent findings from physicist James Spottiswoode suggest that... electromagnetic radiation from our Milky Way galaxy and...solar flares degrade psychic functioning."[40]

A rational explanation for prevalence of theta band brain activity present in children who are telepathically gifted and in those proficient in meditation derives from recognition that the posited unification of Consciousness with self-consciousness through theological space, which enables transmission of thoughts and images over spatiotemporal distances, still requires association with a brain to transform the reception of information into a format usable in a physical three-dimensional universe. John Horgan's statement in *Rational Mysticism* now makes sense: "The shape and timing of an electromagnetic signal, rather than its amplitude, determines its ability to influence a brain."[41] The "shape and timing of an electromagnetic signal" is a reference to its form and frequency, which differ among the four standard brainwave signals. *Alpha* (8–12 Hz), *beta* (over 12 Hz), *delta* (below 2 Hz), and *theta* (4–8 Hz) waves will affect a brain in different ways. It is not difficult to believe theta waveforms are a common denominator for parapsychological and mystical outcomes.

Transmission of reliable, accurate information between Consciousness and self-consciousness, which would occur during a mystical experience, or communication between one self-consciousness and another, which would occur during a telepathic experience, would be incapable of corruption by signal interference or electromagnetic noise because data transfer employs coherent Energy as the medium. When correlated with neural activity in the brain, distracting noise and signal interference *can* develop according to the laws of physics for electromagnetic radiation operating in three-dimensional spacetime. Signal degradation by distortion of the transmitted wave shape or attenuation of the frequency may facilitate explanation of characteristically poor results reported for telepathic communication: information would be harder to interpret correctly, and recipients would display uncertainty in their responses as a direct consequence.

What makes the electromagnetic Energy of Consciousness different than the electromagnetic energy of normal three-dimensional spacetime physics? *The Interconnected Universe*: *Conceptual Foundations of Transdisciplinary Unified Theory* by Ervin Laszlo offered an intriguing possibility: "In their interaction with the zero point field, electrically charged quanta are believed to produce secondary electromagnetic fields... This component consists not of...transverse electromagnetic waves, but...longitudinal propagating scalar ('Tesla') waves.... We envisage the zero-point field as a structured field with a scalar-mediated electromagnetic spectrum."[42]

An object moving in the same direction as a wavefront is said to undergo *longitudinal motion*; objects moving in a direction perpendicular to a moving wavefront are said to undergo *transverse motion*. Scalar waves move with a speed dependent upon surrounding object masses: faster motion is possible the more densely populated the space.[43] This is contrary to the norm: special relativity theory claims electromagnetic radiation in the form of light rays, a transverse waveform, is the measuring stick for speed in a three-dimensional universe, and moves fastest in a vacuum. Laszlo further posited these scalar waves carry information that can be encoded and decoded upon interaction with normal spacetime.[44] His zero point vacuum field has been speculated to be a physical description of theological space.

Taken in combination, these theories suggest the electromagnetic Energy of global consciousness, present in discrete form as the auric field associated with self-consciousness surrounding all living organisms, can interact with classical spatiotemporal electromagnetic energy to convey data interpretable by transformation into neurophysiological and biochemical operations in the brain of an intelligent being. A difference in properties between longitudinal and transverse electromagnetic waveforms provides a necessity for harmonic resonance allowing communication between the two, a notion previously expressed as the means employed to overcome the *activation energy barrier* segregating spatiotemporal energy from the Energy of Consciousness.

If the electromagnetic field surrounding people and objects serves as the force responsible for transmission and reception of information, how does this energy align with the Energy that *is* theological space? Recourse must be made to a proposition offered in chapter five: "Gravity is substantively more significant with respect to the subjective physics for which Consciousness... plays the principal part... Within theological space, gravity is... 10^{42} times more powerful than the electromagnetic force. The product of force strengths in normal spacetime would be equivalent to the product of force strengths in theological space."

The gravitational field of theological space influences SpaceTime and spacetime in a similar fashion as its three-dimensional counterpart does. This force, in conjunction with its weaker three-dimensional counterpart, connects "here" to "there," and "now" to "then," such that all information contained in spacetime is accessible to Consciousness. The equivalence in magnitude of the product of the forces in the two domains allows for synchronization of energy with Energy: the transfer of selected data from Consciousness to self-consciousness or the reverse, or between one person's self-consciousness and another's, is now permissible. If the data concerns issues of principal import to cosmogonic reality, the information transfer is characterizable as mystical in content. If the data concerns issues of principal import to cosmologic reality, the information transfer is characterizable as parapsychological in content.

A statement made by Boethius almost fifteen hundred years ago in *The Consolation of Philosophy* now takes on new meaning with this postulated mechanism of action:

> Since every judgment...comprehends the things lying before it according to its own nature, and since God's status is abidingly eternal and in the present, his knowledge too transcends all movements in time. It abides in...its present, embraces...past and future, and...ponders all things as if they were being enacted in the present. Hence your judgment will be more correct should you seek to envisage the foresight by which God discerns all things not as a sort of foreknowledge of the future, but as knowledge of the unceasingly present moment.[45]

Global consciousness—synonymous with God, Allah, Brahman, Tao, the Command of Heaven, Wakantanka (The Great Mystery), or countless other designators of cosmological reality in its totality—is omniscient. All parts of the manifest universe are spatiotemporally connected into a single cohesive "here-now" due to the force of transcendental gravity.[46] This Energy can be reconfigured into any signal shape, frequency, and amplitude. Information reflecting the conditional state of *any* discrete portion of three-dimensional spacetime can be imparted to *any* individualized self-consciousness localized in *any* region of spacetime. This mechanism is posited to transpire in order to produce telepathy, precognition, clairvoyance, sensory cognition, thought, intuition, and mystical revelation. In sum, it is responsible for the continuum hypothesis.

Mention has been made of the ability possessed by some individuals to alter biophysical and biochemical processes, either their own or another's—the bio-PK phenomenon. Laszlo gave another illustration of this occurrence that is of direct pertinence to the proposed mechanism of action: "Glen Rein proposed...scalar energy is transduced into linear electromagnetic energy in the body... His experiments on nerve cells in tissue culture indicate...scalar energy can modulate basic biochemical communication between nerve cells as mediated by the known neurotransmitters.... Uptake of noradrenaline in cells exposed to an eight hertz scalar wave field was inhibited by 19.5 percent compare to cells in the control dish."[47] The utilization of a *scalar* electromagnetic field of eight hertz—the value mentioned by Bentov, Hunt, and Fraser in their citations—would seem to be beyond coincidence, and suggests coherency among all of these reports in that they are discussing the same type of Energy.

Though the proposal sounds fanciful, a key test of plausibility is whether it can serve as a vehicle explaining known properties of mystical and psychic phenomena. Mystic experiences possess qualities of ineffability, transience, noesis, and passiveness, among others. Ineffability—an inability to articulate experiential details with appropriate language—may be understood as an inevitable result from one's self-consciousness detaching from the physical body while sending and receiving electromagnetic signals bloated with data. By definition, a pure Consciousness event means normal awareness is not available to focus on the event as it occurs. If mind and body are transitorily separate, brain cannot engage.

As was claimed in chapter seven, the process is not unique to mystical endeavors. An analogy would be the intensity of concentration required to solve complex mathematical equations to the point where an individual can simply lose all sense of time, space, and one's immediate environment. For information transmission involving theological space, intensity of personal focus *must* be on maintaining an energetic synchronization with the Energy of global consciousness.

Transience is expected: (a) the purpose of transcendental communication is to convey meaningful data; once this has been accomplished, there is no further justification in maintaining a link to theological space; (b) translation from Space to space is physically fatiguing for a favored individual's *brain* because translation of the scalar waveform into the more common transverse waveform for processing places demands on the body's natural resources. The mystical literature, in recognition of the issue, consistently promotes training as a means to give a person greater endurance during contact with the transcendent.

The practice of mental exercises can improve both overall focus as well as clarity of thought and memory. One can train one's sensory organs to be more sensitive to stimuli: subtle distinctions in similar sights, sounds, smells, tastes, and feels can be easily made. If, as has been claimed, psychic skill is a developmental talent, it is not unreasonable to propose physical stamina will be an important component for creation of practical techniques designed to enhance ability. This idea implies poor results in manifesting psychic ability suggesting its successes are due principally to good guesswork and wanes over the time of one test session occurs not, as Dean Radin asserted, because individuals get *bored*; it occurs because people get *tired*.

Noesis would seem to be self-explanatory, as it is the *raison d'être* for the entire process. Certitude in *initial* correctness of the acquired information necessarily follows, since the source is the unimpeachable absolute of global consciousness. Nonetheless, translation of longitudinal into transverse waves can be accompanied by signal distortion within the brain: interpretation of revealed information can be faulty. The discussion on mysticism revealed the potential for misstatements of meaning even by those to whom the event occurred. Substantive errors in thought or in sensory analysis are common. Precognitive, clairvoyant, or clairaudient experiences are sometimes flawed in details even as the general outlines of the information are proven correct.

The trait of passivity expresses nothing other than the realistic fact that the active agent in the exchange is the transcendent. No information can be transferred absent the allowance of grace. While this fact is emphasized in exploration of mystical phenomena, it is valid for other forms of cognition, for example: (a) aesthetic inspirations in music and art seldom appear on demand—hence, the frequent attribution of creative intuitions as dependent on the muses; (b) efforts of profound thought, no matter how strongly willed, are periodically unproductive; (c) psychics are more attuned, receive more reliable data, on some occasions in comparison to others.

In Robert Forman's *The Problem of Pure Consciousness: Mysticism and Philosophy*, Daniel Matt emphasized the crucial contributory role played by passivity during acts of prayer hopefully leading to revelation: "The mystic's only active role is the decision to pray and the effort to maintain the clarity essential for conveying divine energy: 'One who merits this level is nothing but a channel through which are conducted words from on high'."[48] Valerie Hunt's extension of Matt's assessment of the individual's role in communion with theological space during an act of prayer conforms to the mechanistic proposal: "When a person deeply and quietly prays, his field often reaches its highest...vibration and becomes more coherent with all energy focused in the same direction and in sync. This is a state of immense power.... In the most profound spiritual experiences, the person moves to a state of oneness with divine vibrations, where the identity experiences no separation between the human and divine fields."[49]

What about the characteristic tendencies revealed by parapsychological research? Can these be accounted for through appeal to this mechanism for thought or image transference over inordinately long durations and distances in three-dimensional time and space? Previous declarations presented earlier in the chapter must be revisited in light of the provided physical explanation for psychic transmission. The alternatives considered to be most feasible in explanation of the accumulated data for parapsychological experiments were: (a) "tested subjects...lack authentic psychic gifts," and engaged in random guesswork with variable degrees of success; or (b) "psychic skill" is "an evolutionarily acquired capacity," but "at the infantile stage with respect to human development."

Reconsideration of available data suggests the second contender might be preferred. Like most skills, successful psychic functioning is a product of innate aptitude combined with training to enhance one's natural gifts to their most heightened level. Just as mystics are both born and created—the latter *via* intensive preparation in techniques of contemplation and meditation—those inherently adept at psychic functioning must learn to enhance their skill in achieving resonance with Consciousness. Fatigue during display of ability, or the utility of positive reinforcement by offering timely feedback on good responses, is symptomatic of novices just beginning to learn how to adjust to the arduous exertions of mental and physical discipline needed in exercise of those talents.

This rationale provides a justification for the claim that augmentation of parapsychological skill in functional interaction with the environment is an evolutionary phenomenon. Unclear is whether they are emergent abilities, or constitute a reversion to skills once possessed in abundance, but which have eroded due to increased specialization in brain structure and/or function, particularly in the neocortex, leading to establishment of other talents, such as communication by verbal language.

As a species, *Homo sapiens* are just beginning to acquire the knowledge and technical skill needed to *scientifically* probe the reality of transcendental Energy. Our understanding of physics, chemistry, and biology had to reach a level of sophistication in which mind-brain interactions could be explored. Prior to now, our relative ignorance could only allow for the assertion of a metaphysical self-consciousness and a coincident declaration without proof of its association with a brain-body organism. "Assertions" and "declarations without proof" are now convertible to self-consistent theories amenable to rigorous qualitative, possibly quantitative, examination. The cosmogonic-cosmologic crisis central to speculation about the metaphysical foundations of the universe demands the development of an intentional ability in sentient creatures to skillfully interact with non-physical Energy if those lifeforms are to play a role in resolution.

Can this mechanistic thesis explain other observable features of psychic experience? Consider the implications of comments found in the literature on parapsychology. In *Prayer is Good Medicine*, Larry Dossey remarked: "In the field of parapsychology…it has long been known that 'sheep' (those who believe a distant effect is possible) are more likely to achieve greater results …than 'goats' (skeptics who do not believe in the possibility)."[50] Russell Targ suggested in *Miracles of Mind*: "Seriousness of purpose, trust among participants, and acceptance of psi all enhance remote viewing."[51] Sylvia Fraser mentioned "personality tests indicating that extroversion, openness, confidence and an ability to recall one's dreams correlated with psychic talent."[52] Also, in reference to telepathic communication, "the most striking incidents happened when a loved one was in danger and other forms of communication were blocked. Most often this took place between parent and child or deeply bonded mates."[53]

What all of these anecdotal statements have in common is their utility as *psychological* motivational enhancements. The task of gaining information by nonsensory modes is an arduous one. Strengthening personal commitment to making the effort is a desirable, perhaps even necessary prerequisite. If a person has a deep emotional connection to someone in distress, or simply has an intellectual acceptance of the reality of atypical channels of knowledge attainment, he or she is more likely to attempt to attune their personal energy with the Energy of global consciousness. Strong emotions causally intensify the amplitude of electromagnetic radiation generated by an individual's auric field.

Mental acceptance and acknowledgment of the possibility of expanding one's connection to another out of profound need or desire will stimulate the will's *power*. The latter enhances the attractive gravitational force existing between theological space and ordinary spacetime, precipitating activation of the "push me, pull you" process hopefully culminating in the bestowal of grace. Psychology-based mechanisms have been invoked for a successful life of contemplative prayer leading to mystical insight.[54] Ann and Barry Ulanov offered a summary comment on this theme in *Primary Speech: A Psychology of Prayer*: "Part of building this extraordinarily intimate relationship with the divine is finding what wishes and fears, hopes and desires impel us toward God."[55]

Of immense importance to the metaphysical speculation grounding this proposal for the mechanism of psychic functioning, Dean Radin presented evidence for increases in successful outcomes in precognition studies among multiple individuals working in consort: "Same-sex pairs...tended to achieve null or slightly negative outcomes, whereas opposite-sex pairs produced an effect that was approximately twice that of individuals.... When the pair was a 'bonded' couple, such as spouses or close family members, the effect size was more than four times that of individuals."[56] It should be expected some people would perform better than others. This means that just as aptitude for sports or music or scientific creativity is not uniformly spread across a given population, neither is skill in accessing information through nontraditional methods.

A ready explanation for the gender-based pairings success rate is not at hand, assuming it is not an artifact of data collection, and suggests the need for further investigation to elucidate relevant explanatory factors. The results are significant: resolution of the cosmologic-cosmogonic crisis underlying formation of manifest three-dimensional spacetime requires the cooperation of massive numbers of sentient beings. Evidentiary support for the feasibility of cooperative interaction in traversing the Energy/energy barrier segregating metaphysical reality from physical reality is directly pertinent to the overall speculation.

Remarks made by R. B. Roberts on parameters constituting a theory of parapsychology were called "naïve" because he did not include any demand for a rational mechanism for the operation of psi. In *The Conscious Universe*, Dean Radin prescribed several attributes of a good theory. It would have to: (a) coherently mesh with information available from relevant scientific fields on the nature of Consciousness or self-consciousness; (b) offer a mechanism of operation unconstrained by supposed limits of space and time; (c) explain the variable sensitivity to receipt of information experienced by individuals so that "we are not overwhelmed with information all the time"; (d) include an understanding of the role of phenomenological intentionality in the outcomes of psychic events.[57] Radin also seemed insistent its details be able to respond to specific questions as an indicator of the theory's power:

> How does psi information "get into" a living organism? Are there secret senses we have overlooked? What are the limits of...mental interactions on living systems? Is psi an "invisible" carrier of information...? Is psi interconnectedness related to the unitary sense of self in human beings...? Does psi imply the existence of an even larger unity among human beings, among all sentient creatures, or among all life? Does it serve any evolutionary purpose? Is it an ability, a talent, a throwback to a more primitive sense, or a glimpse of our future?[58]

How do the set of proposals argued for in these pages measure up against these requirements for any theory on the mechanism of operation for psychic phenomena? Self-consciousness involves a harmonious integration of a non-physical Consciousness with a physical brain whose structure and function, the combination of specific tissue groups and neural interconnections, must reflect with precision what is known through investigation into the subjects of neuroanatomy, neurophysiology, and brain chemistry. Ideas developed in these pages take the relevant facts wholesale without alteration: whatever is accepted as fact by the specialists in these areas is admitted into evidence.

The disciplines of psychology and physics, as separately constituted and explored, are currently ill-suited for the vision offered here. They need to be interwoven into psychophysics in order to provide proper grounding for the interaction of mind with brain. Both subjects need expansion: physics by inclusion of a fourth spatial dimension allowing for a mechanism by which material and non-material events could be coherently conjoined; psychology through research into parapsychological and mystical phenomena proffering evidence for the multitudinous ways mind and brain can interact seamlessly.

All investigations into Consciousness appear in agreement: spacetime is not a constraint on operations pertaining to information transfer. The force of gravity is strongest in theological space, allowing for all information to be accessed as though it were confined to "here and now." The electromagnetic force, in the form of both scalar longitudinal and vectorial transverse wave fields, serves as the conduit for transmission of data from the fourth spatial dimension to normal spacetime and back.

The process demands great effort by human intelligence to access global consciousness. Mystics and psychics—by virtue of sustained effort at prayer, meditation, contemplation, and energy concentration—have developed skills allowing for occasional, at this time in history non-commanding, entrance into that vast storehouse of information. It is not a constant and continuous availability because human development has not yet reached the stage where ultimate reliance on grace, on the acceptance by global consciousness for intrusion into theological space, is rendered unnecessary. Therefore, we are unlikely to be "overwhelmed with information all the time." The elaboration of all pertinent principles, laws, and mathematical formulae producing an integrated science of psychophysics must be diligently pursued in order to comprehend how transcendence combines with immanence. Once coherently assembled, it will become clear how mind-matter operations occur such that none are random.

How does information passing through the invisible fourth dimension enter living creatures? The short answer is, "electromagnetically." The ever-present Energy field surrounding living beings—humans, animals, and plants have all been observed to possess an electromagnetic shield extending out from all parts—serves as a sort of lightning rod for information retrievable by entry into theological space.[59] For humans, the most direct path is from this field to the surface of the skin, from there into the sense receptors, from the receptors along the neural net into the brain.

Upon arrival in the brain, information encoded as scalar longitudinal waves of electromagnetic energy gets relayed to the pertinent organizational area for processing. Processing consists initially of decoding: conversion into transverse electromagnetic waveforms. That step is followed by enactment of the three-stage theory of judgments, articulated in chapter six, whereby self-consciousness generates appropriate intellectual, emotional, and behavioral responses formulated as a consequence of applying personal values to the data. Are there limits to the information retrieval process? Not physical ones, since spacetime as experienced is irrelevant to the mechanism of transfer. The limits are psychological: the extent of personal discipline and effort willfully expended in attempting to penetrate theological space.

Is parapsychological ability related to mystical experience? Absolutely. Both are attempts to access the fourth spatial dimension. Mysticism currently proceeds along a more organized, painstakingly diligent course than does the random, immature modality of psychic functioning. This situation will alter over time as human beings make greater systematic efforts at development of psychic abilities. It is interesting to realize children have greater capacity in this regard than adults: in a sense, adults need to re-learn what children know intuitively.[60] Psychic talents are either evolutionary advances or regressions, but either way, they allow for greater coordination of knowledge and effort, vital factors in resolving the cosmologic-cosmogonic crisis.

In *The Self-Aware Universe*, the Goswami's reported on an experiment offering insight into the mechanistic consequences of telepathy. Two people are allowed to establish what they mutually regard as a state of telepathic communication. They are then placed in separate *Faraday cages*, enclosures designed to block penetration by transverse electromagnetic radiation, and individually wired to an electroencephalograph recording their personal brain wave patterns. One of the pair is then shown a flickering light. The result?

The person who is *not* observing the stimulus registers "electrophysiological activity...quite similar in shape and strength to the...stimulated brain."[61]

The significance of the Faraday cages relates to the two available options for comprehending the mechanism of interaction: either transfer of data from stimulated brain to recipient brain is non-electromagnetic in nature, or it is and electromagnetic transfer has an extra-dimensional aspect not inhibited by Faraday cages. Consider the following premises: (a) Faraday cages can block three-dimensional transverse waveform transmissions, but are not effective against transmission of scalar longitudinal waveforms; (b) self-consciousness is not external to either cage given its interactive connection to each human brain; (c) self-consciousness unites with self-consciousness in theological space during telepathic acts; (d) thought transmission occurs through use of longitudinal waveforms. *If* all these conditions hold, *then* the propositional claim that electromagnetic transmissions through theological space provide the conduit for information transfer is plausible.

Do they hold? Criterion (a) is a scientific fact; criterion (b) must be true if self-consciousness is real and other than brain; criteria (c) and (d) are core components of the proposed mechanistic theory of action.[62] Detectable signal transfer by standard electroencephalograph instrumentation is conclusive for rationality of the proposal. Otherwise, one would need to posit one medium to allow for signal transfer between the cages, and a separate mechanism for this medium's conversion into a recordable electromagnetic waveform upon completion of the transfer. It is far more reasonable to postulate the process is entirely electromagnetic. Furthermore, a lack of conscious awareness of the signal by either participant means it was below the requisite intensity level for awareness followed by verbal report, but detectable by the personal omnipresent alertness that *is* self-consciousness.

It is a tacit presupposition of the theory that all parapsychological skills operate according to a single mechanism. Justification for this notion rests partially on application of Ockham's razor. It is far simpler to theorize about a single, encompassing mechanism to explain all aspects of an issue than to increase the level of complexity by invoking separate mind-brain processes for telepathy, precognition, and clairvoyance. Aside from reliance on this epistemological tenet, there are suggestions by other theorists conforming to this idea.

Sylvia Fraser revealed in *The Quest for the Fourth Monkey: A Thinker's Guide to the Psychic and Spiritual Revolution*: "In 1960...Ninian Marshall theorized that our brains' neural circuits might, through resonance, pick up the mixture of possibility and actuality in probability waves as they sweep toward a future. This would be experienced...as precognition."[63] Michael Talbot's *The Holographic Universe* paraphrased a proposal voiced by Robert

Jahn and Brenda Dunne: "Psychokinesis actually involves an exchange of information between consciousness and physical reality, an exchange that should be thought of less as a flow between the mental and the material, and more as a *resonance* between the two."[64]

Each citation invoked "resonance" as a partial mechanistic explanation for precognition or psychokinesis. Disputes over results in parapsychological experiments as being due to operation of telepathy rather than clairvoyance, or psychokinesis instead of precognition, are inconsequential because there is no difference in outcomes dependent upon which skill set is being employed. The fact resonance is deemed a plausible element in comprehension of both psychic forms is evidence in support of a common mechanism. The theses of Marshall, or of Jahn and Dunne, can be subsumed within the present theory, since it also employs the notion of resonance in elaboration of a mechanism of action, unless it can be shown each author's concept of resonance innately prohibits any possibility of overlap.

Chapter Nine: Psychophysics

> We can say that mind is a psychophysical process that encodes information, developing over time.
>
> —Max Velmans, *Understanding Consciousness*

If the nascent field of psychophysics is to fulfill its potential, it cannot be content with qualitative mechanistic descriptions explaining epistemological actions of transcendent pure Consciousness, or of its immanent forms herein characterized as global consciousness and individualized self-consciousness. Independent of the self-consistency underlying all such procedures, or their compatibility with data obtained from sources covering a range of thematic areas, a mature psychophysical epistemology must incorporate quantitative elements where possible. Any offer of pertinent mathematical formulas must be considered tentative at this stage of development; as a relatively new field of investigation, psychophysics will likely go through an extensive growth process requiring many revisions of applicable numerical relationships.

One presupposition will be taken as reasonable *a priori*: the mathematics involved is expected to be relatively noncomplex in form; a suitable example is provided by the famous $E = mc^2$ describing the interconversion of matter with energy. Justification derives from characterization of Consciousness as a coherent field of Energetic alertness. It has no parts; it is a whole whereby its "thoughts" *are* its "actions." Energetic transformations are expected to be complex only when brain operations are central.

If, upon investigation, this assumption turns out to be false, there are two alternatives to consider: (a) the formulas needed for detailing the processes pertaining to Consciousness involve many variables, or (b) mathematics is an entirely inappropriate tool with respect to understanding its operations. In the latter case, psychophysics would be relegated to qualitative declarations of relationships between mind and mind, or mind and brain. It is possible to eventually prove or disprove the proposal's validity only by first making the assumption of formulaic simplicity.

Some parameters commonly found in physics formulas will have no role in symbolic representations of laws pertaining to Consciousness. Distance has no metaphysical meaning, for example: Consciousness is not composed of discrete parts capable of differentiating "this piece" from "that one." On the contrary, Consciousness is monolithically fungible because distinction is not even theoretically possible. With the absence of the notion of Distance, standard concepts—kinetic energy as a function of velocity and forces acting on objects represent two examples—intrinsically associated with distance are inappropriate. Nonetheless, it is meaningful to use "gravitational force" or "electromagnetic force" for events related to theological space. Why?

The role of forces is to signify the mechanism for passage of information between four-dimensional spacetime and transcendent Conscious Space. As both forces have meaning for physical laws, their employment as shorthand descriptions of events in theological space simplifies comprehension through familiarity. What should be understood by these notions is Conscious Energy acts *as if* it were a conventional force of the electromagnetic or gravitational type even though distance is involved in their respective physics formulas and the inverse square law is not pertinent to Consciousness.

As the opening epigraph from Max Velmans indicates, the two principal elements most relevant to any psychophysical scheme pertain to *information transfer* as a function of *time passage*. Four stages are necessarily operative in an epistemological sequence leading to the movement of information: (a) between Consciousness and self-consciousness; (b) between one person's self-consciousness and another's; (c) between self-consciousness and brain; (d) within the singular brain. An accounting of the detailed mechanistic steps for each level must be accessible if psychophysical theorizing is to lead to a pragmatically explanatory endpoint. In addition, a thorough understanding of the nature of Time/time must allow for the three-dimensional experience of time directionality from past to present to future while also incorporating a viable mathematically-based rationale for the hypothesis that the operations of theological space are, in a relativistic sense, Time/time insensitive.

A preliminary simplification of the psychophysics of Consciousness can be offered in the claim that the only two parameters affecting transmission of information in theological space are Energy and Time. Time, when meant in the usual sense to mark the movement of physical bodies, also doesn't apply to Consciousness, but can be re-interpreted to possess significant meaning in terms of knowledge gained. Hence, a statement to the effect that Time moves forward means there is knowledge gain for Consciousness. In what manner, or for what purpose, could this condition be regarded as both plausible and important? Crisis resolution induced by Interpassage will largely depend on information amount and type possessed by Consciousness. Will cosmogonic teleology result in Union or Separation? Will cosmologic teleology result in an open or closed universe?

The knowledge gain for Consciousness is with respect to what sentient creatures in the three-dimensional universe will contribute to the issue. As discussed in the chapter on metaphysics, sentient lifeforms inherently have free will because, as bodies associated with self-consciousness, a derivative combination of dual Conscious Spaces, those "pieces" of self-consciousness can choose to facilitate cosmic openness or closure as an *independent* choice. Consciousness is therefore ignorant as to what will transpire. If knowledge gain becomes an actual condition for Consciousness, then Time can be said to move forward, as it does in the existential universe, because it can only become more apparent *over time* what the likely outcome will be.

It is clear from this thesis Time will move slowly in the conventional sense: millennia in conventional time will be needed to ascertain whether lifeforms will become developmentally capable of involvement in teleology; in which direction their intentionality to act *as a community* will lie; and in the real-time likelihood of successful accomplishment of the intended effort. So sporadically does Time move it is easy to allege that, relatively speaking, Time doesn't move, or that theological space activity occurs with zero Time passage. Thus, like time, Time moves absolutely, but for different underlying causes; unlike time, however, Time does not move relatively.

If Time refers to knowledge gain by Consciousness, and the acquisition of knowledge is a function of information transferred from a transmitter to a recipient, then the conversion of Energy into meaningful information content lies within the domain of Consciousness as its sole activity. Message content and its transmission is said to be an *electromagnetic* Energy transformation because within spacetime the Energy is converted into the normal movement of electrons within the neural network of a brain in a manner of execution no different than in standard electrical circuitry.

Exchange of messages between global consciousness and any particular self-consciousness, or between two person-associated self-consciousnesses, occurs through *gravitational* Energy unifying the two agencies involved in the transaction into a continuum so that transmission of information content involves zero distance. João Magueijo remarked on the unusual nature of the gravitational force in *Faster than the Speed of Light*: "Gravity is different from all other forces in our daily lives in one very important aspect. All other forces seem to be contact forces."[1] Magueijo was wrong. He judged gravity was a force acting on distant objects because he assumed three-dimensional space was the only reality. *Gravity's novelty resides in the fact it is the one concept with an inherent ability to offer an in-the-universe mechanism for continuity of process.*

In sum, Conscious Energy is capable of adopting several atypical forms in addition to its presentation as radiation usually imaged as light waveforms: (a) as alertness, or what might be designated *potential* Energy; (b) as packets of meaningful content conveying a discrete message; (c) as unifying agent for "pieces" of self-consciousness within theological space.

Gravity provides a potent symbol for the cosmogonic-cosmologic crisis best exemplified by the central role for mass density of matter in space as an indicator of universal expansion or contraction in time. In this connection, a history of the mathematics relating to cosmology becomes germane. Einstein tried to make his then-new Theory of General Relativity compatible with the dominant cosmological theory of his time, in which it was accepted that the size of the universe had to be constant because it was assumed to be spatially closed and incapable of alteration.

In *Science and Human Transformation*, William Tiller revealed the most significant implication for this then-accepted view: "According to Einstein's General Theory of Relativity, the geometrical properties of space are related to the density of energy in the universe.... A constant proportional to the vacuum energy density has been defined. It is...the *cosmological constant*, $\Omega = 8\pi\rho G/c^4$."[2] Tiller acknowledged, "The calculated cosmological constant is many orders of magnitude larger than we know it to be." To explain this passage, it should be understood Einstein *invented* the cosmological constant as a means to force-fit astronomical data on mass density of the universe to the *Steady State Universe* theory.

An explanation for the errant magnitude is readily at hand, yet it is one which every physicist or cosmologist has failed to recognize. *Quantitatively, gravity is a weak force in spacetime because it is a potent transcendental force.* In the formula for the cosmological constant, the value of the constant for gravity, 'G,' is taken as its standard spacetime magnitude. In actuality, it is necessary to include its magnitude in theological space as well. When this is done, 'Ω' is more in alignment with known values for 'ρ,' the mass density in universal space, and 'c,' the speed of light in the relative vacuum of space.

The relationship between electromagnetism and gravity is transparent in the fourth spatial dimension—analogous to fourth-Dimensional theological space—where their respective field equations have been theoretically unified into a seamless whole. As has been previously stated, their combined force strength there is equal to their sum in three dimensions. Alternatively stated, gravitation is proportionately stronger than electromagnetism in theological space to precisely the same degree electromagnetic force is stronger than its complementary partner in three dimensions:

$$Gm_1m_2/d^2 + Kq_1q_2/d^2 \text{ (fourth dimension)} = Gm_1m_2/d^2 + Kq_1q_2/d^2 \text{ (three dimensions)}$$

$$Gm_1m_2/d^2 : Kq_1q_2/d^2 \text{ (fourth dimension)} :: Kq_1q_2/d^2 : Gm_1m_2/d^2 \text{ (three dimensions)}$$

By this formulation, a cosmological constant whose magnitude is larger than anticipated indicates gravity's involvement in the psychophysics of the transcendental theological space. Besides establishing global consciousness as a "here-now" with respect to manifest spacetime, gravity facilitates self-arrangement of Consciousness into bundles of Energy in the form of symbol messages.

At least six distinct cognitive modes require mechanistic explanation by elucidation of operative psychophysical principles. These major varieties are: ratiocination, intuition, revelation, telepathy, precognition, and clairvoyance. *Ergotropic* and *trophotropic* were introduced to classify techniques useful for inducing episodes of mystical revelation should the institution of grace allow for such insight. Ergotropic methods are energy intensive; trophotropic modalities are energy minimization routes.

If this classification scheme were expanded to incorporate the remaining cognitive modalities, considered primarily ergotropic would be ratiocination, revelation by contemplation, and clairvoyance. Primarily trophotropic would be intuition, revelation by meditation, and precognition. Telepathy could be classified under either heading with respect to intentionality: sometimes one actively tries to send or receive a message; on other occasions, especially as recipient, an exchange of thought happens without volition. It is reasonable to presuppose this categorical distinction may impact psychophysical factors involved in their use.

The concept of *resonance* was employed to indicate an activation energy barrier seemingly existent between the energy familiar to manifest spacetime and Conscious Energy. The most effective psychophysical means to explain the intrinsic nature of the barrier and its means of being overcome to allow transmission of data invokes the concept of *phase*. When two energy waves intersect, they either augment or diminish the total amplitude, the combined strength of both. These options are known as *constructive*, for enhancement, or *destructive*, for reduction, *interference*.

The same variations exist in theological space when Energy is converted into energy, or the reverse, for transmission into or from three-dimensional spatiotemporal reality. For Energy/energy intersection to be constructive, a prerequisite for message transfer, the radiant fields must be aligned, *in* phase. Since the innate coherent Energy of Consciousness is motionless, the energy of self-consciousness must adjust its phase to the transcendental standard. The *resonance condition* can be said to be attained when: $E = e + \varphi$, where 'E' is the Energy frequency of global consciousness, 'e' is the energy frequency of personal self-consciousness, and 'φ' is the frequency phase adjustment to be performed. *The smaller the correction factor*, 'φ,' *the more likely grace is granted* to the seeker by global consciousness.

Why are the two forms *necessarily* discordant, thus requiring adjustment of the phase relationship? Explanation is provided by reference to a principle derived from the theory of optics. *Snell's Law* relates the directional changes of light moving in different media—through air, water, solid objects, or in a vacuum—to a property of the medium called *index of refraction*. Light alters its path through different media resulting in increases or decreases in speed depending upon resistance encountered as it traverses them. The quantitative ratio of the speed of light through diverse substances has been found to be inversely proportional to the degree of path alteration, the index of refraction, measured relative to a standard, movement in a vacuum, given a refractive index of $n = 1$. Snell's Law is: $\sin \theta_1/\sin \theta_2 = n_2/n_1$, where '$\theta$' is the change in angular direction of light measured in degrees in different media and 'n_i' are the respective indices of refraction of the two media. In descriptive terms, light entering a system with a higher index of refraction changes its direction of motion to a greater degree compared to its direction of movement before entering that medium, and therefore slows down.

Three-dimensional spacetime and theological space are different media to the extent one is physical while the other is metaphysical. Some properties of spacetime are detectable through the senses; SpaceTime is not sensorially detectable. Self-consciousness associated with physical organisms must alter its Energetic properties to compensate for the parameters needed to transmit information in spacetime. When compared to the invariant Energy of global consciousness, they can be said to possess different indices of refraction.

The adjustment in phase undertaken by the Energy of self-consciousness when attempting to penetrate theological space is a measure of the difference in their refractive index. The auric field Energy attributable to every lifeform is uniquely dependent upon the total field strength established by that entity. Every being's refractive index directly correlates with the Energetic field of its self-consciousness, which means every entity must adjust its phase to a different degree before attempting to interact with theological space for the purpose of information transfer.

Aside from total field strength of the Energy of self-consciousness, the principal characteristic needed to merge with global consciousness in order to make feasible information transfer is the precise degree of *coherency*, or organizational orderliness, of that self-conscious Energy. Coherency refers to the intrinsic entropy associated with a being. *TOPP* claimed the development of each sentient creature required an enhancement of their basal energy level and a diminution of internal mental and emotional chaos: "Evolution defined in scientific terms, an apparent disposition of the universe towards lower enthalpy and higher entropy, is in opposition to evolution in mystical terms, an individualized, self-disciplined attempt to gain higher enthalpy and lower entropy."[3]

Enthalpy, or basal Energy, can be correlated with the index of refraction of the auric self-consciousness field. Entropy refers to the focused ability to direct that self-conscious Energy as efficiently as possible in order to possess maximal strength with little distortion due to conflicting internal motivations and goals. For any organism, entropy 'S_o,' and coherency, 'C_o,' are indirectly proportional: $C_o \alpha 1/S_o$.

A reasonable first estimate of relationship is: $\varphi = n_o/C_o$, where 'n_o' is the refractive index attributable to the activity of the self-conscious Energy of an organism, which would be empirically determined relative to the refractive index reference standard attributable to global consciousness *aka* theological space, $n_{TS} = 0$. Larger 'n_o' and smaller 'C_o' values indicate a greater phase adjustment is needed to coalesce and interact with theological space. In such circumstances, the ability of that self-consciousness to send or receive data is lower in comparison with those individuals whose self-consciousness can be said to possess smaller 'n_o' and larger 'C_o' values. The potential variance in both parameters explains the deviation in skills existing among individuals independent of mode used in the apprehension of new information.

How can the variables, refractive index and coherency, be evaluated? If an individual's ability to interact with global consciousness is refractory due to mental strength, refractive index is a measure of *willpower*: one's ability to act and react quickly, decisively, competently, and effectively. The greater a being's strength of will, the lower their refractive index relative to global consciousness.

Coherency, an ability to focus one's mental efforts without distraction or distortion by internal or external influences, is a measure of *integrity*. This term is not meant to be restricted to an ethical-moral sense: adherence to an ethical code expressed by moral behavioral acts without self-contradiction or inconsistency. Integrity should be interpreted to mean that all personality and character traits interact cooperatively, and with unified purpose, without the manifestation of any *quasi*-neurotic tendencies capable of undermining self-expression. Beings with greater integrity innately possess higher quantitative coherency.

Willpower is the fuel for the brain engine allowing creatures to function as they do. It will be shown to be correlatable with 'P/N,' which is the signal-to-noise ratio in message transmission. Refractive index is the reciprocal of this value, the noise-to-signal ratio for any organism: $n_o = N_o/P_o$. Using this definition, a justification why $n_{TS} = 0$ is at hand: no noise is associated with global consciousness, hence $N/P = 0$. It is important to note both power and noise are *scalar* quantities—values are omnidirectional, the same throughout the affected volume of Space/space—a fact consistent with their application to the homogeneous coherency of Consciousness.

Coherency is the antithesis of entropy, and is a function of psychological rather than physical factors. The best way to convey the essence of integrity as a description for personal coherency is to modify a statement concerning the virtue of courage made by Aristotle in *Nichomachean Ethics*: "He who faces and fears the right things for the right motive and in the right way and at the right time, and whose confidence is similarly right, is courageous; for the courageous man in his emotions and actions has a sense of fitness and obeys the law of reason."[4] Adaptation of this perspective for integrity would result in its characterization as proper interplay of values, attitudes, thoughts, emotions, and behavior: the right thing is believed, felt, and done in the right way, to the right degree, at the right time, for the right reasons. Conceivably, it is measurable by test coupled to observation: using psychological profiles to determine whether any given lifeform will act in a manner promoting life and individual development or act in self-inhibiting, self-defeating ways to retard growth in knowledge and ability.

Total coherency 'C_T' would be the arithmetic sum of its parts: individual personality and character traits each assigned empirical values based upon a numerical scale and added together to provide an overall determinant. If they can be placed on a relative scale for all tested subjects, the values can then be normalized to a maximum score indicative of flawless coherency attributable to theological space, $C_{TS} = 1$. Individual scores would then be some decimal value less than one.

Given $\varphi = n_o/C_o$ and reference standard values for the Conscious Energy of theological space, $n_{TS} = 0$ and $C_{TS} = 1$, the phase adjustment for global consciousness is, as anticipated, zero: $\varphi = 0/1 = 0$. For any sentient creature, $n_o = N_o/P_o$ will be a decimal value above zero, and with 'C_o' being a decimal equivalent less than one, the phase 'φ' can take on all values above zero. *The transcendental significance of zero and one become discernible as necessary characterizations of Consciousness* with this formulation.

These equations have as an implicit presupposition the idea that all the cognitive techniques for acquiring information not already available in three-dimensional spacetime have equal accessibility to theological space, a notion theoretically plausible, but not pragmatically supportable by evidence. More likely *a priori*—without accumulating statistics to confirm the contention—individuals engaged in contemplation of an answer for solution of a problem probably obtain insight by gaining rapport with global consciousness more frequently than those who meditate receive revelations of significance. This hypothesis suggests 'φ' is *not* the same for all modalities, but must vary for every individual according to some property, or combination of properties.

One possible parameter is where the cognitive mode appears along the ergotropic-trophotropic energy scale. This is nothing other than a restatement of an assertion expressed previously pertaining to the relevance of this scale in classifying different cognitive modalities: "This distinction would impact all psychophysical factors involved in their use." Why should this continuum be significant? Energy intensive modes, such as deep contemplation, would demand greater resource application by an individual compared to a passive technique, meditation for example. Therefore, more power, 'P,' needs to be generated by an act of contemplation than an act of meditation in order to create a resonance condition with theological space. This was referred to as a "me push" versus "you pull" attitude: it takes more energy to push than to be pulled; hence, $P_{ergotropic} > P_{trophotropic}$.

On the other hand, the likelihood of distraction preventing completion of the goal is greater for passive techniques compared to the focused willpower generated by energetically aggressive methods. In fact, the ease of distraction is a lesson taught to all novices engaged in meditation independent of the meditative technique employed. Recognition of this fact is to say the noise level potential, 'N,' is greater for trophotropic than for ergotropic modalities: $N_{trophotropic} > N_{ergotropic}$.

Since $n_o = N_o/P_o$, and these variables are oppositely affected by location of the cognitive mode along the energetic continuum, the prior formulation, $\varphi = n_o/C_o$, must be amended: $\varphi_{ratiocination} \neq \varphi_{intuition} \neq \varphi_{revelation} \neq \varphi_{telepathy} \neq \varphi_{clairvoyance} \neq \varphi_{precognition}$. Moreover, in general, $N_{ergotropic}/P_{ergotropic}$ will be less than $N_{trophotropic}/P_{trophotropic}$—small number divided by large number *vs.* large number divided by small number—suggesting phase adjustments are smaller for ergotropic compared to trophotropic modes. This conclusion is internally consistent with the hypothesis that information transfer by contemplation, an ergotropic technique, will be successful more often than the attainment of meaningful data by a trophotropic method like revelation.

Although coherency does not change with respect to cognitive modality used for gaining information because it is a function of the person and not the technique, it is recognized and acknowledged there are currently severe pragmatic difficulties with the positing of coherency as an element capable of quantification. First is the precise selection of appropriate personality and character traits. Which ones, among the vast number of theoretical options, most plausibly need to be taken into account to obtain an accurate value for coherency? Second is the methods to be used in testing and observation to properly assess a being's level of attainment in each category. How can these be determined to be useful and reliable indicators? Third is the issue of who is qualified to compile the list and perform the evaluations. Who would make a viable judge possessing the psychological insights required? Fourth is the obvious question as to whether these determinations are, in fact, quantifiable. It is currently unclear whether all, or even some, of the pertinent component parts will lend themselves to numerical assignment.

The concept of compiling personality and character traits for the purpose of assessment for their degree of appropriate coherence within an individual is not new. Psychologists have made such efforts for decades, usually though not always for the purpose of uncovering deviancy from some arbitrary norm as indicators of potential or actual psychopathology. A review of some of the common evaluation instruments reveals none are suitable for employment in the present circumstances: (a) there are qualitatively descriptive tools, such as the Briggs-Myers Personality Indicator; (b) there are instruments designed to be psychopathology detection oriented, such as the Minnesota Multiphasic Personality Indicator; (c) there are catalogs that initially look promising, such as the Five Factor Model, but they omit numerous relevant factors needed for precision measurement of a completely integrated personality and character.[5]

The first-named assessment tool, the Myers-Briggs Personality Indicator (MBPI) developed by Isabel Myers, categorizes individuals through use of generic descriptions of personality types based on the work of Carl Jung. From self-assessment questionnaires demanding forced choice preferences between two options, test takers are classified as predominantly possessing a set of four traits in contrast to their presumptive opposites: (a) *extroverted* or friendly *vs. introverted* or withdrawn; (b) *intuitive* or relationship oriented *vs. sensate* or fact oriented; (c) *thinking* or logical *vs. feeling* or empathic; (d) *judging* or organized *vs. perceiving* or spontaneous. The diagnostic's absence of quantitative data indicates an impossibility for the detection of potential or actual personality composition differences among people having the same group of four characteristics.

The second method, the Minnesota Multiphasic Personality Indicator-2 (MMPI-2), is a revision of the original MMPI, statistical in its output, and also based on self-assessment from prompts on a wide array of topics in a questionnaire format. Its original design intent was as a diagnostic for people having severe personality disorders requiring incarceration in locked ward facilities. Hence, selection of the proper composition for any control group of "normal people" is not only of critical importance, but results in a numerical grading scale innately relative to an average, not single, benchmark standard. Its specialized usage does not succeed in uncovering many of the requisite traits for establishing a degree of overall functional coherency for mentally healthy human beings.

The third technique listed, the Five Factor Model (FFM), was formalized as catchall categorical labels taken from numerous independently created and used evaluation instruments. Apparently, both psychologists and those being assessed frequently employed the same terms descriptive of traits considered pertinent to coherent personality. The Five Factor Model takes its name from the groups of qualities emerging as all-inclusive upon reductive analysis of linguistic tendencies, and diverse elements of character are subsumed within each class. The five broad headings are: *Agreeableness*, *Conscientiousness*, *Extraversion*, *Neuroticism*, and *Openness*. Due to the ubiquity of reference to this theory in the psychology literature as an indicator of the attention it has received, both supportive and critical in type, it will prove useful to display the Model's strengths and weaknesses by tabulating representative qualities included in each class.

Based upon simultaneous observation by ten psychologists over a single weekend of 140 males and 140 females of different age ranges, desirable and undesirable traits were obtained from statistical analysis of their reports. A listing of these qualities provides an illustration of the perceived utility of the Five Factor Model:[6]

Agreeableness	**Conscientiousness**	**Extraversion**
Desirable:	Desirable:	Desirable:
sympathetic	organized	bossy
generous	efficient	sociable
affectionate	reliable	active
Undesirable:	Undesirable:	Undesirable:
stingy	frivolous	reserved
cruel	undependable	quiet
unfriendly	forgetful	withdrawn

Neuroticism	**Openness**
Desirable:	Desirable:
anxious	curious
fearful	clever
emotional	artistic
Undesirable:	Undesirable:
calm	shallow
stable	narrow interest
contented	unintelligent

It is plausible this tabulation reflects the personalities of the observers at least as much as those of the subjects: (a) the reference indicated there were a number of possibly misclassified traits; (b) the questionnaires constructed by different observers measuring "Big Five" factors resulted in more consistent evaluation for characteristics denominated Extraversion, Conscientiousness, and Neuroticism compared to those traits included under the Agreeableness and Openness categories.

Any possible miscategorization of traits may be considered to necessitate a minor refinement of the Model. A more serious charge was presented by critics Walter Mischel and Yuichi Shoda in "Integrating Dispositions and Processing Dynamics within a Unified Theory of Personality: The Cognitive-Affective Personality System (CAPS)," in which they claimed: "The...goal is to characterize...people in terms of a comprehensive but small set of... stable 'behavioral dispositions' or factors on which they differ. These factors are assumed to remain invariant across situations and to determine a broad range of important behaviors."[7]

The presumed stability for expression and continuous usage of traits best characterizing each person professed by adherents of the Five Factor Model goes directly *counter* to the idea of coherency contained herein. If coherency means doing the right thing at the right time, it implies a person's method of expression will be situationally dependent: at one moment anger may be the appropriate response, while in another setting cool detachment is needed, for example. Any static list of favorable or unfavorable traits must fail to capture the notion of integrity underlying coherency of personality and character.

Daniel McAdams rendered the most potent of all criticisms of the Five Factor Model, the essence of which was conveyed by Annie Murphy Paul in *The Cult of Personality*: "An awareness of where a person stands on the five factors is...'crucial information in...evaluation of strangers and others about whom we know very little.' This aspect of the model's utility...is precisely its limitation. Once it has yielded its...speed-read of personality, it cannot offer any deeper...knowledge of an individual's character."[8] It would appear from a subsequent attribution by Paul that McAdams has an appreciation for the inherent complexity of integrity and coherency, and is aware of many of the informational needs required for proper assessment of any human being. She wrote:

> Several additional types of information are required... One of these is *contingent* information: the way an individual responds in different moods or situations.... Other...information cited by McAdams include knowledge about how people's personalities have changed and developed over their lifetimes; the cultures—ethnic, religious, social, professional—in which they are immersed; the current concerns and anxieties they harbor; the political beliefs and value systems to which they subscribe; the desires and goals they hold for the future.[9]

In response to the Five Factor Model and its deficiencies, Mischel and Shoda developed the Cognitive-Affective Personality System and claimed:

> The theory resolves the consistency paradox by conceptualizing...personality...in ways that allow one to model its interaction with situations, and...makes variability of behavior across situations an essential reflection of the stable personality system

> and...its distinctive signature. This...model not only allows an explanation of cross-situational inconsistencies...it...opens a route to predict...how individuals should vary meaningfully as a function of the subjective cues...different situations contain for them.[10]

The last sentence is key to unraveling the complexities of integrity, and reveals why CAPS also represents an invalid view of coherency. Coherency means that although there are behavioral differences displayed by a given individual in diverse settings, these variants are *not* revealing fundamental "cross-situational inconsistencies." Instead, they are actually exemplars of a comprehensive philosophy of values. McAdams' call for knowledge of one's current value system as well as their future projections of desires and goals is a more encompassing, and potentially more fruitful, approach to the problem.

From this recitation of some of the more common systems of personality assessment, one can deduce none are suitable as measures of coherency on their own merits. While traits contained within each instrument may facilitate the eventual compilation of a comprehensive list, the ranking scale employed and the persons able to best serve as assessors are still unanswered questions. Another factor unaccounted for by any system or method is an analysis of the decision making procedures utilized by each individual in determination of the manner of expression of any given set of traits in any particular context. This element must be delineated prior to the fine-grained construction of an integration scale for personality and character coherency.

The inclusion by McAdams of the idea of *future* orientation introduces another vital aspect of psychophysical theory. If accessing information from global consciousness is intended for the purpose of gaining knowledge upon processing and integrating the new data with information already on hand, and if such a sequence is accomplished, it is likely a person's personality and character—expressed inwardly by his or her value system and outwardly by one's behavior—will be modified to a lesser or greater degree as the new information is influential with respect to the old. This is why the construct of personal development—central to the metaphysical teleology permeating this work—is a function of time. Put another way, *individual knowledge grows as time passes*, suggesting one's coherency does also, *resulting in decreases in* 'φ' *values over time.*

The time rate of change of psychophysical parameters must be analyzed and, if possible, subjected to mathematical formulation. Self-consciousness does not undergo development over time; it is always invariant because it is always Consciousness. Mind-brain interaction becomes more efficient as the brain capabilities for distributing and processing data expand in order to keep pace with greater integration of accumulated knowledge. In other words, the rate-determining step for growth in knowledge over time for each individual is a function of his or her ability to organize data into as low an entropic state as possible. Superior organization allows improved retrieval, which in turn produces more efficient assimilation with new incoming information.

One example of how psychophysical parameters change over time is by consideration of *intensity* as it applies to electromagnetic waveforms. This notion is defined in physics as the energy of the wave per unit area over time. Intensity can be related to a statement made in chapter seven: "An increase in concentration over a smaller area of actual or potential activity allows for an accumulation of mental power." Combining the two ideas means intensity is a measure of mental concentration per unit time. As an individual develops superior knowledge organization skills, which reduces entropy and increases coherency, their ability to concentrate at a high level for longer time periods, to become more intense, leads to greater mental power with less distractive noise, with the result that a diminution of refractory power is attained. In this way, both 'C_o' and 'n_o' change in directions necessary to decrease the phase adjustment, 'φ,' needed to resonate with theological space.

Turning to consideration of the content of messages rather than focusing on the parameters affecting the self-consciousness of sender or recipient of the information, it should be noted the coherency of a transmitter or receiver is distinct from the entropy contained in messages. Information entropy is a central construct for epistemological psychophysical formulations. Its utility becomes clear after one comprehends the meaning of entropy as it applies to message content. Jeffrey Mishlove offered, in *The Roots of Consciousness*, a mathematical expression for the energetic requirements associated with data movement from source to destination:

> The number of bits of information in a given situation is defined to be equal to the logarithm to the base two of the total number of possibilities for that situation ($I = \log_2 P$). One bit refers to two possibilities, two bits refer to four possibilities... If knowledge in the form of ΔI bits of information is gained from a system, S, in any measurement or observation, then a minimum amount of free energy, ΔF, is lost from S as given by the psycho-energetic formula: $-\Delta F = kT\Delta I$, where $k = 1.38 \times 10^{-23}$ joules/Kelvin, and T is the temperature, in Kelvin, for S.[11]

Since the central tenet for the metaphysical speculation undergirding this volume postulates formation of the universe from an Energetic interaction of two Conscious Spaces in a process called Interpassage, it is significant that information requires, at minimum, an inherent duality: "One bit refers to two possibilities." This statement confirms the propriety of a prior representation likening the brain to a binary digital computer: both are capable of operation dependent upon whether specific circuits are turned "on" or "off" allowing signal transmission. "Information" is *form-in-action*, conversion of Energy into a cognizable *form* able to *act in* a four-dimensional spacetime universe.

The citation's second important aspect is its interpretation of knowledge as a function of system thermodynamics. Thermodynamics is the branch of chemistry concerning systemic energy changes associated with collections of molecules in a confined space coinciding with, or introducing, alterations in properties such as pressure, volume, temperature, heat capacity, and so on. The passage claims a corresponding loss of free energy occurs in conjunction with any gain in knowledge by a system.

Free Energy, an invention of chemist J. Williard Gibbs at the start of the twentieth century, is a quantity generally symbolized as 'G,' not 'F.' It is a bookkeeping term representing the amount of energy remaining in a system after the molecules have absorbed as much energy as they can, and after they have used some of the energy to reorganize the system contents into a more or less ordered arrangement in comparison to the condition of the collection prior to the information gain. *Information gain increases personal coherency and is vital to any eventual decrease in one's refractive index*. The origin of the aphorism "knowledge is power" is revealed to be as an intuition derived from psychophysical doctrine.

As mentioned in chapter three, thermodynamics is intimately involved in the concept of time directionality. There is time reversibility *only* if there is process reversibility occurring between initial state and final state for any system of molecules irrespective of what occurs on an individual molecular level. Contrarily, time has an inherent irreversible directionality experienced as past to present to future *only* if there is irreversible systemic change.

Acquisition of information is necessarily an irreversible process barring system breakdown, such as occurs with debilitating conditions like strokes or the onset of Alzheimer's disease. Advances in knowledge imply creation of continuously larger groupings of contingent information, lead to truth and, eventually, to Truth. Irrespective of the likelihood of temporal gain or loss of information by a single organism, sentient beings *as a group* irreversibly retain received information as *history*. It is the transition from discreteness/individuality to continuity/community that is needed for time directionality; manifest time, the domain of discreteness, demands transcendental Time, the domain of continuity, for its experiential unfolding.

Cosmogonic continuity and symmetry for the whole—Conscious Spaces as Energetic fields of intelligent alertness engaged in mutually-dependent Interpassage—leads to *cosmologic continuity and asymmetry for the whole*—the manifest universe displays asymmetry in the kinds of matter it contains and in its time-related operations—leads to *chemical-biological discreteness and asymmetry for the part*—every lifeform's individualized character traits and capacities—leads to *subatomic discreteness and symmetry for the part*—the Standard Model's mirror-image redundancy of particle properties—leads back to *cosmogonic continuity and symmetry for the whole*—the equivalence of matter and Energy of Consciousness according to special relativity theory.

Dependency upon Boltzmann's constant and temperature for free energy change in a system upon receipt of information is of immense importance for elucidation of psychophysical principles. The combination 'kT' is a measure of molecular kinetic energy, the energy of motion, a concept definable from another formula, $E = ½\ mv^2$, containing the parameters of particle mass and velocity. Since velocity is defined as the distance traveled per unit time, $v = d/t$, then the energy of particle motion is a function of time. In fact, Einstein's special relativity formula, $E = mc^2$, is a specialized version of the classical expression. Equating the two formulas for kinetic energy gives: $E = 3kT/2 = ½\ mv^2$, or $3kT = md^2/t^2$. This expression supports the contention claiming time attains irreversibility status only for an ensemble of particles—single particles do not possess a temperature, it is solely an attribute of systems—and time has been defined as innately inclusive of the concept of motion.

With respect to the epistemological psychophysical operations involving global consciousness, the implications to be derived from Mishlove's citation confirm the idea standard physical parameters are inappropriate descriptors of activity. Why? It has been posited traditional concepts of space and time do not apply to theological space: it is a fourth-dimensional Spatial domain best characterized as "here-now." If distance and time have no quantitative meaning, motion doesn't occur: d/t is indefinable. This also implies kinetic energy is *not* the nature of global consciousness; it is *static, potential Energy*: self-transformable into information packets accessible to self-consciousness under conditions typical for sensing, thinking, intuiting, psychic functioning, and mystical revealing.

Since the background radiation of normal spacetime has been found to be less than three Kelvin, according to Mishlove's formula 'kT' is less than 4×10^{-23} joules. If this background radiation is a remnant of the amount of Conscious Energy present in global consciousness, a microscopically small amount of Energy is lost in theological space per bit of data transferred. It is uncertain how to determine, even approximately, the total Energy of global consciousness. Could it be done, one would obtain an upper bound limit for its potential information content.

When $-\Delta F = kT\Delta I$ is rewritten as $\Delta G = -kT\Delta I$, it can at once be observed that its formalism precisely mirrors another key thermodynamic relationship, $\Delta G = -nq\Delta E$. This relationship states the change in free energy of a system is proportional to the electrical energy change, 'ΔE,' of the system multiplied by the number of electrically charged particles, 'n,' in motion in that system, where $q = 1.6 \times 10^{-19}$ joules, the amount of charge on a single electron. Taken together, the formulae demonstrate *information transmission is equivalent to electron motion*: information moves from location to location by a process mathematically indistinguishable from one that is electrical. This discovery supports earlier postulation of electromagnetic force as central to knowledge acquisition. It also reifies the positing of the electromagnetic composition of the auric field surrounding the physical body of all organisms as intimately involved in accessing data.

The centrality of thermodynamics for time directionality and information transfer demands two statements from chapter three be revisited. The first, from *The Arrow of Time* by Peter Coveney and Roger Highfield, expressed the assertion, "increasing entropy coincides with time's forward movement"; the second, taken from John Pierce's *An Introduction to Information Theory*, affirmed, "the more we know about what message the source will produce, the less uncertainty, the less the entropy, and the less the information."

In thermodynamics, the definition of entropy, usually symbolized as 'S,' was given as "the energy to reorganize the system contents into a more or less ordered arrangement." Its relation to coherency, with respect to the self-consciousness associated with any sentient lifeform, was expressed. In information theory, entropy also refers to the organizational state of systems; it indicates the maximal number of permutations for symbol characters to be incorporated and interpreted in messages: "Entropy increases as the number of messages among which the source may choose increases."[12]

Pierce provided two related formulations for the calculation of entropy. Using the thermodynamic symbol for entropy—to avoid confusion as well as the unnecessary proliferation of equivalent representations—the first, more generalized form was $S = n \log_2 s$, where 'n' is the number of symbols used in the message, and 's' is the total number of different symbols available.[13] The second representation offered a specific example: "The simplest choice we can imagine is...where there are...two possible symbols...between which the message source chooses repeatedly, each choice uninfluenced by any previous choices." The entropy was defined as $S = -(p_0 \log_2 p_0 + p_1 \log_2 p_1)$ bits per symbol, where 'p_0' is the probability of choosing the first symbol, 'p_1' is the probability of choosing the second symbol, and $p_0 + p_1 = 1$.[14]

The second expression is a subset of a more generalized form with major importance for the metaphysical speculation at the foundational core of this volume. It has been claimed all sentient lifeforms must work in harmony in order to affect the teleology of cosmogonic and cosmologic reality. Critical is the construct of *harmony*: a community of discrete individuals working in consort. Pierce explained why this element was vital: "If separate messages are sent...the total amount of information must be the sum of the amounts of information sent.... To get the entropy of several message sources operating simultaneously...add the entropies of the separate sources."[15]

In other words, $S_T = S_1 + S_2 + \ldots$ or, in compact form: $S_T = \Sigma S_n$, where 'S_T' is the total entropy of transmission, and 'n' the total number of messages sent. If message entropy increases due to transmission of "mixed messages" in attempts to resolve the cosmogonic-cosmologic crisis, total coherency of self-consciousness Energy will decrease even if the individual coherencies are maximal. Symbolically, $S_o \propto 1/C_o$ for the Energy of a single person acting as transmitter, $S_T \propto 1/C_T$ for the total Energy of the community of senders. The combination of 'S_T' (transmitting agents) and 'S_T' (messages sent) will determine whether the agents would: (a) be able to establish resonance with global consciousness to transmit usable information, and (b) be successful in communicating their influence *vis à vis* teleological resolution.

The first formula is *identical* to that found for thermodynamic entropy, $S = K \log_{10} W$, where 'W' is number of possible arrangements for a collection of molecules, and 'K' is an arbitrary mathematical constant—with allowance made for binary notation, $\log_2$, in information theory in contrast to decimal notation, $\log_{10}$, for chemistry. The thermodynamic and information science representations for entropy are equivalent in form to the first psychophysical equation promulgated, *Fechner's Law*, created by Gustav Theodor Fechner, the founder of psychophysics. Ken Wilber gave the form of Fechner's Law in *Integral Psychology* as: "Fechner's law...is stated as $S = K \log_{10} I$ (the mental sensation varies as the logarithm of the material stimulus)."[16]

For the interpretive parallelism to hold, mental sensations, 'S,' must be understood as entropic: capable of causing order or disorder for the system subject to the stimuli. The material stimulus, 'I,' must reflect the number of distinct possible combinatorial states to which the mind-brain may be subject upon receipt of a specific sensation. This analysis suggests an increase in the complexity of the stimulation introduced into the mind-brain leads to a more disordered response unless an ordering principle, *which is neither material stimulus nor mental sensation*, is imposed. This conclusion is consistent with the notion of coherency previously suggested. The ordering principle is the third-order judgment instituted by self-consciousness upon introspection of the meaning of experience.

If a being's set of rules for evaluating experience is sufficiently ordered to induce the *correct* combinatorial state—"the interplay of values, attitudes, thoughts, emotions, and behavior such that the right thing is believed, felt, and done in the right way at the right time for the right reason"—then, since $S = K \log_{10} I$, and $I = 1$ in this circumstance, $S = 0$. This means *the mental sensation doesn't vary once the stimulus is received.* In other words, the sensation appropriately corresponds to the stimulus and the entity's integrity or coherency is perfectly harmonized with the rest of that lifeform's values and attitudes.

The ordering principle was declared to be the domain of the third-order judgment: a functional operation of self-conscious introspection. Chapter six described this process as one involving the accumulation of "all secondary level experiences so as to be able to discern general patterns, the purpose of which is to establish personal values capable of serving as guideposts for future experiences deliberately sought." If introspection on a mental stimulus serves to organize it, to facilitate its coming into alignment with, previously held values, attitudes, ideas, and feelings, the resultant integration lowers the intrinsic entropic state of the entity and provides knowledge leading to the development of what constitutes Truth.

If the incoming stimulus fails to cohere with dispositions already held, entropic disorder increases and the system, the physical organism, will begin to act erratically; its behavior will become more uncertain, less predictable. These ideas are in perfect accord with a statement uttered by Pierre Teilhard de Chardin presented in chapter seven: "A consciousness is that much more perfected according as it lines a richer and better organized material edifice."

What has now been constructed is an epistemological cycle leading to knowledge acquisition in the service of the eventual attainment of Truth: (a) minimized states of personal internal entropy facilitate communication with global consciousness due to grace; (b) the consequence of grace is the receipt of new information by means of some combination of ratiocination, intuition, parapsychological skill, and mystical revelation; (c) greater complexity in the possible meaning obtainable from the incoming data—the more entropy the transmission contains—induces greater urgency in assimilation of the added information, lest the influx destabilize the welfare of the being by exceeding his, her, or its current entropic limit; (d) efficient activation of the third order judgmental process allows for higher levels of integration among the existing corpus of knowledge and the newly arrived information so that the resulting accumulation can be handled in a predictable and useful manner for future activities; (e) any failure to adequately merge old and new data in a timely fashion—inefficient activation of the third order judgmental process—leads to neurotic or, in the extreme, psychotic pathologies; (f) enhancements in coherency between old and new data bits results in a more organized mental, emotional, and physical state; (g) greater internal coherency means decreased internal entropy allowing for enhanced access to theological space in search of new information.

The end is the beginning at a new level, the primary characteristic of a helix: four-dimensional spatiotemporal repetition of a cycle from an elevated perspective. An ongoing iteration of cycles is called *evolutionary progress*. *If* it is true the ability to organize incoming data is a necessity for increasingly rational behavior in response to a wide spectrum of external influences; *if* individualized advancement to the level designated *perfection* is a reasonable criterion for participation in resolution of the cosmogonic-cosmologic crisis underlying creation of the universe; *then* factors of pertinence to absorption, distribution, integration and extrapolation of information are important.

One essential component is the speed at which data is sent from source to recipient. Remarkably, the *same formulation* already encountered becomes germane: "If we send symbols...at a constant rate, the speed of transmission, W, is related to m, the number of different...current values available, by $W = K \log_2 m$.... K is a constant whose value depends on how many successive current values are sent each second."[17] If the source is global consciousness or personal self-consciousness, there are two ways to view the values for 'W' and 'K' independent of how much information, 'm,' is being sent. The first alleges there is no passage of time within theological space—that is, $t = 0$—and both the overall speed of transmission and the rate of transfer become indefinable: the equation is meaningless. The second perspective claims that *because* there is no passage of time, overall speed of transmission and rate of transfer must be considered instantaneous, since the net result is the actual conveyance of data from one agent to another.

If the transfer is from self-consciousness to associated brain or within the brain between neurological structures, time *does* pass during exchange and the mathematical expression is of import. Under these conditions, additional parameters become relevant, as Pierce noted: "The amount of information which can be transmitted is proportional to the bandwidth times the time of transmission.... The number of...different current values which can be sent over a circuit per second is twice the...bandwidth of frequencies used."[18]

The bandwidth enabling information exchange between brain and self-consciousness, or intracranially, has been discussed: mystic revelation and psychic skills both appear to operate in the *theta* band range of four to eight Hertz. As a first approximation, it is reasonable to assume intuitive cognition utilizes the same frequency range. Why? Revelation and intuition both claim immediate clarity and certitude as distinguishing characteristics. Although human psychic functioning does not at present appear to endow the recipient individual with either clarity or certitude, this is not considered inherent: as people mature in their use of these skills, it is likely they will also come to view these qualities as coincidental attributes of the received information.

This assumption must be distinguished from use of *alpha* waves, those in the eight to twelve Hertz range, and *beta* waves, those above twelve Hertz, routinely employed to think about information already present in spacetime, on data not *immediately* originating from global consciousness. Are any facts available bearing on transmission time between self-consciousness and brain, or even within the brain? The authors of *The Self-Aware Universe* reported the results of experiments performed by Benjamin Libet and coworkers:

> Libet and Feinstein measured the time it takes for a touch stimulus on a patient's skin, traveling as a spiked electrical activity along the neuronal pathway, to reach his brain. It was about 0.01 seconds. What Libet and Feinstein found is that the patient did not verbally report being consciously aware of the stimulus for close to half a second. In contrast, a behavioral response by such subjects (such as pushing a button or saying the word "go") takes only 0.1–0.2 seconds.[19]

A time of ten milliseconds needed to transmit an environmental stimulus to the brain, which then takes an additional 100–500 milliseconds for the person to become aware of that influence and articulate one's awareness of it, is consistent with data provided by Evan Harris Walker in *The Physics of Consciousness*: "We write, for the minimum consciousness time interval, $\Delta t = 0.04$ seconds."[20] The brain functions of awareness and report operate at a fraction of the overall speed, as measured by time passage, of self-conscious alertness to a stimulus, which occurs instantaneously with no time delay. The efficiency ratio, defined as $t_{awareness}/t_{alertness}$, is impossible to calculate because $t_{alertness} = 0$ and division by zero is an impermissible arithmetic operation.

The only time estimate available is the time necessary for a synapse to fire, which Walker claimed to be 0.3 milliseconds. Therefore, the temporal gap between awareness in the brain and cognition of stimulus as reported by the extremity involved—eye, ear, nose, *et cetera*—involves a delay from 0.3 milliseconds to somewhere between 10 and 40 milliseconds, depending upon whether Libet's or Walker's data is used. This means awareness at the site of the stimulus precedes awareness localized in the brain by a factor of (10–40 milliseconds/0.3 milliseconds)=33.33–133.33. Since it is certain one can act in a temporal sense only subsequent to a prior cognition, it can be said willful response is a fractional portion of awareness, and awareness is a fractional portion of alertness.

The calculations of Libet and Walker refer to common external stimuli serving as the trigger for self-conscious alertness to transform into awareness associated with brain activity. William Tiller's measurement of the speed of telepathic signals sent between human beings and plants produced a value of 5–25 milliseconds, as recorded in chapter two. One would think a telepathic signal transmitted between an individual's self-consciousness and correlated brain should occur faster than one transferred between two indirectly related objects, like a human being and a plant. The time of transmission internal to a being should have an upper limit of five milliseconds. Can this analysis be refined?

Walker's value of 0.3 milliseconds required to induce the first synaptic firing once a stimulus is perceived at the skin's surface does not directly allow an estimate for transit time between self-consciousness and the skin, but it is plausible the value represents an upper bound limit. If transfer time on a metaphysical level between Consciousness and self-consciousness is so immeasurably fast as to regard it as effectively instantaneous, and transfer time on a physical level between two components involved in triggering a synaptic firing takes 0.3 milliseconds, one can reasonably estimate a transfer time between a metaphysical Dimension and a physical dimension, such as between self-consciousness and skin surface, between the two extremes.

Using values of four to eight Hertz for bandwidth and 0.3 milliseconds for transmission time between self-consciousness and surface sense receptor, "the amount of information which can be transmitted is proportional to" the product of the two numbers after conversion of Hertz—a unit equivalent to one cycle per second—to milliHertz, where one milliHertz is one thousandth of one Hertz: (0.004–0.008 cycles/millisecond)(0.3 milliseconds) = 0.0012–0.0024 cycles of scalar longitudinal electromagnetic Energy transit between self-consciousness and sense receptor before the latter is triggered and sends a signal to the brain that information is incoming.

The miniscule amount of data represented by this calculation supports the proposal of highly efficient coordination between self-consciousness and physical organism: the brain-body system is thrust into awareness by self-conscious alertness at the merest hint of reception of a noteworthy stimulus. In addition, since "the number of distinct, different current values...is twice the...bandwidth," between eight and sixteen data bits of information can be sent per second from self-consciousness to brain, or intracranially to different functional neural networks. This is the value for 'K' in $W = K \log_2 m$. Can an estimate be made for 'm,' the number of symbols capable of transmission?

In the English language, there are 26 letters, 10 numerical digits, and an assorted number of symbols such as punctuation marks, dollar signs, and so on that are regularly used in written or verbal communication. In computer programs written for the English language, a data set of 256 so-called ASCII characters has been established as a standard compilation of frequently used symbols. Each character requires eight current values, on/off states or bits of data, because $2^8 = 256$.

This means (256 characters)(8 bits/character) = 2048 data bits would be used to transmit each ASCII character once. If messages were sent consisting simply of each ASCII symbol, the needed transmission speed would be: W = (8–16 bits/second)($\log_2 2048$) = 88–176 bits/second. Since a message can be any arbitrary length in terms of the number of symbol characters conveyed and the number of times any particular symbol is transmitted, it is impossible to estimate a value for 'm' satisfying every conceivable condition.

It is possible to approach the issue from another direction. If the speed of transmission exceeds a recipient's capacity to absorb, distribute, or integrate it, important parts of a message can be lost entirely causing a fragment to be received and subsequently misinterpreted through assumption the entire data stream was captured. In *The Physics of Consciousness*, Evan Harris Walker defined *consciousness channel capacity* as "the rate at which information flows into consciousness."[21] Based upon the visual perception of pictures and the amount of data each picture typically contains, he estimated this capacity as fifty million bits per second.[22]

The estimate refers to awareness, not to basal self-consciousness. True channel capacity will likely be larger: we select out of a vast quantity of input a much smaller subset as objects for awareness. Once data is received, it must be distributed across the various brain tissues specialized to handle distinct types of data, where it is analyzed at that regionally localized level. Walker devoted several pages in his book to the functional biochemistry of neurons and gave a detailed description of how electrical currents flow in the central nervous system.[23] For technical reasons not requiring elaboration, he selected soluble RNA, *ribonucleic acid*, molecules as the agent conducting electrical signals to the targeted tissues and organs.[24]

After a lengthy explanation of the factors involved in the calculation, summarized in the Notes, Walker revealed the formula he used to determine the consciousness channel capacity: $C = in^2tNf/\tau M$, where 'i' represents the information one synapse sends; 'n' the number of electrons traveling from one synapse; 't' the time a synapse usually pauses before firing; 'N' the total number of synapses in the brain; 'f' the frequency of synaptic firing; 'τ' the time for the electron to make one movement across a single synapse; 'M' the number of molecules of soluble RNA.[25] Substituting calculated values for each variable, Walker concluded C = 47.5 million bits per second, which was in agreement with his initial estimate of fifty million bits per second.

If Walker's value estimates and accounting of relevant factors is taken as complete and reasonably accurate—in other words, if one accepts his value for the channel capacity of human beings—the maximal message length self-consciousness can send to the brain for processing *with a guarantee of not making any errors and without any information loss* is: $m = \text{antilog}_2 (47.5 \times 10^6$ bits/second)/(8–16 bits/second).[26] After doing the arithmetic: $m = \text{antilog}_2 (3.0\text{–}5.9 \times 10^6) = 21.5\text{–}22.5$. At that transmission speed, the human brain can absorb only 22 symbols flawlessly every second.

As has been mentioned, every symbol character requires eight data bits for transcription into electrical signals. Hence, (22 symbols)(8 bits/symbol)

= 176 bits maximally transmitted each second without error. The reciprocal, 1/176, is 0.00568 seconds/bit, or 5.68 milliseconds/bit. We can handle one information bit every five to six milliseconds at maximum efficiency, *which is in agreement with Tiller's experimental determination of the transmission time for a telepathic signal between human and plant.* The time required for self-consciousness to transfer data is *independent of the physical recipient*: human being or plant makes no difference.

This result: (a) indicates the rationale underlying the presupposition of a transmission speed faster than the 0.3 milliseconds needed to induce synaptic firing was incorrect; (b) "proves" self-consciousness is not only distinct from body, but as a form of Consciousness, can interchangeably associate with any creature; (c) demonstrates that Walker's determination for the Conscious awareness channel capacity is likely to be accurate, and that theta band brain waves are involved in information transmission from theological space, since these factors were used in the calculation employing Pierce's formula; (d) means the transit time of five milliseconds is only 2–8 times faster than the ten to forty millisecond time needed for an electrical signal to travel to the brain for processing, if Walker's or Libet's timing data are accurate.

Irrespective of how slowly or how simply constituted a message is when sent by self-consciousness, factors for successful data transfer intrinsic to brain functional operation must be considered. William Tiller, in *Science and Human Transformation*, and John Pierce, in *An Introduction to Information Theory*, gave another formula for channel capacity radically different from the relationship used by Walker. Tiller's notation varied only slightly from Pierce's: $C = \Delta V \log_2 (1+P/N)$, where 'C' is channel capacity, or reception speed, in bits per second; 'ΔV' is channel bandwidth in cycles per second, which must be the same as the effective frequency range of the transmitter; 'P' is the signal power of the receiver in watts, which is equivalent to joules per second, because power equals energy per unit time; 'N' is the noise level of the channel for the receiver, also calibrated in watts.[27]

If the receiver is global consciousness or self-consciousness, its Power must be indefinable, the ratio of an inexhaustible amount of Energy to zero time passage within theological space. The Noise inherent in Consciousness must also be zero: it is defined as a coherent field of alertness, meaning there is an absence of Energy scattering or distortion during transformative acts. Hence, the ratio 'P/N' is another indefinable expression, due to the arithmetic impermissibility of division by zero. The ratio 'P/N" can make no discernible contribution to the formula, which is reduced to $C = \Delta V \log_2 1$.

It has been asserted $n_{TS} = N_{TS}/P_{TS} = 0$ despite an inability to determine an explicit value for 'P' because Consciousness is a noiseless channel. Since the logarithm of 1 is zero independent of the base—$x^0 = 1$ regardless of the value of 'x'—the expression reduces to C = 0. Does this result make any sense? The claim of *zero reception speed for Consciousness is to state motion does not exist within theological space*: a conclusion previously obtained from the idea that "here-now" constitutes the best SpaceTime characterization for this Dimension. The transformation of Conscious Energy to content-rich symbol messages, or the reverse occurs instantaneously; there are *no moving parts*.

Once again, conditions change if the movement of data is confined to the central nervous system. Given physiological sensitivity to local alterations in chemical ion balances in the body as the carrier of electrical charge to, from, and within the brain, it is not guaranteed one will always reach the necessary signal power levels to ensure receipt of an accurate message. There is also an issue concerning occurrence of potential noise distortion of message content during information processing.

Evidence has been presented 'ΔV' refers to theta band brain wave energy operating in the range of four to eight Hertz. Are there any data concerning power and noise levels related to movement of electromagnetic currents in the brain allowing for evaluation of channel capacity reception rate? As an alternative, is anything known about reception rates providing insight into quantifiable power and noise levels? It is plausible self-consciousness selects and transmits to the brain only those parcels of data upon which it is intended we respond to and act on in some manner. Walker's channel capacity would be the channel capacity referred to by Tiller and Pierce in their formula; that is, C = 47.5 million bits/second. Substituting into the Pierce-Tiller equation, $C = \Delta V \log_2 (1+P/N)$, the known values for 'C' and 'ΔV,' and rearranging the terms, gives:

$$\log_2 (1+P/N) = (47.5 \times 10^6 \text{ bits/second})/(4\text{–}8 \text{ cycles/second}) = 5.9375\text{–}11.875 \times 10^6, \text{ or}$$
$$(1+P/N) = \text{antilog}_2 (5.9375\text{–}11.875 \times 10^6) \sim 22.5\text{–}23.5, \text{ and } P/N \sim 21.5\text{–}22.5$$

With respect to energy, *the signal level must be approximately twenty-two times the noise level in order to minimize processing errors*. Any ratio less than this would render the incoming data stream susceptible to potential loss, fragmentation, or misinterpretation during subsequent analysis. *This is the same value as was obtained when the operation was considered from the perspective of the transmitter*. This is to be expected: to ensure an error-free message, sender and receiver must be aligned such that the rate of data being sent is no greater than the capability of the receiver to process it.

Another way to state this criterion is: $n_o = N_o/P_o = 1/21.5\text{–}1/22.5$, or $n_o = 0.0444\text{–}0.0465$, is the *minimum* possible index of refraction a human being can possess. That value would allow him or her to attain a state of resonance with global consciousness at a pre-specified level of internal coherency 'C_o'

given the harmonization condition, $\varphi = n_o/C_o$, while still allowing perfect—that is, no inadvertent error—processing of received information in the brain. Any index of refraction greater than this amount may mean harmonization is still attained, albeit with a greater phase adjustment needed to attain it, but access to theological space now comes with an intrinsic cost: the probability of information loss or mismanagement is increased.

This result explains James Spottiswoode's observation in chapter eight: "Electromagnetic radiation from our Milky Way galaxy...degrade psychic functioning." Atmospheric disturbances alter the electromagnetic conditions of our environment, increase one's noise level, and might lead to a lack of access to theological space due to failure to attain the resonance condition. Even if resonance is still achieved, extraneous electromagnetic radiation can act as baseline noise and lead to distortion of the incoming data. As a result, while information is, as always, flawlessly transmitted between source and recipient self-consciousness *via* theological space, an ability to process and interpret the information obtained through parapsychological techniques is compromised by the brain's reliance on bioelectrical pathways to conduct signals.

The significance of a comment made by William Tiller in *Science and Human Transformation* can now be placed into an appropriate context: "To maximize an entity's capacity to act as a channel...he/she must grow in consciousness to increase ΔV; he/she must increase their ability to generate or conduct a high-power signal, P; and he/she must quiet their fears and uncertainties to reduce the noise power, N."[28] While 'ΔV' cannot be altered—theta band frequency is fixed at four to eight Hertz—increasing 'P' and/or reducing 'N' improves the 'P/N' ratio, or alternatively stated reduces the $n_o = N_o/P_o$ ratio. The result is a lessening of refractory power of brain linked to self-consciousness, and improvement of the probability of establishing a resonance condition through diminishment of phase adjustment, 'φ.' Any technique capable of enhancing signal to noise ratios up to twenty-two would naturally improve information interpretation.

Improvement of 'P/N' ratios is important not only for the inward bound arc of message transfer, but for the outward bound direction as well: the realm of active response to stimuli. Acting in combination with a body, self-consciousness has three epistemological functions: (a) be continuously alert to, and track, all stimuli produced by the surroundings inclusive of sources both external and internal to that organism; (b) from the set of incoming data transmissions, decide which influences the brain should become aware; (c) determine which of the latter should be responded to in some overt manner. In terms of a previous statement: "Willful response is a fractional portion of awareness, and awareness is a fractional portion of alertness."

The role of will, as one component element comprising the ensemble of activities for self-consciousness, was assessed by Evan Harris Walker in *The Physics of Consciousness* in a manner compatible with its characterization in chapter six: "The concept of will demands...the physical laws...must specify a range of potentialities as to what the body could do, and the process that selects...must be clearly outside...the physical laws.... Brain and body will then do one of those things...physics permits so that the thing the mind wills becomes the state of the brain and the action of the body."[29] Walker defined the *will channel capacity* in a manner analogous to that for consciousness channel capacity—as the rate at which information flows into the will—and found it to be 58,666 bits/second.[30]

It was calculated that awareness at the initial site of a stimulus, based upon data for the timing of synaptic firing, occurs at a speed 33.33–133.33 times as rapidly as awareness reaching the brain does, with the latter based upon the time needed for movement of electrical signals through the nervous system. Can the temporal relationship between awareness and willfulness to act be quantified? The ratio of the conscious awareness channel capacity to the will channel capacity is: (47.5 x10^6 bits/second)/(58,666 bits/second) = ~810. If it takes 0.3 milliseconds for data to begin to travel to the brain after a stimulus is received at an outlying sense receptor, it will take 810 times that for data to be acted upon with willful intentionality: (810 x 0.3 milliseconds) = 0.243 seconds. This result is in agreement with Libet's data: "The patient did not verbally report being consciously aware of the stimulus for close to half a second. In contrast, a behavioral response by such subjects...takes only 0.1–0.2 seconds."

The transmission or reception of information through theological space occurs in four stages with different transfer times: (a) zero time passes during data transfer between global consciousness and self-consciousness; (b) five milliseconds pass during data transfer between self-consciousness and sense receptor at the skin's surface; (c) ten to forty milliseconds pass during data transfer between sense receptor and brain; (d) ~250 milliseconds pass during data transfer between skin sense receptor and willful response manifesting in external behavioral display.

Given this data for typical human mind-brain functional interaction, it is additional evidence anthropocentrism is a bankrupt philosophical position, as we have little or no information about comparable values for other species. Do animals take in more or less information per unit time than we do? Do they proportionately act on more or less of that amount than *Homo sapiens*? It is not at all inconceivable that the refinement of their senses enabling them to survive allows them to process more relevant data, and act on it in a more efficient manner, than do ours. This is *not* meant as a justification for a claim that *all* animals are capable of mind-brain interactions superior to our own. It is to posit *a continuum* upon which some species process more data and act on a higher percentage of their capacity than human beings do, while others characteristically are less efficient in their ability to contemplate and respond to everything about which they are aware. Given the substantial number of unknowns, it *cannot* be rationally alleged, "man is the measure of all things."

An individual must accept two doctrinal theses in order to conclude substantive merit is present within the pages of *Discreteness, Continuity and Consciousness: An Epistemological Unified Field Theory* or in its companion volumes—*The Other Perennial Philosophy: A Metaphysical Dialectic* and *Going Against the Flow: An Exercise in Ethical Syncretism.* These are: (a) Consciousness is transcendental reality; (b) its subordinate representation as global consciousness or self-consciousness are important idea constructs for elaboration of a rigorous, self-consistent, comprehensive philosophy.

It is recognized a reader might choose to disregard entirely, or reject as unconvincing: (a) the extensive compilation of evidence accumulated from a large array of authoritative sources disseminated in all three books; (b) the arguments invoked against philosophical materialism dispersed over more than thirteen hundred cumulative pages of text; (c) the expression of novel propositions in favor of the fundamental metaphysical scenario grounding the entire project. Attempting to leave no stone unturned, this work will close with a detailed analysis of the approach taken by Evan Harris Walker, whose *The Physics of Consciousness*, cited numerous times, may be the most well-developed materialist philosophy ever published; one built on contemporary scientific ideas and a rigorous mathematical formalism.

Walker believes the physical universe is causally closed, and constructed his model on the premise that self-consciousness must be completely derived from the operation of conventional physical principles. In his book, several pages were devoted to justification of this position. He considered alternative theses, offered explanations as to how and why these were inadequate, and concluded the Schrödinger wavefunction equation was the only remaining plausible option capable of serving as the foundation for the development of self-consciousness. It is worth reviewing this line of thought because several notable errors and misconceptions will be revealed in the process.

Walker began by investigating whether it was sensible to associate self-consciousness with the manifest universe itself, and posited that if this were so, he would expect to have immediate and direct first-hand experience of all actions occurring in the universe on all scales down to the subatomic level.[31] He believed his assessment to be a self-evident *reductio ad absurdum.* The problem with his reasoning is he never conceived the possibility of existence of a meaningful distinction between the continuity of global consciousness and the discreteness of personal self-consciousness. Any proposal suggesting individualized self-consciousness is identifiable with the existential universe in its entirety is contradictory, and to this extent his argument was valid. It has been an ongoing contention that *global consciousness* is associated with the extant three-dimensional reality. Walker's fixation on the *physics* of self-consciousness apparently did not permit him to contemplate the *metaphysics* of Consciousness.

The previous suggestion was intended to show self-consciousness could not be identified with the continuity of the whole. His next proposition went to the other extreme in alleging self-consciousness could not be equated to the discreteness of the part. In other words, Walker pondered the likelihood of self-consciousness being intrinsically related to subatomic particles, such as a single electron or a quark, and concluded this could not be the case. Why? Each particle was "totally isolated" and could not experience all that a human individual experiences.[32] Isolation in three-dimensions, however, does not lead to the conclusion there is lacking all connection, for it is the function of global consciousness to unify the manifold discrete into an all-embracing continuity. Evidence consistent with existence of a Spatial Dimension/spatial dimension—theological space bridging transcendence with immanence—has been provided. Walker's insistence on material causal closure again did him a disservice.

One of his own statements undermined the notion that Consciousness is fundamentally physical: "Only if...some field...spans the brain and extends over the data-processing events...will we have an acceptable candidate with which consciousness could be identified."[33] *No physical anatomical feature "spans the brain." When objects are dissected, everything is discrete.* It is the nature of reality: *continuity is not a property of three-dimensional spacetime.* An aura has been claimed to "span the brain" and the rest of the physical organism: an Energy field representing non-material self-consciousness. The auric field is directly involved in collection, distribution, and processing of information—those processes Walker attributed solely to the brain.

In contrast to previous statements, Walker expended many words on his argument against electromagnetism as foundational for self-consciousness. It offered several elements of importance for any reasoned response:

> The various regions of the brain are not electrically interconnected... The presence of these local currents produces an electromagnetic field...that...involves the whole brain in a continually pulsing field.... When we go to sleep...consciousness...goes away.... If consciousness resided in the electromagnetic fields associated with the brain's activity...consciousness could ...be modified by subjecting the individual to electromagnetic radiation.... We are bathed in electromagnetic radiation all the time without any effect on our conscious experience.[34]

It is indisputable electromagnetic phenomena play a critical role in the interplay between self-consciousness and brain. Through restriction of self-consciousness to awareness of experience, Walker's argument was flawed. Self-consciousness as identical to, or primarily identified by, awareness is a frequently claimed, though thoroughly erroneous conjecture. Alertness—the anticipatory readiness to acquire information prior to cognitional awareness of its receipt—is *the* inherent feature of all forms of Consciousness.

Walker's assertion was also contrary to another statement written in his book: self-consciousness is not "thinking," not "self-reflection," needs "no words" and "no things," is not "becoming aware," not "perception," not "wakefulness," and not "attention."[35] This earlier passage had it *exactly right*. Temporally prior to acts involving the brain is the metaphysical condition of stillness, alertness, or preparedness to act. *This condition alone* is indicative of Consciousness, and inherent in global consciousness or personalized self-consciousness as well.

Second, his charge, "we are bathed in electromagnetic radiation all the time without any effect on our conscious experience," is both logically faulty and pragmatically untrue. If *always* exposed to electromagnetic radiation, how would we ever be able to discern its effects? We might, with effort, be able to *hypothesize* its absence and *imagine* life without that exposure, but how could we develop a sense of certainty about our conclusions? Walker's confidence in his conviction about the impact of electromagnetic radiation is misplaced.

Moreover, during the exploration of psychic phenomena conducted in chapter eight, and the point was restated earlier in these pages, evidence was cited proclaiming electromagnetic phenomena interfere in parapsychological function. Operating over vast distances from a three-dimensional perspective involves an expenditure of concentration and energy necessary to harmonize with theological space to gain access to the desired information. It is to be expected that sensitivity to atmospheric electromagnetic radiation would be heightened in comparison with the degree of mind-brain sensitivity to stray signals during acquisition of localized stimuli typically involved in personal experience.

Walker, consistent with his commitment to physics as both the origin and terminus of Consciousness, finally arrived at his preference for the key to the point of initiation for self-consciousness. He claimed the locus of mind-brain connectivity was the "intersynaptic cleft" between an axon, the head, of one neuron and a dendrite, the tail, of another.[36] Walker may be correct about the level at which the brain-mind interface might be sought, but the implication drawn was too restrictive. A quantum mechanical formulation for synaptic transmission of data along the neural network in the brain is pertinent to the operational mechanism of mental processes on a physical level *if and only if* quantum mechanics is a viable reference frame for conceptualization. This contention has been called into question based on several lines of analysis.

Reliance on the Schrödinger equation's wavefunction as the quantitative entryway between mind and brain is problematic. Extensive recitation of the philosophic and arithmetic difficulties posed by quantum mechanical theory was presented in chapter five. One consequence for dependence on collapse of the quantum mechanical wavefunction as inevitably leading to induction of resolution of all potential probabilistic results into one realistic outcome is its relationship to time. Peter Coveney and Roger Highfield made the point in *An Arrow of Time* that the abstract form of the Schrödinger wavefunction is time symmetric, whereas physical events invariably display directionality for time:

> When a particular measurement is made, the electron will be recorded as having arrived at one, and only one, position.... This removes the symmetry between the past state of the system (potentiality) and the present state (actuality).... Thus the very act of measurement introduces an arrow of time into the phenomena described by quantum mechanics.... Neither the collapsed wavefunction nor its precursor says anything about the direction of time.[37]

John von Neumann and Eugene Wigner proposed Consciousness causes wavefunction collapse to a single mathematical value for any quantifiable property.[38] Walker agreed: "State vector collapse must have already occurred or...occurs at the time of...observation.... *Observation* is...a euphemism for consciousness."[39] If wavefunction collapse is precipitated by Consciousness, either in the guise of global consciousness or self-consciousness, it must be through a time-dependent Consciousness wavefunction interacting with the Schrödinger wavefunction to introduce a time direction, a history, for all of the particles involved in any manifest event.

Given: (a) the enormous number of particles involved in any action; (b) mathematical approximations must be employed in solving any Schrödinger wavefunction; (c) the precision fit of Newtonian mechanics to most everyday phenomena; (d) the Consciousness wavefunction must do the "heavy lifting" when it comes to explaining experiential directionality for the nature of time; the question to be asked is: Why propose physical operations of particles to abide by quantum mechanical theory? It would seem to be extraneous to the operation. Walker could still use electron movement along neural synapses as the physical mechanism underlying data transfer to/within/from the brain by evoking classical laws of electromagnetism.

Furthermore, the brain's linkage to self-consciousness does not depend solely on physical parameters, but on psychological elements also. It is these latter that are most pertinent to the actual process of acquiring data before it can be transformed and be capable of transmission throughout the brain. To the degree the relevant psychological factors can be quantified is the degree to which psychophysics attains maturity as a sophisticated discipline.

Walker saw no necessity in professing an order of Consciousness beyond human self-consciousness. Consequently, his implicit belief in the potency of quantum mechanics in conjunction with self-consciousness was overstated. It is clear a number of cosmic events unfold in the absence of human observers. Hence, "the only observer who could collapse a conventional wavefunction of the universe must be God," as Coveney and Highfield noted.[40] Little did they realize how correct they were.

Of course, the Newtonian-Einsteinian hybrid is also time symmetric, so the problem of time directionality is not resolvable on the level of physics.[41] In *Critique of Judgment*, Immanuel Kant obliquely defined the solution to the conundrum that is time: "A transcendental principle is one by means of which is represented, *a priori*, the universal condition under which...things can be in general objects of our cognition."[42] Irreversible time must derive from a transcendent domain of irreversible Time, which was instituted with the onset of cosmogonic Interpassage. *Consciousness is the transcendental principle*.

Notes

Chapter One: Skepticism and Knowledge

1. (a) Toshihiko Izutsu, *Sufism and Taoism*, Berkeley: University of California Press, 1983, p.423: "Chuang-tzu often speaks of 'what cannot be evaded' or 'that which cannot be made otherwise.' Everything is necessarily fixed and determined by a kind of Cosmic Will...called the Command of Heaven.... Complete freedom is obtained only when man identifies himself with Necessity itself...the natural course of things and events." (b) Wing-tsit Chan, *A Sourcebook in Chinese Philosophy*, Princeton: Princeton University Press, 1963, p.22–23: "What *t'ien-ming* (Mandate of Heaven) is depends upon one's own philosophy.... Confucianists before the T'ang dynasty (618–907 C.E.) understood it to mean either the decree of God, which determines the course of one's life, or the rise and fall of the moral order, whereas Sung (960–1279 C.E.) scholars...took it to mean 'the operation of Nature which is endowed in things and makes things be as they are'." (c) Lao Tzu, *Lao Tzu*, chapter 16, from Chan, *A Sourcebook in Chinese Philosophy*, p.147–148: "To return to destiny is...eternal Tao. To know the eternal is...enlightenment.... Being one with Nature...is in accord with Tao. Being in accord with Tao...is everlasting."
2. (a) Śaṅkara, *Crest Jewel of Discrimination* (Swami Prabhavananda and Christopher Isherwood, trans.), Los Angeles: Vedanta Press, 1947, p.42: "The sages have spoken of four qualifications... First is...discrimination between the eternal and the non-eternal. Next comes renunciation of the enjoyment of the fruits of action.... Then come the six treasures of virtue, beginning with tranquility. And last...is the longing for liberation." The list of the "six treasures of virtues" on p.43 include: (1) tranquility: detachment of one's mind from objective things by continually seeing their imperfection; (2) self-control: detachment of one's senses from objective things; (3) mental poise: no reaction to stimuli; (4) forbearance: endurance of affliction without complaint or lament; (5) faith: having conviction that teachings of scriptures are true; (6) self-surrender: concentration of intellect upon pure Brahman and keeping it fixed there always. (b) Dwight Goddard, *The Laṅkāvatāra Sūtra* (D. T. Suzuki, trans.), San Rafael: Dawn Horse Press, 1983, p.151: "Transcendental Intelligence rising with the attainment of enlightenment is a permanent nature. The Truth-essence...is discoverable in the enlightenment of all who are enlightened, is realizable as the regulative and sustaining principle of Reality, which forever abides."
3. (a) Rajmani Tigunait, *Seven Systems of Indian Philosophy*, Honesdale: Himalayan Institute of Yoga, 1983, p.104: "*Nyāya* philosophy accepts four independent sources of knowledge—perception, inference, comparison, and testimony—but *Vaiśeṣika* accepts only two—perception and inference." (b) Ibid, p.137: "*Sāmkhya* philosophy accepts only three independent sources of valid knowledge: perception, inference, and testimony." (c) Ibid, p.199-200: "Sources of...knowledge according to *Mīmāmsā*—perception, inference, comparison, testimony, postulation, non-perception." *Perception* is sense-based cognition of obvious qualities; *inference* involves analysis of perceptions to elucidate non-obvious characteristics; *testimony* derives solely from the *Vedas*; *postulation* is the hypothesizing of information used to resolve discrepancies or conflicts in knowledge; *non-perception* means reasoning by virtue of the *absence* of a thing or fact.

4. (a) Ronald H. Isaacs, *Everyperson's Guide to Jewish Philosophy*, Northvale: Jason Aronson, 1999, p.62–63: "Torah is the word of God and is the source of all knowledge. The biblical narratives are not only records of the past but portents of the future.... The commandments are all for the good of man, either to keep man from something hurtful, to remove man from evil habits, or to teach mercy and goodness." (b) Paul Szarmach (ed.), *Introduction to the Medieval Mystics of Europe*, New York: State University of New York at Albany Press, 1984, p.100: "The *Bahir* [*Book of Clarity*] and all Kabbalists after it claim that the true subject of scripture is God himself, that revelation is essentially an act of self-disclosure... Only in the exoteric, public sense is revelation primarily a matter of divine will, teaching the commandments that man is to follow in order to lead the good life."
5. R. C. Zaehner, *The Dawn and Twilight of Zoroastrianism*, New York: G.P. Putnam's Sons, 1961, p.60: "The main doctrines preached by the prophet Zoroaster can, then, be summed up as follows: (1) There is a supreme God who is creator of all things both spiritual and material.... (2) The world as we know it is divided between Truth and the Lie.... (3) The creatures of the Wise Lord are created free—free to choose between Truth and the Lie." The pantheon of gods and goddesses in Greek or Egyptian religion—Zeus, Hera, Apollo, Athena, Set, Isis, Horus, Osiris, Amon-Ra, *et cetera*—are too well known to require authentication by literature citation, as is the building of numerous temples devoted to the paying of homage to one or more of these gods and goddesses by specially trained priests or priestesses serving as intermediaries between the divinities and the general public.
6. Nikolai Berdyaev, *The Beginning and the End* (R. M. French, trans.), New York: Harper and Row, 1957, p.71.
7. Friedrich Nietzsche, *The Birth of Tragedy*, from Walter Kaufmann (trans.), *Basic Writings of Nietzsche*, New York: Random House, 1992, p.96.
8. (a) Eduard Zeller, *Outlines of the History of Greek Philosophy* (L. R. Palmer, trans.; Wilhelm Nestle, ed.), New York: Dover Publications, 1980, thirteenth edition, p.101: "He was a 'philosopher' in the original, modest sense of a man who...seeks after the truth." (b) Warren Ashby, *A Comprehensive History of Western Ethics* (W. Allen Ashby, ed.), Amherst: Prometheus Books, 1997, p.38–39: "Socrates' confession of ignorance had a double meaning. Socrates intended to say that he was not in the possession of truth which he could give to any man.... It was no use to tell a man what he was to find, for each...must understand with his own mind what to accept and reject.... He attempted to convince men of their ignorance... Once aware of their limited knowledge they...would be driven by the desire for truth."
9. Richard Popkin, *The History of Skepticism from Erasmus to Spinoza*, Berkeley: University of California Press, 1979, p.47.
10. Julia Annas, *The Morality of Happiness*, New York: Oxford University Press, 1993, p.201.
11. Mary Beth Ingham and Mechtild Dreyer, *The Philosophical Vision of John Duns Scotus*, Washington: The Catholic University of America Press, 2004, p.39.
12. Richard Popkin and Avrum Stroll, *Skeptical Philosophy for Everyone*, Amherst: Prometheus Books, 2002, p.57.
13. The last possibility—the equation is, in fact, false and the proponent *knows* it to be false, but proclaims it to be True in any case—is not viewed as a contention that anyone would deliberately invoke.
14. Mo Tzu, *Mo Tzu*, chapter 35, from Wing-tsit Chan, *A Sourcebook in Chinese Philosophy*, p.222.

15. Hannah Arendt, *The Life of the Mind*, one volume edition, New York: Harcourt Brace and Company, 1971, volume II: Willing, p.37–38.
16. Richard Popkin and Avrum Stroll, *Skeptical Philosophy for Everyone*, p.55–56.
17. (a) Julia Annas, *The Morality of Happiness*, p.207–208. (b) Sextus Empiricus, *Outlines of Pyrrhonism*, I, 25–26, from Julia Annas, *The Morality of Happiness*, p.209: "For Skeptics began to do philosophy…to decide between appearances and to apprehend which are true and which false, so as to become untroubled…and being unable to decide this they suspended judgment. And when they suspended judgment, untroubledness in matters of opinion followed fortuitously." All the writings of Sextus Empiricus (second century C.E.) have been translated into English: R. G. Bury, Cambridge: Loeb Classical Library, four volumes, 1933-1947.
18. (a) David Hume, *A Treatise of Human Nature* (Ernest C. Mossner, ed.), London: Penguin Books, book I, part iv, section 4, 1985, p.275: "The fundamental principle of that philosophy is the opinion concerning colors, sounds, tastes, smells, heat and cold; which it asserts to be nothing but impressions in the mind, derived from the operation of external objects, and without…resemblance to…qualities of…objects." (b) David Hume, *An Enquiry Concerning Human Understanding* (Eric Steinberg, ed.), Indianapolis: Hackett Publishing Company, 1977, p.10-11: "By the term *impression*…I mean all our more lively perceptions, when we hear or see or feel or love or hate or desire or will…. Impressions are distinguished from ideas, which are the less lively perceptions, of which we are conscious, when we reflect on any of those sensations or movements above mentioned."
19. George Santayana, "Reason in Art," *The Life of Reason*, Amherst: Prometheus Books, 1998, p.343: "Knowledge transcends sensation by rising to a supersensuous plane, the plane of principles and causes. These principles and causes are…the intelligible or the real world; and the sensations, when they have been so interpreted and underpinned, are what we call appearance."
20. (a) Richard Tarnas, *The Passion of the Western Mind*, New York: Ballantine Books, 1991, p.118. (b) Daniel C. Matt, "Ayin: The Concept of Nothingness in Jewish Mysticism," from Robert Forman (ed.), *The Problem of Pure Consciousness: Mysticism and Philosophy*, New York: Oxford University Press, 1990, p.123–124: "*Creatio ex nihilo* first appears in second-century Christian literature, evoked by the confrontation with Gnostic heresy and Greek philosophy. It represents a denial of the prevailing Platonic notion that creation was out of eternal primordial matter, a notion that compromises the sovereignty of God…. Its denial was tantamount to the undermining of revealed religion. Though…*creatio ex nihilo* was not indigenously Jewish, under the influence of Christian and Moslem thinkers, it penetrated Jewish philosophical and religious circles."
21. Nikolai Berdyaev, *The Beginning and the End*, p.43: "Truth has two meanings. There is truth in the sense of knowledge of reality and there is truth which is reality itself. Truth is not only an idea, and a value, it is also an entity, something which exists."
22. (a) Aristotle, *Metaphysics* (Richard Hope, trans.), Ann Arbor: University of Michigan Press, 1960, 1005b, p.67–68: "And the surest principle of all is that about which it is impossible to be mistaken; for such a principle does not rest on assumptions and must be known…. Such a principle is the most certain of all and we can formulate it thus: 'It is impossible for the same thing at the same time to belong and not to belong to the same thing and in the same respect'." (b) Ibid, 1011b, p.83: "And the possibility of a middle between contradictories is excluded; for it is necessary either to assert or deny one thing of another. This is clear from the definition of truth and falsity; for to deny what is or to affirm what is not is false, whereas to affirm what is and to deny what is not are true; so that any

judgment that anything is or is not states either what is true or what is false. Hence either what is, is affirmed or denied, or else what is not is affirmed or denied."

23. René Descartes, *Discourse on Method*, from René Descartes, *Philosophical Essays* (Laurence J. LaFleur, trans.), New York: Macmillan Publishing Company, 1964, p.24–25. In addition to *Discourse on Method*, *Philosophical Essays* includes Descartes' *Meditations Concerning First Philosophy* and *Rules for the Direction of Mind.*
24. Sören Kierkegaard, *Concluding Unscientific Postscript* (Howard Hong and Edna Hong, ed. and trans.), Princeton: Princeton University Press, 1992, p.317: "If the I in *cogito* is understood to be an individual human being, then the statement demonstrates nothing: *I am* thinking ergo I am, but if *I* am thinking, no wonder...that I *am*... What, then, is the concluding formula supposed to mean? Indeed, there is no conclusion, for then the statement is a tautology."
25. Abū Hāmid Muhammad al-Ghazālī, "Deliverance from Error and Attachment to the Lord of Might and Majesty," from W. Montgomery Watt (trans.), *The Faith and Practice of al-Ghazālī*, Oxford: Oneworld Publications, 1998, p.19–20: "Sure and certain knowledge is that knowledge in which the object is disclosed in such a fashion that no doubt remains along with it, that no possibility of error accompanies it, and that the mind cannot even entertain such a supposition. Certain knowledge must...be infallible; and this infallibility or security from error is such that no attempt to show the falsity of the knowledge can occasion doubt or denial."
26. Hans Küng, *Does God Exist?* (Edward Quinn, trans.), Garden City: Vintage Books, 1981, p.7.
27. Francis Bacon, *Novum Organum*, book I, aphorisms 19 and 22 from Robert M. Hutchins (editor-in-chief), *Great Books of the Western World*, Chicago: Encyclopedia Brittanica, 1952, volume 30.
28. (a) M. Friedlander, introduction to Moses Maimonides, *The Guide for the Perplexed* (M. Friedlander, trans.), second edition, New York: Dover Publications, 1956, p.xxv: "According to his [Maimonides'] opinion, man should only believe what he can grasp with his intellectual faculties, or perceive by his senses, or what he can accept on trustworthy authority." (b) Roy Perrett, *Hindu Ethics: A Philosophical Study*, Honolulu: University of Hawaii Press, 1998, p.67: "The Cārvāka admitted only perception as a valid means of knowledge... Vaiśeṣika and the Buddhists admitted both perception and inference as *pramānas* [valid means of knowledge]. Sāmkhya allowed testimony as a third means. Vaiśeṣika added comparison. Mīmāmsā added postulation to these four. Advaita Vedānta added...a sixth source, noncognition." As a method, "noncognition" allows for reasoning by virtue of the *absence* of a thing or fact. Both comparison and noncognition are dependent upon perception. Note the absence of intuition or revelation as acceptable modalities, perhaps surprising omissions from religious epistemological theory. (c) Chapter one, note #3, p.291.
29. Augustine, *On Free Choice of the Will* (Thomas Williams, trans.), Indianapolis: Hackett Publishing Company, 1993, book 2, chapter 12, p.55.
30. Edmund Husserl, *Formal and Transcendental Logic* (D. Cairns, trans.), The Hague: Nijhoff, 1969, section 106, p.281, from Dermot Moran, *Introduction to Phenomenology*, London: Routledge, 2000, p.96.
31. David Hume, *An Enquiry Concerning Human Understanding*, p.102–103.
32. (a) Bernard J. F. Lonergan, *Insight: A Study of Human Understanding*, New York: Harper and Row, 1978, p.79: "Empirical method involves four distinct elements, namely (1) the observation of data, (2) insight into the data, (3) the formulation of the insight or set of insights, and (4) the verification of the formulation." (b) Alan M. Laibelman, *The Other Perennial Philosophy: A Metaphysical Dialectic*, Lanham: University Press of America,

200, p.xxi: "A four step process that transforms facts into knowledge: data acquisition, data classification, data analysis, data extrapolation."

33. René Descartes, *Discourse on Method*, from René Descartes, *Philosophical Essays*, p.25.
34. Ibid, p.25–26.
35. Paul Weiss, *Being and Other Realities*, Peru: Open Court Publishing Company, 1995, p.88.
36. Abraham Joshua Heschel, *Man is not Alone*, New York: Farrar, Straus, and Giroux, 1951, p.12.
37. Donald Rothberg, "Contemporary Epistemology and the Study of Mysticism," from Robert K. C. Forman, *The Problem of Pure Consciousness*: *Mysticism and Philosophy*, p.175–178.
38. I write as a scientist with a doctoral level degree in chemistry and over twenty years of research in organic chemistry, principally in the pharmaceutical industry. I am personally acquainted with hundreds, and indirectly with several thousands, of active scientists and know how they think about the significance of their work.
39. Charles Hartshorne, *The Divine Relativity*, New Haven: Yale University Press, 1964, p.6.
40. John Locke, *An Essay Concerning Human Understanding* (Peter H Nidditch, ed.), New York: Oxford University Press, 1975, book IV, chapter I, sections 2, p.525.
41. John Stuart Mill, *On Liberty*, from John Stuart Mill, *On Liberty* and *Utilitarianism*, New York: Bantam Books, 1993, p.22–23.
42. William James, *The Meaning of Truth*, from William James, *Writings* 1902–1910, New York: Library of America, 1987, p.959.
43. Dermot Moran, *Introduction to Phenomenology*, p.188: "Not many philosophers have wanted to follow Husserl's claim that intuition...validates all other kinds of knowing.... We can in fact be deeply wrong about what we think is immediately evident to us."
44. (a) Bernard J. F. Lonergan, *Insight: A Study of Human Understanding*, p.430: "The contribution of science and scientific method to philosophy lies in a unique ability to supply philosophy with instances of...heuristic structures which a metaphysics integrates into a single view of the concrete universe." (b) Abraham Joshua Heschel, *God in Search of Man*, New York: Farrar, Straus, and Giroux, 1955, p.10: "Philosophy of religion has to be pursued in two ways: as radical understanding of religion in terms of its own spirit and as a *critical reassessment of religion from the point of view of philosophy*. It represents an effort of religion to justify its claims; to set forth its *validity*, not merely its *relevance*." (c) Jos DeCorte, "Saint Anselm of Canterbury on Ultimate Reality and Meaning," *Ultimate Reality and Meaning*, 1989, 12(3): 177–191, p.179: "If the main principles of faith were taken as axioms from which deductions were made according to logical rules, a religious philosophy was conceivable." Disciplines primarily creatively aesthetic—art, music, and dance—are derivative from a combination of science, religion, and philosophy unique to each.
45. Abraham Joshua Heschel, *God in Search of Man*, p.34.
46. In ancient India, these three paths were also known, and were designated as *karma yoga* [the path of action] for science, *bhakti yoga* [the path of emotion] for religion, and *jnāna yoga* [the path of knowledge] for philosophy.
47. Teresa of Avila, *The Life of Saint Teresa of Jesus*, fifth edition, London, 1916, chapter xii, p.191–192, from Joseph Maréchal, *The Psychology of the Mystics* (Algar Thorold, trans.), Mineola: Dover Publications, 2004, p.191–192: "For when our Lord suspends the understanding and makes it cease from its acts, he puts before it that which astonishes and occupies it; so that, without making any reflections, it shall comprehend in a moment more than we could comprehend in many years with all the efforts in the world."

48. Hans Küng, *Does God Exist?*, p.18: "The two cognitive powers (the natural light of reason and the light of faith stemming from divine grace) and the two spheres of knowledge (the realm of clear and distinct ideas and the realm of revelation) are absolutely distinct: there is neither mingling of the two nor any mutual opposition, but a parallelism and a proper distance between them." Descartes tried to resolve this issue, improperly in my view, by steering a middle course. I believe these two modes of learning are portions of a continuum of methodologies.
49. Dermot Moran, *Introduction to Phenomenology*, p.106. Husserl formally launched the method of phenomenology through the publication of *Logical Investigations* in 1900. He intended it to study features of common, everyday experience so as to link the workings of the internal conscious mind with the external environment available through sensory input. Phenomenology was later adapted to investigate mystical experiences.
50. Rom Harré, *One Thousand Years of Philosophy*, Oxford: Blackwell, 2000, p.256–257: "How is what is immanent in consciousness related to what is transcendent to it?... The *noetic* investigation of the contents of the mind is a study of those forms of consciousness in which objectivities are presented experientially. By 'bracketing'...we may be able to discern how the objectivity of...a thing...is presented in consciousness. A *noematic* investigation is a study of the objectivities themselves, as they are re-presented in reflection. Experiencing...and reflecting...are two different acts.... The contents of these acts are different.... Whatever serves as the unanalyzed ground...is *doxetic*.... The second group of terms to grapple with includes 'epoché,' which is closely related to the idea of phenomenological reduction.... To grasp how objectivities present themselves in consciousness as such, I must 'bracket' the fact...I unreflectingly take them as objective. The phenomenological reduction abstracts the marks of each taken-for-granted 'reading' of experience. After a sequence of reductions, these marks are wholly bracketed, and the pure doxetic phenomena remain." Husserl invented the terms *noetic*, *noematic*, and *doxetic*, and borrowed terminology for the major steps. By *bracketing* is meant extracting some chosen component of an actual experience from its context so one could focus the point of analysis. *Reduction* is meant in much the same way as it is commonly used in science and mathematics: to continuously remove the more complex elements of an abstraction until the core is reached incapable of being reduced further; that is, arrival at the lowest common denominator (to use mathematical language), or at its subatomic level (to employ the language of physics). *Epoché*, in the original radical skeptical sense of "maintaining neutrality without judgment" is crucial to the reductive procedure: to be able to continuously remove the higher levels of complexity, no matter how difficult they may be to discern, in order to reach the absolute core of the experience, one must remain uncommitted to the retention of any particular aspect of that experience on some *a priori* grounds.
51. Charles Sanders Peirce, "How to Make Our Ideas Clear," *Popular Science Monthly*, volume xii, 1878, p.286.
52. William James, *Pragmatism*, from William James, *Writings 1902–1910*, p.512.
53. Johann Gottlieb Fichte, *Foundations of Transcendental Philosophy* (Daniel Breazeale, ed. and trans.), Ithaca: Cornell University Press, 1992, second introduction, p.90.
54. Heinrich Dumoulin, *Zen Buddhism: A History—India and China* (James W. Heisig and Paul Knitter, trans.), New York: Macmillan Publishing Company, 1988, p.260: "The Rinzai school aims at a sudden experience [*satori*] that affects a profoundly reorienting conversion.... For the Soto school the practice of zazen is primary. In addition it speaks of a quiet experience of depth that is carried through one's daily activities and is in no way inferior to sudden satori."

55. Even the most ardent admirer of Moses, Buddha, Jesus, Muhammad, or other similarly esteemed personages would be hard-pressed to credibly claim any of these learned Truth in exhaustive detail in every conceivable domain of information-gathering during his or her encounters with an omniscient transcendent source.
56. Richard Popkin and Avrum Stroll, *Skeptical Philosophy for Everyone*, p.131.
57. (a) Immanuel Kant, *Critique of Pure Reason* (F. Max Muller, trans.), Garden City: Doubleday and Company, 1966, p.28. (b) Ibid, p.39–40: "As the external sense gives us nothing but representations of relations, that sense can contain in its representation only the relation of an object to the subject, and not what it is inside the object by itself."
58. Nikolai Berdyaev, *The Beginning and the End*, p.29: "In Kant's philosophy a turn is definitely made towards a philosophy of the ego, of the subject, as against a philosophy of the world, of the object. This presupposes to begin with a dualism between appearance and the thing-in-itself, between the order of nature and the order of freedom." Much discussion has ensued about this "turn"—generally referenced in conformance to Kant's own designation as a "Copernican Revolution." It is difficult to refrain from commenting upon the absurd irony of this label: Copernicus demonstrated the Earth is not the center of the universe, but simply one luminary among a vast number orbiting our sun; Kant, however, proclaimed reality as dependent on human knowledge. Human epistemology, according to Kant, figuratively speaking "makes the world go round."
59. (a) Charles Hartshorne and William L. Reese, *Philosophers Speak of God*, Amherst: Prometheus Press, 2000, p.58: "Aristotle contends...substance is the sole basic category, while...nine subcategories—quantity, quality, relation, place, time, position, state, action, and affection—are attributes of substance." (b) Aristotle, *Metaphysics*, book delta, 1017a, p.98: "The varieties of essential being are indicated by...categories; for in as many ways as there are categories may things be said to 'be.' Since predication asserts sometimes what a thing is, sometimes of what sort, sometimes how much, sometimes in what relation, sometimes in what process of doing, or undergoing, sometimes where, sometimes when, it follows that these are all ways of being."
60. Immanuel Kant, *Critique of Pure Reason*, p.262–263.
61. Richard Tarnas, *The Passion of the Western Mind*, p.398–399.
62. John Locke, *An Essay Concerning Human Understanding*, book II, chapter VIII, sections 9-10, p.135: "These I call *original* or *primary qualities* of body, which I think we may observe to produce simple *ideas* in us, solidity, extension, figure, motion, or rest, and number. Secondly, such *qualities*, which in truth are nothing in the objects themselves, but powers to produce various sensations in us by their *primary qualities*...as colors, sounds, tastes, etc. These I call *secondary qualities*."
63. (a) Plato, *Parmenides*, from Benjamin Jowett (trans.), *The Works of Plato*, four volumes, New York: Tudor Publishing Company, undated, volume IV, p.322: "But may not the ideas, asked Socrates, be cognitions only, and have no proper existence except in our minds, Parmenides?... The more probable view, Parmenides, of these ideas is, that they are patterns fixed in nature, and that other things are like them and resemblances of them; and that what is meant by the participation of other things in the ideas, is really assimilation to them." (b) Richard Tarnas, *The Passion of the Western Mind*, p.11: "The exact degree to which the Ideas are transcendent rather than immanent—whether they are entirely separate from sensible things, with the latter...imperfect imitations, or whether they are in some manner present in sensible things, with the latter essentially sharing in the Ideas' nature—cannot be determined from the...references in the different dialogues. Generally speaking, it seems that as Plato's thought matured, he moved towards a more transcendent interpretation."

64. Lenn E. Goodman, *Avicenna*, London: Routledge, 1992, p.56: "To answer Aristotle's objections to the self-subsistence of the Forms, the Neoplatonists placed them within *Nous*, the Aristotelian principle of intelligence...now identified with Platonic Being, and hence with the eternal Forms. Since the Ideas are now identical with the mind that knows them, they are no longer isolated but alive, in and as divine Intelligence."
65. (a) John Locke, *An Essay Concerning Human Understanding*, book II, chapter XIV, section 3, p.182: "'Tis evident to any one who will but observe what passes in his own mind...there is a train of *ideas*, which...succeed one another in his understanding, as long as he is awake." (b) David Hume, *A Treatise Of Human Nature*, p.261: "The only existences of which we are certain are perceptions, which being immediately present to us by consciousness, command our strongest assent and are the first foundation of all our conclusions."
66. Richard Tarnas, *The Passion of the Western Mind*, p.402.
67. David Hume, *An Enquiry Concerning Human Understanding*, p.103: "There is another species of skepticism, *consequent* to science and enquiry, when men are supposed to have discovered...the absolute fallaciousness of their mental faculties, or their unfitness to reach any fixed determination on all those curious subjects of speculation, about which they are commonly employed. Even our very senses are brought into dispute...and the maxims of common life are subjected to the same doubt as the most profound principles or conclusions of metaphysics and theology."
68. David B. Burrell, *Knowing the Unknowable God*, Notre Dame: University of Notre Dame Press, 1986, p.32: "Truth, then, is not a property of things or of propositions, but results from an activity called judging; a second-order activity in the realm of knowing which completes the inquiry by ascertaining whether what we maintain is in fact the case."
69. Kurt Gödel, "Über Formal Unentscheidbare Sätze der *Principia Mathematica* und Verwandter Systeme," *Monatshefte für Mathematik und Physik*, 1931, 38: 173–198. The title translates as: "On Formally Undecidable Propositions in *Principia Mathematica* and Related Systems."
70. (a) Aristotle, *Nichomachean Ethics* (J. E. C. Welldon, trans.), Amherst: Prometheus Books, 1987, book VI, chapter vi, p.193–194: "Science is a mode of conceiving universal and necessary truths.... Demonstrable truth or science in general implies first principles... The only possible conclusion is...these principles are apprehended by intuitive reason." (b) Friedrich Nietzsche, *On the Genealogy of Morals*, third essay, section 24, from Walter Kaufmann, *Basic Writings of Nietzsche*, p.587–588: "There is no such thing as science 'without any presuppositions'; this thought does not bear thinking through, it is paralogical: a philosophy, a 'faith,' must always be there first of all, so that science can acquire from it a direction, a meaning, a limit, a method, a *right* to exist." (c) Ronald H. Isaacs, *Everyperson's Guide to Jewish Philosophy*, p.74: "In the preface to *Or Adonai* [*Light of God*]...Hasdai Crescas (1340-??) argued that it is absurd to speak of a divine commandment to believe in the existence of God, since such a belief cannot be a commandment itself, but rather a presupposition for any commandment.... Crescas considers as presuppositions God's existence, unity, and incorporeality." (d) Abraham Joshua Heschel, *God in Search of Man*, p.20: "Some of the basic presuppositions of Judaism cannot be completely justified in terms of human reason. Its conception of the nature of man as having been created in the likeness of God, its conception of God and history, of the election of Israel, of prayer and even of morality, defy some of the realizations at which we have honestly arrived at the end of our analysis and scrutiny."
71. (a) Robin George Collingwood, *An Essay in Metaphysics*, Oxford: Clarendon Press, 1940. (b) Arthur O. Lovejoy, *The Great Chain of Being*, Cambridge: Harvard University Press,

1964, p.7: "It is the beliefs which are so much a matter of course that they are rather tacitly presupposed than formally expressed and argued for, the ways of thinking which seem so natural and inevitable that they are not scrutinized with the eye of logical self-consciousness, that often are most decisive of the character of a philosopher's doctrine, and still oftener of the dominant intellectual tendencies of an age."

72. (a) Paul Trainor, "History and Reality: R. G. Collingwood's Theory of Absolute Presuppositions," *Ultimate Reality and Meaning*, 1984, 7(4): 270–287, p.283. (b) John Locke, *An Essay Concerning Human Understanding*, book I, chapter II, sections 2–4, p.49: "There is nothing more commonly taken for granted, than that there are certain principles both *speculative* and *practical* universally agreed upon by all mankind: which therefore they argue, must needs be the constant impressions, which the souls of men receive in their first beings, and which they bring into the world with them, as necessarily and really as they do any of their inherent faculties. This argument, drawn from *universal consent*, has this misfortune in it, that if it were true in matter of fact, that there were certain truths, wherein all mankind agreed, it would not prove them innate, if there be any way shown, how men may come to that universal agreement, in the things they do consent in.... But, which is worse, this argument of universal consent, which is made use of, to prove innate principles, seems to me a demonstration that there are none such: because there are none to which all mankind give a universal assent."
73. Alfred North Whitehead, *Adventures of Ideas*, New York: Macmillan Company, 1933, p.287.
74. Hans-Georg Gadamer, "The Universality of the Hermeneutic Problem," from D. Linge (ed.), *Philosophical Hermeneutics*, Berkeley: University of California Press, 1977, p.9.
75. Plato, *Charmides*, from Benjamin Jowett, *The Works of Plato*, volume IV, p.37.
76. Paul Trainor, "History and Reality: R. G. Collingwood's Theory of Absolute Presuppositions," p.283. The direct quote is from: Robin George Collingwood, *An Essay on Metaphysics*, p.33.
77. Ervin Laszlo, *The Interconnected Universe: Conceptual Foundations of Transdisciplinary Unified Theory*, River Edge: World Scientific, 1995, p.8: "[Ilya] Prigogine spoke with good reason of a 'divergence property' basic to the evolutionary process. But he would also need to speak of a 'convergence property,' for systems not only diversify in space and time, they also converge within higher-level systems."
78. Aristotle, *Metaphysics*, book gamma, 994a, p.36.
79. Vivekananda, "Practical Vedānta, Part IV," *Jnāna Yoga*, New York: Ramakrishna-Vivekananda Center, revised edition, p.275.
80. David Hume, *An Enquiry Concerning Human Understanding*, p.11.

Chapter Two: Anthropocentrism

1. Immanuel Kant, *Critique of Pure Reason* (F. Max Muller, trans.), Garden City: Doubleday and Company, 1966, p.32.
2. Alan M. Laibelman, *The Other Perennial Philosophy: A Metaphysical Dialectic*, Lanham: University Press of America, 2000, p.151–152: "Time is to Form as Space is to Content. Or alternatively...Time relates to Content the same way...Space relates to Form. Content Exists...Form Changes. Space reveals Essence as Absolute, while Time reveals Essence as Relative. Space regards Reality as Being...Time regards Reality as Becoming.... The principle of Inherent Duality claims...proper understanding of Space would demonstrate the Absolute Being of two transcendent Realities as the Content of universal existence. In a similar vein, Inherent Duality posits that a proper understanding of Time would verify the Relative Becoming of two transcendent Realities as the Form of universal existence."

3. Kathleen Freeman, *Ancilla to Pre-Socratic Philosophers*, Cambridge: Harvard University Press, 1966, p.125, from Warren Ashby, *A Comprehensive History of Western Ethics* (W. Allen Ashby, ed.), Amherst: Prometheus Books, 1997, p.37.
4. Aristotle, *Nichomachean Ethics* (J. E. C. Welldon, trans.), Amherst: Prometheus Books, 1987, book II, chapter v, p.52–53.
5. Livia Kohn, *Early Chinese Mysticism*, Princeton: Princeton University Press, 1992, p.74. These are Kohn's words to describe Kuo Hsiang's (died 312 C.E.) commentary on a passage from the *Chuang Tzu*, which was written, at least in part, by the author of the same name.
6. Nikolai Berdyaev, *The Beginning and the End* (R. M. French, trans.), New York: Harper and Row, 1957, p.11.
7. Chapter one, note #58, p.297.
8. Hans Küng, *Does God Exist?* (Edward Quinn, trans.), Garden City: Vintage Books, 1981, p.389.
9. (a) Pierre Teilhard de Chardin, *The Phenomenon of Man*, New York: Harper and Row, 1965, p.165–166. (b) Augustine, *On Free Choice of the Will* (Thomas Williams, trans.), Indianapolis: Hackett Publishing Company, 1993, book 1, chapter 7, p.12: "We often see animals that have been tamed by human beings.... Do you think...there is any way...a wild animal...could...attempt to subdue a human being?... So what is it...human beings have in virtue of which they are superior, so...no animal can subdue them, but they can control many animals? Is it not...what is usually called 'reason' or 'understanding'?" (c) Paul Weiss, *Being and Other Realities*, Peru: Open Court Publishing Company, 1995, p.127: "What is not in nature are beings with purposes. Only humans have these.... Because only humans have purposes, moreover, only they could be creative."
10. P. D. Ouspensky, *Tertium Organum*, revised translation, Garden City: Vintage Books, 1982, p.74.
11. John Andrew Fisher, "The Myth of Anthropomorphism," from Marc Bekoff and Dale Jamieson (ed.), *Readings in Animal Cognition*, Cambridge: MIT Press, 1999, p.4.
12. Ibid, p.7–8.
13. Temple Grandin and Catherine Johnson, *Animals in Translation*, New York: Scribner, 2005, p.52-54.
14. Donald R. Griffin, *Animal Minds*, Chicago: University of Chicago Press, 1992, p. 3–4.
15. Roger Crisp, "Evolution and Psychological Unity," from Marc Bekoff and Dale Jamieson, *Readings in Animal Cognition*, p. 311–312: "The brains of all multicellular animals...are made of the same matter. The fundamental characteristics of neurons and synapses are roughly the same. It is the numbers and structures that differ: the brain of a human being contains billions of neurons, while those in an ant brain run only into the hundreds or thousands.... The brains of small whales, dolphins, and porpoises are close in size to those of human beings, both absolutely and in relation to size of body.... All vertebrate brains can generally be described in terms of three main parts: the hindbrain, midbrain, and forebrain.... The cerebral hemispheres of lower vertebrates have various physiological properties in common with those of mammals."
16. John Andrew Fisher, "The Myth of Anthropomorphism," from Marc Bekoff and Dale Jamieson, *Readings in Animal Cognition*, p.8.
17. Donald R. Griffin, *Animal Minds*, p.253.
18. Roger Crisp, "Evolution and Psychological Unity," from Marc Bekoff and Dale Jamieson, *Readings in Animal Cognition*, p. 314–315.
19. C. Lloyd-Morgan, *An Introduction to Comparative Psychology*, London: Walter Scott, 1894, p.53.

20. Hugh Wilder, "Interpretive Cognitive Ethology," from Marc Bekoff and Dale Jamieson, *Readings in Animal Cognition*, p.35.
21. Ibid.
22. G. Matthews, "Animals and the Unity of Psychology," *Philosophy*, 1978, 53: 437–454, p.437.
23. Roger Crisp, "Evolution and Psychological Unity," from Marc Bekoff and Dale Jamieson, *Readings in Animal Cognition*, p.309.
24. Ibid, p. 310.
25. C. G. Gallup, "Chimpanzees: Self-Recognition," *Science*, 1970, 167: 86–87.
26. Güven Güzeldere, introduction to Ned Block, Owen Flanagan, and Güven Güzeldere (ed.), *The Nature of Consciousness: Philosophical Debates*, Cambridge: MIT Press, 1997, p.25.
27. Michael Philips and Steven N. Austad, "Animal Communication and Social Evolution," from Marc Bekoff and Dale Jamieson, *Readings in Animal Cognition*, p.260–261.
28. (a) Temple Grandin and Catherine Johnson, *Animals in Translation*, p.273. (b) C. N. Slobodchikoff, C. Kiriazis, C. Fischer, E. Creef, "Semantic Information Distinguishing Individual Predators in the Alarm Calls of Gunnison's Prairie Dogs," *Animal Behavior*, 1991, 42: 713-719.
29. Temple Grandin and Catherine Johnson, *Animals in Translation*, p.275: "Prairie dogs are *super-prey*—there's almost no meat eater in the vicinity of prairie dog burrows that *doesn't* eat them.... To make things worse, prairie dogs live in the same burrows for hundreds of years. That means every single predator in the vicinity knows exactly where to find them.... It's likely prairie dogs are so vulnerable they had to develop a really good system of communication to survive as a species."
30. (a) Carolyn Ristau, "Aspects of the Cognitive Ethology of an Injury-Feigning Bird, the Piping Plover" from Marc Bekoff and Dale Jamieson, *Readings in Animal Cognition*, p.79–80. (b) Carolyn Ristau (ed.), *Cognitive Ethology: The Minds of Other Animals*, Hillsdale: Lawrence Erlbaum Associates, 1991.
31. Carolyn Ristau, "Aspects of the Cognitive Ethology of an Injury-Feigning Bird, the Piping Plover" from Marc Bekoff and Dale Jamieson, *Readings in Animal Cognition*, p.83–85: "The data derive from nineteen different experimental sessions and from ten birds.... In forty-four out of forty-five cases, the bird's direction of display would have caused an intruder who followed it to get further from the young at the end of the period of injury-simulating display than at the beginning.... With respect to location, forty-four of the forty-five broken-wing displays were made in front of the intruder...within a 180° arc of the intruder's visual field.... The bird always moves before displaying. Sometimes the bird moves by flying... In all thirteen cases of flying, the bird's new position was closer to the intruder than...its position before flight.... Furthermore, in eleven of those thirteen cases...the bird...was closer to the front of the intruder than it had been...more directly in the center of the intruder's visual field and/or the path of the moving intruder.... In seventeen of thirty-one instances when the intruder did not follow the display, the bird stopped its display and reapproached the intruder by either flying or walking closer. In nine instances, the bird...increased the intensity of the display...by flapping its wings more vigorously or vocalizing raucously while displaying. Of the remaining five cases, after displaying, the bird flew to the location of the young in three instances." The experiments were performed in 1982 on a Virginia beach using human intruders whose movements were intentionally randomized. The birds, both adults and their young, were continuously videotaped for reactions to the intruders; the investigators also recorded in-the-moment human commentaries on the events.

32. (a) Donald R. Griffin, *Animal Minds*, p.156–157: "One of the clearest examples of natural animal communication that suggests conscious thinking stems from studies of the alarm calls and other vocalizations of vervet monkeys.... When they see dangerous predators they emit at least three types of alarm call... This differentiation of alarm calls...leads to clearly distinct responses.... The best ways to escape from the three principal predators ...are mutually exclusive, and it is very important...the alarm calls inform other members of the group *which* danger threatens." (b) T. T. Struhsaker, *The Red Colobus Monkey*, Chicago: University of Chicago Press, 1967. (c) Louis M. Herman and Palmer Morrel-Samuels, "Knowledge Acquisition and Asymmetry between Language Comprehension and Production," from Marc Bekoff and Dale Jamieson, *Readings in Animal Cognition*, p.298: "One dolphin was specialized in a language in which words were represented by computer-generated sounds. For the second...words were represented by gestures of a signer's arms and hands, as in human sign languages. In each case, the words referred to objects, actions, properties, and relationships." (d) Louis M. Herman, "Cognition and Language Competencies of Bottlenosed Dolphins," from R. J. Schusterman, J. Thomas, and F. G. Wood (ed.), *Dolphin Cognition and Behavior: A Comparative Approach*, Hillsdale: Lawrence Erlbaum Associates, 1986, p.221–252.
33. Dermot Moran, *Introduction to Phenomenology*, London: Routledge, 2000, p.25.
34. Robert K. C. Forman, *Mysticism, Mind, Consciousness*, Albany: State University of New York Press, 1999, p.69–70.
35. (a) Michael L. Rosenzweig, "Do Animals Choose Habitats?," from Marc Bekoff and Dale Jamieson, *Readings in Animal Cognition*, p.192–193. (b) F. Bene, "The Role of Learning in the Feeding Behavior of Black-Chinned Hummingbirds," *Condor*, 1945, 47: 3–22.
36. (a) Donald R. Griffin, *Animal Minds*, p.88–89. (b) G. Pilleri, "Ingenious Tool Use by the Canadian Beaver in Captivity," *Investigations on Beavers*, volume 1, Berne: Brain Anatomy Institute, 1983.
37. (a) Temple Grandin and Catherine Johnson, *Animals in Translation*, p.183. (b) P. Bateson, "Assessment of Pain in Animals," *Animal Behavior*, 1991, 42: 827–839.
38. Temple Grandin and Catherine Johnson, *Animals in Translation*, p.183–184. The authors did not supply a reference source for this data; it may be the same as in the previous citation.
39. (a) Temple Grandin and Catherine Johnson, *Animals in Translation*, p.184. (b) F. C. Colpaert, J. P. Taryre, M. Alliaga, and W. Koek, "Opiate Self Administration as a Measure of Chronic Nociceptive Pain in Arthritic Rats," *Pain*, 2001, 91: 33–34.
40. (a) Donald R. Griffin, *Animal Minds*, p.259. (b) M. Jouvet, "What does a Cat Dream about?" *Trends in Neuroscience*, 1979, 2: 280–282.
41. (a) Dianė Collinson, *Fifty Major Philosophers*, London: Croom Helm, 1987, p.39: "William of Ockham is famous for a principle of economy known as 'Ockham's razor.' The principle states...'What can be explained by the assumption of fewer things is vainly explained by the assumption of more things'." (b) P. Boehner (ed. and trans.), *Ockham: Philosophical Writings*, Edinburgh: Nelson, 1957.
42. (a) John Dupre, "The Mental Lives of Nonhuman Animals," from Marc Bekoff and Dale Jamieson, *Readings in Animal Cognition*, p.335–336. (b) David J. Chalmers, *The Conscious Mind*, New York: Oxford University Press, 1996, p.103: "Once all the physical facts about a mouse are in, the nature of its conscious experience remains an *open question*: it is consistent with the physical facts about a mouse that it has conscious experience, and it is consistent with the physical facts that it does not."

43. (a) Peter Thompkins and Christopher Bird, *The Secret Life of Plants*, New York: HarperCollins, 1973, p.5–7. (b) Cleve Backster, "Evidence of a Primary Perception in Plant Life," *International Journal of Parapsychology*, 1968, 10(4): 329–348.
44. (a) J. H. Rush, "Physical Aspects of Psi Phenomena," from *Parapsychology: Its Relation to Physics, Biology, Psychology, and Psychiatry* (Gertrude R. Schmeidler, ed.), Metuchen: Scarecrow Press, 1976, p.20. (b) R. M. Brier, "PK on a Bioelectrical System," *Journal of Parapsychology*, 1969, 33:187–205.
45. Peter Thompkins and Christopher Bird, *The Secret Life of Plants*, p.40.
46. Ibid, p.55.
47. (a) Max Velmans, *Understanding Consciousness*, London: Routledge, 2000, p.195. (b) Benjamin Libet, "Subjective Referral of the Timing for a Conscious Experience: a Functional Role for the Somatosensory Specific Projection System in Man," *Brain*, 1979, 102: 193–224.
48. William A. Tiller, *Science and Human Transformation*, Walnut Creek: Pavior Publishing, 1997, p.5–12: "The actual detector has a sandwich-like shape with the gas (30% xenon and 70% carbon dioxide) isolated between two parallel planar dielectric surfaces... Thin gold electrodes...were deposited on the outside of the dielectric surfaces. An external voltage, generated by a high voltage power supply...powers the device to a level *below* the electrical breakdown voltage of the gas.... The current output from this device passes through a sensing resistor and can be detected by...an oscilloscope or a pulse counter. With such a pulse counter, any current pulse larger than a preset value is recorded as a single count. The electrical current versus time characteristics of a...pulse was revealed by the oscilloscope.... The human intentionality effect on the device functioning could be determined via an increased number of counts in a five minute test period." For the data cited in the chapter, the external voltage was set at 4325 or 4350 volts, with a constant frequency of 475 Hertz.
49. Peter Thompkins and Christopher Bird, *The Secret Life of Plants*, p. 92–93.
50. (a) Jagadis Chandra Bose, *Response in the Living and Non-Living*, New York: Longmans, Green and Company, 1902. (b) Jagadis Chandra Bose, *Plant Response as a Means of Physiological Investigation*, New York: Longmans, Green and Company, 1906. (c) Jagadis Chandra Bose, *The Nervous Mechanism of Plants*, New York: Longmans, Green and Company, 1926. (d) Jagadis Chandra Bose, *Motor Mechanisms of Plants*, New York: Longmans, Green and Company, 1928.
51. Evan Harris Walker, *The Physics of Consciousness*, Cambridge: Perseus Publishing, 2000, p.43.
52. Ted Bastin, "A Clash of Paradigms in Physics," from Ronald Duncan and Miranda Weston -Smith (ed.), *The Encyclopedia of Ignorance*, New York: Pergamon Press, 1977.
53. Isaac Newton, *Scholium* (Andrew Motte, trans.), revised edition, London, 1803, from Alfred North Whitehead, *Process and Reality* (David Ray Griffin and Donald W. Sherburne, ed.), corrected edition, New York: The Free Press, 1978, p.70–71.
54. Amit Goswami, Maggie Goswami, and Richard E. Reed, *The Self-Aware Universe*, New York: Jeremy Tarcher Press, 1993, p.28–30.
55. Evan Harris Walker, *The Physics of Consciousness*, p.52–53.
56. Newton's Laws apply in inertial reference frames operating at non-relativistic velocities. An inertial reference frame is one where any particle's motion can be measured by a single time clock independent of the position of the particle within the frame. Two or more frames are inertial with respect to each other if they are moving at a constant velocity relative to the other; if one frame is accelerating with respect to the other, they

cannot both be inertial reference frames. Hence, in inertial reference frames, quantities such as force, mass, acceleration, and time are invariant.

57. Paul Davies, "Our Place in the Universe," from John Leslie (ed.), *Modern Cosmology and Philosophy*, Amherst: Prometheus Books, 1998, p.312: "It was from the intellectual ferment brought about by...merging...Greek philosophy and Judeo-Islamic-Christian thought, that modern science emerged, with its unidirectional linear time, its insistence on nature's rationality, and its emphasis on mathematical principles. All the early scientists such as Newton were religious in one way or another. They saw their science as a means of uncovering God's handiwork in the universe."
58. Immanuel Kant, *Critique of Pure Reason*, introduction, p.xxxiii.
59. Amit Goswami, Maggie Goswami, and Richard E. Reed, *The Self-Aware Universe*, p.115.
60. Ibid, p.97–98: "When is a measurement complete? When the transcendent consciousness collapses the wave function by means of an immanent brain-mind looking on with awareness."
61. Ibid, p.78–79.
62. Richard Tarnas, *The Passion of the Western Mind*, New York: Ballantine Books, 1991, p.279.
63. Plato, *The Laws*, from Benjamin Jowett (trans.), *The Works of Plato*, four volumes, New York: Tudor Publishing Company, undated, volume IV, p.462–463.

Chapter Three: Discreteness and Continuity

1. Rom Harré, *One Thousand Years of Philosophy*, Malden: Blackwell Publishers, 2000, p.231–232: "We do not intuit, that is, directly apprehend space and time, only things and events ordered in certain ways. Similarly, we do not apprehend causality as such, but events ordered in such a way that one is a condition for the other. We do not apprehend our selves, our egos, but we experience our minds as unified."
2. The electromagnetic spectrum of light consists of regions listed in order of increasing energy: radio and television waves, microwaves, infrared light, visible light, ultraviolet light, x rays, and gamma rays. Our normal visual range covers wavelengths between 370 and 730 nanometers. Some species of fish can see in the infrared and ultraviolet portions of the spectrum. Our normal hearing range is from 200 to 20,000 cycles per second; other animal species, such as whales, can comfortably and routinely detect sounds up to 200,000 cycles per second. Dogs can smell odors humans simply cannot detect.
3. Jeffrey Mishlove, *The Roots of Consciousness*, New York: Random House, 1975, p.208.
4. Franklin Merrill-Wolff, *The Philosophy of Consciousness Without an Object*, New York: Julian Press, 1973, p.161.
5. This value of 99.999% empty space in an atom is an upper limit. The heavier elements have substantial numbers of protons and neutrons in the nucleus, making the nucleus correspondingly bigger in size. While these larger atoms also have more electrons around them, the volume of space the electrons occupy does not grow as rapidly as the volume of space the nucleus occupies. The ratio of the volumes between nucleus and atom as a whole decreases as the atoms become more massive. The volume of a three-dimensional sphere, the general overall shape of an atom, is calculated from the formula: $V = 4/3\pi r^3$, and the diameter, d, is equal to twice the radius, r. Since we are concerned only with the ratio between two volumes, the transformation of diameters to volumes for both the nucleus and the entire atom mathematically cancel each other, and we can use only the ratio of the diameters to make the statement about the ratio of volumes. With regard to masses, the mass of both a proton and a neutron are each approximately two thousand

times the mass of an electron. These facts explain the statement that 99% of the atom's mass resides in the nucleus.

6. Alfred North Whitehead, *Process and Reality* (David Ray Griffin and Donald W. Sherburne, ed.), corrected edition, New York: The Free Press, 1978, p.61.
7. P. D. Ouspensky, *Tertium Organum*, revised translation, Garden City: Vintage Books, 1982, p.225: "Separateness and combination are not opposites in the real world, but exist together and at the same time, without contradicting each other. We must realize that in the real world the same thing can be both a part and a whole."
8. George Santayana, "Reason in Common Sense," *The Life of Reason*, Amherst: Prometheus Books, 1998, p.20.
9. Arthur O. Lovejoy, *The Great Chain of Being*, Cambridge: Harvard University Press, 1964, p.276.
10. William James, *A Pluralistic Universe*, from William James, *Writings* 1902–1910, New York: Library of America, 1987, p.717: "In the *physical* world there is no real compounding. 'Wholes' are not realities there, parts only are realities.... The 'whole'...is nothing but our vision, nothing but an effect on our sensorium when a lot of things act on it together.... In the mental world, on the contrary, wholes do in point of fact realize themselves *per se*."
11. (a) Masao Abe, "Nonbeing and *Mu*—The Metaphysical Nature of Negativity in the East and the West," from Masao Abe, *Zen and Western Thought* (William R. LaFleur, ed.), Honolulu: University of Hawaii Press, 1985, p.128. (b) Keiji Nishitani, *Religion and Nothingness* (Jan Van Bragt, trans.), Berkeley: University of California Press, 1982, p.123: "The emptiness of *Śūnyatā* is not an emptiness represented as some 'thing' outside of being or other than being. It is not simply an 'empty nothing,' but...an *absolute nothingness*, emptied even of these representations of emptiness. And for that reason, it is at bottom one with being, even as being is at bottom one with emptiness."
12. P. D. Ouspensky, *Tertium Organum*, p.209.
13. (a) Lenn E. Goodman, *Avicenna*, London: Routledge, 1992, p.15–16: "Metaphysics, al-Farabi explains, takes its name from the fact that its themes and questions rise above the questions of physics about motion and change. Mathematics too deals with nonphysical objects, but only by conceptual abstraction, whereas the objects of metaphysics may actually exist entirely beyond the physical realm, divorced form matter not just notionally but in actuality. The primary object of metaphysics is being and thus oneness (since unity or self-identity is the most general mark of being). But, correlatively, metaphysics must study nonbeing and otherness as well.... There are three sciences ultimately: physics, mathematics, and metaphysics." (b) Bernard McGinn, *The Foundations of Mysticism*, New York: Crossroad Publishing Company, 1994, p.306: "The lowest forms of science —that is, 'knowledge perfect in its mode'—are the experimental sciences which form the basis for the first level of speculative or properly philosophical science, physica, or the philosophy of nature, which investigates the universe of the *sensible real*. Both the experimental sciences and the philosophy of nature exist on the first level of abstraction. The second level...is the realm of mathematics, which investigates the universe of the *praeter real*; the third level of abstraction studies the universe of the *transensible*, beings that exist without matter and the principles of metaphysics." While not a direct quotation, the citation is an abstract of a statement by Jacques Maritain in *The Degrees of Knowledge*.
14. (a) João Magueijo, *Faster than the Speed of Light*, London: Penguin Books, 2004, p.52. (b) Sylvia Fraser, *The Quest for the Fourth Monkey: A Thinker's Guide to the Psychic and Spiritual Revolution*, Toronto: Key Porter Books, 1992, p.209: "According to Einstein,

gravity was not a force as Newtonian physics implied but a geometric effect. Specifically, it was the curvature of space caused by any massive object."

15. Brian Greene, *The Elegant Universe*, New York: W.W. Norton and Company, 1999, p.56.
16. John Dobson, *Advaita Vedānta and Modern Science*, New York: Vivekananda Vedānta Society, 1979, p.14.
17. Evan Harris Walker, *The Physics of Consciousness*, Cambridge: Perseus Publishing, 2000, p.33–34.
18. Ibid, p.38.
19. Samuel Alexander, *Space, Time, and Deity*, two volumes, London: Macmillan and Company, 1966.
20. (a) Eduard Zeller, *Outlines of the History of Greek Philosophy* (L. R. Palmer, trans.; Wilhelm Nestle, ed.), New York: Dover Publications, 1980, thirteenth edition, p.52–53: "Still more famous and more significant are the four proofs against motion: (1) In order to cover a certain distance a body must first traverse the half and before that the half of this half and so on...it must in a finite time traverse an infinite number of spaces. Thus the movement could not make a beginning. (2) Achilles can never overtake the tortoise when it has any start on him; since by the time he arrived at the spot A, the tortoise has reached the second spot B, and when he is arrived at B it has gone to C and so on; hence the motion could never come to an end. (3) The flying arrow is at rest since at every moment it is in one and the same space; therefore it is at rest in every moment of its flight and...during the whole time. (4) Equal distances must at equal velocity be traversed in the same time.... A moving body passes by a second when this is moving in a contrary direction at the same velocity twice as quickly as when this is at rest. Thus the laws of motion contradict the facts." (b) Muhammad Iqbal, *The Reconstruction of Religious Thought in Islam* (M. Saeed Sheikh, ed.), Lahore: Institute of Religious Culture, 1999, lecture II, note #9, p.163: "It may be added that our primary source for Zeno's famous and controversial arguments is Aristotle's *Physics* (VI.9.239b), which is...said to have been first translated into Arabic by Ishaq b. Hunain (~845–910/911)."
21. (a) Eduard Zeller, *Outlines of the History of Greek Philosophy*, p.53: "The fallacies of his proofs and in particular the fundamental error, the confusion of the infinite divisibility of space and time with infinite dividedness, he...did not notice himself." (b) Muhammad Iqbal, *The Reconstruction of Religious Thought in Islam*, p.30: "The act of movement, i.e. movement as lived and not as thought, does not admit of any divisibility. The flight of the arrow observed as a passage in space is divisible, but its flight regarded as an act, apart from its realization in space, is one and incapable of partition into a multiplicity. In partition lies its destruction."
22. H. Wildon Carr, *The General Principle of Relativity in its Philosophical and Historical Aspects*, London, 1920, p.36: "Zeno therefore is right in saying that the arrow is at rest at every moment of its flight, wrong in inferring that therefore it does not move, for there is a one-to-one correspondence in a movement between the infinite series of positions and the infinite series of instants."
23. Chapter one, note #22, p.293–294.
24. T. R. V. Murti, *The Central Philosophy of Buddhism*, London: Unwin, 1987, p.34.
25. William James, *Some Problems in Philosophy*, from William James, *Writings* 1902–1910, p.1081–1082.
26. Aristotle, *Metaphysics* (Richard Hope, trans.), Ann Arbor: University of Michigan Press, Book Gamma, 1010a, 1960, p.79.
27. (a) David Hume, *A Treatise of Human Nature* (Ernest C. Mossner, ed.), London: Penguin, 1985, p.130–142. (b) Ibid, p.58: "The qualities...by which the mind is... conveyed from

one idea to another are three: resemblance, contiguity in time or place, and cause and effect." On p.61–62, Hume claimed ideas can be compared by: (1) resemblance to each other, (2) identity to each other, (3) relation to space and time, (4) quantitative number, (5) qualitative degree, (6) contrariety as negation of resemblance and identity, and (7) causality.

28. (a) Dean I. Radin, *The Conscious Universe*, New York: HarperCollins, 1997, p.259. (b) Paul A. Weiss, "The Living System: Determinism Stratified," from A. Koestler and J. Smythies (ed.), *Beyond Reductionism: New Perspectives in the Life Sciences*, London: Hutchinson, 1969, p.13. (c) Paul Weiss, *Being and Other Realities*, Peru: Open Court Publishing Company, 1995, p.106–107: "When...a character, a person, and a body change over the course of a life, it is tempting to suppose that individuals change as well. An individual...is a constant. I remain self-same, while expressing myself in different ways at different times through my character, my person, and my body."
29. William James, "The Stream of Consciousness," from Ned Block, Owen Flanagan, and Güven Güzeldere (ed.), *The Nature of Consciousness: Philosophical Debates*, Cambridge: MIT Press, 1997, p.74.
30. Fiona Steinkamp, "Parapsychological Phenomena and the Sense of Self," from Fiona Steinkamp (ed.), *Parapsychology, Philosophy and Mind*, Jefferson: McFarland and Company, 2002, p. 67–68.
31. (a) Abū Hāmid Muhammad al-Ghazālī, *The Ninety-Nine Beautiful Names of God* (David B. Burrell and Nazih Daher, trans.), Cambridge: Islamic Texts Society, 1995. Two names for Allah are *al-Rahman* [The Infinitely Good] and *al-Haqq* [The Truth]. (b) Alaisdair MacIntyre, *A Short History of Ethics*, New York: Touchstone Press, 1996, p.118–119: "To call God good is to name him as the goal of desire.... So 'God is good' is a synthetic proposition." MacIntyre is paraphrasing the doctrine of Thomas Aquinas. (c) Ronald H. Isaacs, *Everyperson's Guide to Jewish Philosophy*, Northvale: Jason Aronson, Inc, 1999, p.76–77: "Crescas writes of God's positive attributes... For Crescas, God is Goodness." (d) Charles Hartshorne, *The Divine Relativity*, New Haven: Yale University Press, 1964, p.12: "God is the measure of truth, as we are not, because He and only He is able to establish a perfect correspondence between His knowing and what He knows, or between what He knows it to be and what it is."
32. Immanuel Kant, *Critique of Pure Reason* (F. Max Muller, trans.), Garden City: Doubleday and Company, 1966, p.105.
33. David Hume, *An Enquiry Concerning Human Understanding* (Eric Steinberg, ed.), Indianapolis: Hackett Publishing Company, 1977, p.16–17.
34. Richard Popkin and Avrum Stroll, *Skeptical Philosophy for Everyone*, Amherst: Prometheus Books, 2002, p.168–170.
35. David Hume, *A Treatise of Human Nature*, p.307–308: "'Tis still true, that every distinct perception, which enters into the composition of the mind, is a distinct existence, and is different and distinguishable and separable from every other perception... We suppose the whole train of perceptions to be united by identity... The understanding never observes any real connection among objects, and that even the union of cause and effect, when strictly examined, resolves itself into a customary association of ideas."
36. Immanuel Kant, *Critique of Pure Reason*, p.108–109.
37. Chapter one, note #58, p.297.
38. Stephen Palmquist, "Kant's Criticism of Swedenborg: Parapsychology and the Origin of the Copernican Hypothesis," from Fiona Steinkamp, P*arapsychology, Philosophy and Mind*, p.160.
39. Immanuel Kant, *Critique of Pure Reason*, p.198.

40. Nikolai Berdyaev, *The Beginning and the End* (R. M. French, trans.), New York: Harper and Row, 1957, p.14.
41. The Earth's rotation period is not exactly twenty-four hours; more precisely measured, it is 23 hours and 56 minutes. In addition, Earth's simultaneous revolution about the sun as it rotates on its axis is being ignored in the remark that there is a return to "its original position." In actuality, there is a slight day-to-day change in the position for any given spot on the Earth relative to the sun.
42. J. T. Fraser (ed.), *The Voices of Time*, second edition, Amherst: The University of Massachusetts Press, 1981, p.410–411: "It was discovered that hyperfine transitions and inversion frequencies in the spectrum of certain elements depend almost entirely on the substance and only very weakly on its environment. Such transitions...are well suited to form the primary reference source of frequency standards.... The most commonly used such 'clock' or time standard is called the cesium clock."
43. (a) Peter Coveney and Roger Highfield, *The Arrow of Time*, London: W. H. Allen, 1990, p.27. (b) Ibid, p.227: "Within the body, cell division is carried out at a variety of rates. Mature brain cells, for example, do not divide at all. Liver cells divide every year or two, while cells in the lining of the gut divide as often as twice a day.... Taken from a human fetus, they [cells] will go through about fifty cycles of division before they eventually die. Similar cells from a forty-year-old stop dividing after forty cycles and cells from an eighty-year-old grind to a halt after about thirty cycles.... This seems to correspond with the overall aging of the body."
44. Priya Hemenway, *Divine Proportion*, New York: Sterling Publishing, 2005, p.126.
45. The "Divine Proportion" in the book's title refers to the *golden mean*, symbolized by the Greek letter *phi*, Φ, and which has a non-terminating mathematical value whose first few digits are calculated to be: 1.618033. It's significance, both philosophical and practical, has been known for over 2500 years.
46. Peter Coveney and Roger Highfield, *The Arrow of Time*, p.150–151.
47. John R. Pierce, *An Introduction to Information Theory*, second revised edition, New York: Dover Publications, 1980, p.23.
48. (a) Alan M. Laibelman, *Going Against the Flow: An Exercise in Ethical Syncretism*, New York: Peter Lang Publishing Company, 2004, p.305. (b) Ilham Dilman, *Free Will*, New York: Routledge, 1999, p.166: "For Moore, 'we have free will' means 'we could have acted differently if we had chosen to act differently,' and this is perfectly compatible with causal determinism. People fail to see this because they confuse causal determinism with fatalism—'The view that *whatever we will* the result will always be the same; that it is, therefore, *never* any use to make one choice rather than another'." (c) George Edward Moore, *Ethics*, Oxford: Oxford University Press, 1947.
49. Augustine, *On Free Choice of the Will* (Thomas Williams, trans.), Indianapolis: Hackett Publishing Company, 1993, p.xi, xiii.
50. Johann Gottlieb Fichte, *Foundations of Transcendental Philosophy* (Daniel Breazeale, ed. and trans.), Ithaca: Cornell University Press, 1992, section 12, p.270.
51. (a) George Edward Moore, "Free Will," from George Edward Moore, *Principia Ethica* (Thomas Baldwin, ed.), revised edition, Cambridge: Cambridge University Press, 1993, p.305–306. "Free Will" was originally published as chapter VI within Moore's *Ethics*. (b) Vivekananda, *Karma Yoga and Bhakti Yoga*, revised edition, New York: Ramakrishna-Vivekananda Center, 1955, p.85: "Everything that we know, or can possibly know, must be subject to causation, and that which obeys the law of causation cannot be free. It is acted upon by other agents and becomes a cause in its turn."

52. René Descartes, *Meditations Concerning First Philosophy*, from René Descartes, *Philosophical Essays* (Laurence J. LaFleur, trans.), New York: Macmillan Publishing Company, 1964, p.132.
53. J. B. Robinet, *De La Nature*, five volumes, 1761–1768, volume IV, p.4–5, from Arthur O. Lovejoy, *The Great Chain of Being*, p.276.
54. Immanuel Kant, *Critique of Pure Reason*, p.171.
55. Aurobindo, *The Life Divine*, Pondicherry: Sri Aurobindo Ashram, 1982, p.144.
56. Bernard McGinn, "Ibn Gabirol: The Sage Among the Schoolmen," from Lenn E. Goodman (ed.), *Neoplatonism and Jewish Thought*, Albany: State University of New York Press, 1992, p.82–83.
57. (a) G. W. F. Hegel, *Phenomenology of Spirit* (A. V. Miller, trans.), Oxford: Oxford University Press, 1977, section 419, p.252: "Spiritual essence is, in its simple being, *pure consciousness* ...*this self*-consciousness. The...*determinate* nature of the individual has lost its positive meaning of being *in itself* the element and the purpose of its activity; it is...a superseded moment, and the individual is a *self* in the form of a universal self." (b) Robert L. Perkins, "Ultimate Reality and Meaning in the Thought of Kierkegaard," *Ultimate Reality and Meaning*, Toronto: University of Toronto Press, 1983, 6(4): 283–289, p.289: "Consciousness, conceived as spirit, is...also self-consciousness."
58. David Buchman, introduction to Abū Hāmid Muhammad al-Ghazālī, *The Niche of Lights* (David Buchman, trans.), Provo: Brigham Young University Press, 1998, p.xxxiii.
59. Muhyi al-Din ibn al-'Arabi, *Meccan Openings*, volume IV, p.252, from William C. Chittick, *Imaginal Worlds*, Albany: State University of New York Press, 1994, p.110: "God did not create creation for the sake of creation itself. Rather, He created it as a similitude for Himself... Hence God brought creation into existence in His own form."
60. Moses de Leon, *Sefer Zohar* [*The Book of Splendor*], 3:31b from Daniel C. Matt, introduction to Daniel C. Matt (trans.), *Zohar: The Book of Enlightenment*, Mahwah: Paulist Press, 1983, p.24: "Through an action below, an action above is aroused."
61. Alice A. Bailey, *A Treatise on Cosmic Fire*, New York: Lucis Press, 1962, p.1066.
62. John Beloff, "Body and Soul: Another Look at the Mind-Brain Problem," from Fiona Steinkamp, *Parapsychology, Philosophy and Mind*, p.9.
63. (a) Güven Güzeldere, introduction to Ned Block, Owen Flanagan, and Güven Güzeldere, *The Nature of Consciousness*: *Philosophical Debates*, p.42. (b) P. D. Ouspensky, *Tertium Organum*, p.120: "Positivism asserts that the phenomena of life and consciousness are... functions of physical phenomena and are no more than a certain complex combination of the latter...further, that all the three kinds of phenomena are...the same, and the higher, i.e., the phenomena of life and consciousness, are nothing but different manifestations of the lower, i.e., of one and the same physico-mechanical or electromagnetic energy."
64. George Berkeley, *Three Dialogues Between Hylas and Philonius* (Robert Merrihew Adams, ed.), Indianapolis: Hackett Publishing Company, 1979, second dialogue, p.44 (first sentence) and p.46 (remainder).
65. Samuel Angus, *The Mystery Religions*, New York: Dover Books, 1975, p.219.
66. Although it has been claimed light energy has particle-like character, and the existence of a particle, a *gravitino*, corresponding to the gravitational force has been proposed, we will see in chapter five how neither concept stands up to scrutiny.
67. Ken Wilber, *The Spectrum of Consciousness*, Wheaton: Theosophical Publishing House, 1977, p.82.
68. (a) G. W. F. Hegel, *Phenomenology of Spirit*, section 590, p.359. (b) F. W. J. Schelling, *System of Transcendental Idealism* (Peter Heath, trans.), Charlottesville: University Press of Virginia, 1978, p.32: "Take away the limits of the empirical, and you have the absolute

self... Pure self-consciousness is an act lying outside time, and by which all time is first constituted; empirical consciousness...arises...in time and the succession of presentations."

69. Alice A. Bailey, *From Intellect to Intuition*, New York: Lucis Press, 1978, p.161.
70. René Descartes, *Rules for the Direction of Mind*, from René Descartes, *Philosophical Essays*, p.185.
71. (a) Mary Beth Ingham and Mechtild Dreyer, *The Philosophical Vision of John Duns Scotus*, Washington D. C.: The Catholic University of America Press, 2004, p.27. (b) Ibid, p.30–31: "Scotus does not hold that intuitive cognition is foundational to our understanding of the world. Rather, he holds that our first acts of cognitive awareness are based...upon the process of sense intuition and the activity of abstraction that begins there. Only subsequently does intuition provide us with that immediate awareness of our own acts of sensing, intellection, and willing."
72. Gustav Theodor Fechner, *Elements of Psychophysics*, New York: Holt, Rinehart and Winston, 1966, from G. Stanley Hall, *Founders of Modern Psychology*, New York: D. Appleton and Company, 1924, p.151.
73. David J. Chalmers, *The Conscious Mind*, New York: Oxford University Press, 1996, p.126–127.

Chapter Four: Metaphysics

1. Bernard J. F. Lonergan, *Insight: A Study of Human Understanding*, New York: Harper and Row, 1978, p.390: "Just as the notion of being underlies and penetrates and goes beyond all other notions, so...metaphysics is the department of human knowledge that underlies, penetrates, transforms, and unifies all other departments."
2. G. W. Leibniz, *Discourse on Metaphysics*, section XVIII, from G. W. Leibniz, *Discourse on Metaphysics* and *Monadology* (George Montgomery, trans.), La Salle: Open Court Publishing Company, 1997, p.33.
3. James Wayne Dye, "Nikolai Berdyaev and His Ideas on the Fundamental Nature of All Entities," *Ultimate Reality and Meaning*, Toronto: University of Toronto Press, 1979, 2 (2): 109–134, p.131.
4. Saadia Gaon, *The Book of Beliefs and Opinions* (Samuel Rosenblatt, trans.), New Haven: Yale University Press, 1948, p.25.
5. (a) Edmund Husserl, *Logical Investigations* (J. N. Findlay, trans.), London: Routledge and Kegan Paul, 1970, investigation 5, section 6, p.545. (b) Dermot Moran, *Introduction to Phenomenology*, London: Routledge, 2000, p.169: "In *Cartesian Meditations*, Husserl says the transcendental ego is responsible...for the being of the world." The reference to *Cartesian Meditations* is section 28, p.62.
6. Johann Gottlieb Fichte, *Foundations of Transcendental Philosophy* (Daniel Breazeale, ed. and trans.), Ithaca: Cornell University Press, 1992, section 17, p.395: "Our thinking always involves a connection, a movement of thought in a certain direction, an act of synthesis, without which there can be no thinking at all. For this reason, the content of our thoughts is never anything taken by itself, but is always a relationship between two things."
7. Immanuel Kant, *Critique of Pure Reason* (F. Max Muller, trans.), Garden City: Doubleday and Company, 1966, p.306–307.
8. João Magueijo, *Faster than the Speed of Light*, London: Penguin Books, 2004, p.80–81.
9. (a) Jean-Paul Sartre, *Being and Nothingness* (Hazel E. Barnes, trans.), New York: Washington Square Press, 1992, p.48–49: "Nothingness is...subsequent to being since it supposes being in order to deny it, since the irreducible quality of the *not* comes to add itself to that undifferentiated mass of being in order to release it. That does not mean only that we should refuse to put *being* and *non-being* on the same plane, but also that we must be careful

never to posit nothingness as an original abyss from which being arose." (b) Alexandra David-Neel and Lama Yongden, *Secret Oral Teachings in Tibetan Buddhist Sects* (H. N. M. Hardy, trans.), San Francisco: City Lights Books, 1967, p.111: "Void is...a state in which the elements do not produce any combination, give rise to non-phenomenon, a state in which only forces, latent and non-manifest, exist." (c) Aurobindo, *The Life Divine*, Pondicherry: Sri Aurobindo Ashram, 1982, p.28: "*Sat* [Being] and *Asat* [Non-Being], if they have both to be affirmed, must be conceived as if they obtained simultaneously." (d) Wing-tsit Chan, *A Sourcebook in Chinese Philosophy*, Princeton: Princeton University Press, 1963, p.357: "The only reality is Emptiness itself, which is the Absolute, Ultimate Void, Original Substance, or in Chinese terminology the correct principle (*cheng-li*)." (e) Rajmani Tigunait, *Seven Systems of Indian Philosophy*, Honesdale: Himalayan Institute of Yoga 1983, p.58: "*Śūnya* may be considered to be a void, emptiness, or nothingness. But those who have attained Nirvana and establish themselves in perfect wisdom describe this void or nothingness as the potential for everything—whatever exists or does not exist."

10. Lucretius, *On the Nature of the Universe* (John Godwin, trans.), London: Penguin Books, 1994, book I, lines 918, 922–924, p.33: "Learn...that *the universe is not bounded in any direction*.... Since you must admit...there is nothing outside the universe, it can have no limit and is accordingly without end or measure."
11. (a) Seyyed Hossein Nasr, *Knowledge and the Sacred*, Albany: State University of New York Press, 1989, p.136: "Since metaphysics as developed in the Occident has almost always been related to ontology, it is important to...discuss the relation of Being to the Principle or Ultimate Reality." (b) Nicholas Rescher, *Process Metaphysics*, Albany: State University of New York Press, 1996, p.171: "Philosophizing is a teleological process... Philosophy is a purposive enterprise aimed at problem-solving, at providing answers to our questions regarding the world's scheme of things and our place within it." (c) Ludwig Wittgenstein, *Tractatus Logico-Philosophicus*, from Hans Küng, *Does God Exist*? (Edward Quinn, trans.), Garden City: Vintage Books, 1981, p.94: "The solution of the riddle of life in space and time lies outside space and time."
12. Immanuel Kant, *Critique of Pure Reason*, p.341.
13. Ibid, p.413–414.
14. Jean-Paul Sartre, *Being and Nothingness*, p.203.
15. Moses de Leon, *Sefer Zohar*, I, 240b, from Gershom Scholem, *Major Trends in Jewish Mysticism*, New York: Schocken Books, 1995, p.222.
16. Rudolf Otto, *Mysticism East and West*, Wheaton: Theosophical Publishing House, 1987, p.217.
17. (a) Julius Lipner, *The Face of Truth*, Albany: State University of New York Press, 1986, p.54. (b) Franklin Merrill-Wolff, *The Philosophy of Consciousness Without an Object*, New York: Julian Press, 1973, p.189: "At certain stages in the analysis of consciousness it appears as though the invariant element were the pure Subject or the Self, but at this stage the analysis has not isolated the subtle distinction between pure Subjectivity and Consciousness... Ultimately, it is found that the Self is derivative as well as the objective pole of Consciousness."
18. Rajmani Tigunait, *Seven Systems of Indian Philosophy*, p.216.
19. (a) Anselm of Canterbury, *Proslogion* (M. J. Charlesworth, trans.), Notre Dame: University of Notre Dame Press, 1979, chapter II, p.117. (b) Thomas Aquinas, *Summa Theologiae* (Thomas Gilby, general editor), Garden City: Doubleday and Company, 1969, question 2, p.62–71.
20. Hans Küng, *Does God Exist*?, p.538.

21. (a) Edward Conze, *Buddhism: Its Essence and Development*, New York: Harper and Row, 1975, p.39: "While...the Buddhists adopt an attitude of agnosticism to the question of a personal creator, they have not hesitated to stress the superiority of the Buddha over Brahma, the god who, according to Brahminic theology, created the universe." (b) Meister Eckhart, "Commentary on Genesis," from Edmond Colledge and Bernard McGinn (trans.), *Meister Eckhart*, Ramsey: Paulist Press, 1981, p.84–85: "So when someone once asked me why God had not created the world earlier, I answered that he could not because he did not exist. He did not exist before the world did.... It is false to picture God as if he were waiting around for some future moment in which to create the world."
22. Nikolai Berdyaev, *The Beginning and the End* (R. M. French, trans.), New York: Harper and Row, 1957, p.102.
23. (a) Meister Eckhart, Sermon 27, from Raymond B. Blakeney, *Meister Eckhart's Doctrines*, New York: Harper and Row, 1941, p.225–226: "God and the Godhead are as different from each other as heaven and earth... Creatures speak of God—but why do they not mention the Godhead? Because there is only unity in the Godhead and there is nothing to talk about. God acts. The Godhead does not. It has nothing to do and there is nothing going on in it." (b) Samuel Alexander, *Space, Time, and Deity*, two volumes, London: Macmillan and Company, 1966, volume II, p.370: "Deity is located only in a portion of the infinite whole of Space-Time, and therefore God, though infinite both in respect of his body and his deity, is only in respect of his body co-extensive with the absolute whole of Space-Time, while his deity is empirical and belongs only to a part of the Absolute."
24. Hans Küng, *Does God Exist?*, p. 649.
25. Alan M. Laibelman, *The Other Perennial Philosophy: A Metaphysical Dialectic*, Lanham: University Press of America, 2000, p.84.
26. Sarvepelli Radhakrishnan, *An Idealist View of Life*, London: George Allen and Unwin, 1932, p.342–343.
27. William C. Chittick, *Imaginal Worlds*, Albany: State University of New York Press, 1994, p.25.
28. Aurobindo, *The Life Divine*, p.545.
29. Charles Hartshorne, *The Divine Relativity*, New Haven: Yale University Press, 1964, p.90.
30. Aurobindo, *The Life Divine*, p. 80.
31. Priya Hemenway, *Divine Proportion*, New York: Sterling Publishing, 2005, p.156 and 159: "Two-, three-, four-, and six-sided shapes are the only ones that will pack a space with regularity and with no gaps.... In a two-dimensional plane, there are three, and only three, regular shapes that can be arranged to completely fill...the plane. These are the equilateral triangle, the square, and the regular hexagon." The first sentence explores the three-dimensional figure possibilities; the second sentence pertains to the corresponding two-dimensional figure possibilities.
32. F. W. J. Schelling, *Ages of the World* (Judith Norman, trans.), Ann Arbor: University of Michigan Press, 1997, p.171.
33. Hazel E. Barnes, introduction to Jean-Paul Sartre, *Being and Nothingness*, p.xxv–xxvi: "Being-in-itself is logically prior to Being-for-itself... Nothing external to being caused the rupture in...self-identity of Being-in-itself. It occurred somehow in Being." Sartre gave an explanation for the distinction between being-in-itself and being-for-itself on p.119–126: Being-in-itself is primordial transcendent being without the appearance of consciousness, while Being-for-itself is being with the presence of consciousness. The latter occurs when being, as subject, introspects on itself as if it were an external object. It can be said that Being-for-itself is self-consciousness. Note the "rupture," the transition from Being-in-itself to Being-for-itself occurred "somehow." Sartre, as was true for other

metaphysicians positing a transcendent unity source, cannot provide a logical explanation for the appearance of duality.

34. F. W. J. Schelling, *Ages of the World*, p.172.
35. Ibid, p.176.
36. William James, *A Pluralistic Universe*, from William James, *Writings 1902–1910*, New York: Library of America, 1987, p.664: "'Can a plurality of reals be possible?' asks Mr. Bradley, and answers, 'No, impossible.' For it would mean a number of beings not dependent on each other, and this independence their plurality would contradict. For to be 'many' is to be related, the word having no meaning unless the units are somehow taken together." The internal reference is to the well-known Idealist philosopher, Francis Herbert Bradley, author of *Appearance and Reality*.
37. Arthur Hyman, "From What is One and Simple Only What is One and Simple May Come to Be," from Lenn E. Goodman (ed.), *Neoplatonism and Jewish Thought*, Albany: State University of New York Press, 1992.
38. Ilham Dilman, *Free Will*, New York: Routledge, 1999, p.165.
39. Carolyn Ristau, "Aspects of the Cognitive Ethology of an Injury-Feigning Bird, the Piping Plover," from Marc Bekoff and Dale Jamieson (ed.), *Readings in Animal Cognition*, Cambridge: MIT Press, 1999, p.85.
40. Joseph Goldstein and Jack Kornfield, *Seeking the Heart of Wisdom*, Boston: Shambhala Publications, 1987, page unstated, from Jean Smith (ed.), *Radiant Mind*, New York: Riverhead Books, 1999, p.290.
41. (a) Edward A. Jarvis, "Josiah Royce's Idea of Ultimate Reality and Meaning," *Ultimate Reality and Meaning*, 3(3): 168–186, 1980, p.170: "The essence of intention is knowledge of what one intends and one cannot fail to agree with what one knows." (b) Moses Maimonides, *Guide for the Perplexed* (M. Friedlander, trans.), second edition, New York: Dover Publications, 1956, p.272: "An agent that acts with intention must have a certain ulterior object in that which he performs.... It is...evident that that which is produced with intention has passed over from nonexistence into existence." (c) Robert Audi, *The Architecture of Reason*, New York: Oxford University Press, 2001, p.62: "Rational action is based on motivating elements, above all desire—in the widest sense of the term, which includes not only intention but any kind of wanting." (d) Colin Allen and Marc Bekoff, "Intentionality, Social Play, and Definition," from Marc Bekoff and Dale Jamieson (ed.), *Readings in Animal Cognition*, p.232: "We will consider the attribution of intentional states significant only if they correspond to actual changes in the organism's cognitive machinery...and these changes are causally responsible for the behavior in question."
42. Hannah Arendt, *The Life of the Mind*, one volume edition, New York: Harcourt Brace and Company, 1971, volume I: Thinking, p.46: "Objectivity is built into the very subjectivity of consciousness by virtue of intentionality. Conversely and with the same justness, one may speak of the intentionality of appearances and their built-in subjectivity. All objects because they appear indicate a subject, and just as every subjective act has its intentional object, so every appearing object has its intentional subject."
43. Aristotle, *Metaphysics* (Richard Hope, trans.), Ann Arbor: University of Michigan Press, 1960, 1026b, p.126.
44. Ibid, 1065b, p.237.
45. (a) The mass of all objects in the universe is estimated to be $\sim 10^{53}$ kilograms, and the volume is believed to be close to 2.25×10^{86} cubic centimeters. (b) B. J. Carr, "Origin, Evolution, and the Purpose of the Universe," *The Irish Astronomical Journal*, March 1982, 15(3): 237–253, from John Leslie (ed.), *Modern Cosmology and Philosophy*, Amherst: Prometheus Books, 1998, p.149.

46. Nikolai Berdyaev, *The Beginning and the End*, p.162–163.
47. Pitirim Sorokin, *Social and Cultural Dynamics*, Boston: Extending Horizons Books, 1957, p.645.
48. Adin Steinsaltz, *The Thirteen Petalled Rose*, New York: Basic Books, 1980, p.155.
49. Isaiah Tishby, "The Doctrine of Man in the Zohar" from Lawrence A. Fine (general editor), *Essential Papers on Kabbalah*, New York: New York University Press, 1995, p.118.
50. (a) Charles Hartshorne and William L. Reese, *Philosophers Speak of God*, Amherst: Prometheus Books, 2000, p.225: "The clearest early instance of temporal theism we have found is Fausto Socinus and his immediate followers in the Socinian movement.... If God is really temporal, then for him there must be past and future as well as present. It follows that this 'eternity' means, not absence of change, but only absence of birth or death or possibility of destruction.... Wisdom cannot consist in a survey of all time in a single moment and...for him, too, there must be...new knowledge—knowledge of events which until they happened were unknown to him." (b) Aurobindo, *The Synthesis of Yoga*, Pondicherry: Sri Aurobindo Ashram, 1976, p.48: "We know the Divine and become the Divine, because we are That already in our secret nature.... Self-attainment is the secret; self-knowledge and an increasing consciousness are the means and...process." (c) Pierre Teilhard de Chardin, *The Phenomenon of Man*, New York: Harper and Row, 1965, p.288–289: "Are we to foresee man seeking to fulfill himself collectively upon himself, or personally on a greater than himself? Refusal or acceptance of Omega? A conflict may supervene. In that case the noosphere...will, when it has reached its point of unification, split into two zones each attracted to an opposite pole of adoration." This passage, replete with terminology invented by de Chardin begs for explanation: *Omega* is the postulated endpoint of evolutionary history; *noosphere* is his term for the generation of a planetary intelligence conceptually no different from group mind or Consciousness. This selection may be interpreted as reflective of the cosmogony incorporated into *TOPP*. Substitute *Separation* for "refusal of Omega," and *Union* for "acceptance of Omega"; superimpose de Chardin's premise of "a conflict" and "a splitting into two poles," and, you have produced the metaphysical speculation posited herein. (d) Abu Bakr Muhammad ben Zakariya al-Razi, *Opera Philosophica* (P. Kraus, ed.), from Majid Fakhry, *A History of Islamic Philosophy*, New York: Columbia University Press, 1983, p.101: "God created man and imparted Reason to him from the 'essence of His divinity' so that Reason might eventually rouse the Soul from its earthly slumber in man's body and remind it of its genuine destiny as a citizen of the higher (intelligible) world and of its duty to seek that world through the study of philosophy.... When this ultimate goal has been reached, and the human Soul, guided by Reason, has been restored to its true abode, this 'lower world' will cease, and matter, which had been forcibly chained to form, will return to its original condition of absolute formlessness and purity."
51. Charles Hartshorne, *The Divine Relativity*, p.141.
52. João Magueijo, *Faster than the Speed of Light*, p.81–82: "The baby universe is therefore fragmented into tiny regions that cannot see each other.... Objects become homogeneous by allowing their different parts to come into contact and...acquire common features.... But the horizon effect prohibits precisely such a process. It tells us ...the vast regions of the universe that we observe to be so homogeneous could not have been aware of each other at first."
53. Paul Davies, "What Caused the Big Bang?" from John Leslie, *Modern Cosmology and Philosophy*, p.234–235.

54. B. J. Carr, "Origin, Evolution, and the Purpose of the Universe," *The Irish Astronomical Journal*, March 1982, from John Leslie, *Modern Cosmology and Philosophy*, p.146: "At one second, the temperature of the universe is 10^{10} Kelvin, hot enough for pairs of electrons and positrons to be created. Before one second...the number of electrons and positrons should be roughly equal to the number of photons.... Before 10^{-5} seconds, the present photon to proton ratio of 10^8 should be manifested as a slight excess of protons over antiprotons, one extra proton for every 10^8 proton-antiproton pairs.... The tiny asymmetry between matter and antimatter at 10^{-5} seconds is crucial for our existence, because all the matter which we observe in the universe today derives from it. If there was no asymmetry, all the matter and antimatter would have annihilated at 10^{-5} seconds to produce a universe of pure light."
55. Heinrich Dumoulin, *Zen Buddhism: A History—India and China* (James W. Heisig and Paul Knitter, trans.), New York: Macmillan Publishing Company, 1988, p.219: "The most important dialectical formula in the House of Lin-chi, 'the four alternatives,' describes four positions regarding the subject-object relationship.... The formula is based on the well-known four propositions of Indian Buddhist logic: being, nonbeing, neither being nor nonbeing, both being and nonbeing.... In the first and second stages illusion is overcome first by the subject and then by the object. The third stage negates both subject and object, but differentiation still obtains. Only in the fourth stage, which affirms the transcendence of...opposition between subject and object, does all confrontation between subject and object cease. Reality is comprehended in its ultimate oneness."
56. Stephen F. Mason, *Chemical Evolution: Origin of the Elements, Molecules, and Living Systems*, Oxford: Oxford University Press, 1991, p.276.
57. Peter Coveney and Roger Highfield, *The Arrow of Time*, London: W. H. Allen, 1990, p.139–140.
58. Paul Weiss, "God and the World," *Science, Philosophy, and Religion: A Symposium*, New York: Conference on Science, Philosophy, and Religion in Their Relation to the Democratic Way of Life, 1941, p.424–425, from Charles Hartshorne and William L. Reese, *Philosophers Speak of God*, p.317–318.
59. Ibid, p.425–426.
60. Ervin Laszlo, *The Interconnected Universe: Conceptual Foundations of Transdisciplinary Unified Theory*, River Edge: World Scientific, 1995, p.16–17.
61. Mauro Ceruti, "Philosophical Perspectives on Quantum Vacuum Interaction Cosmology," from Ervin Laszlo, *The Interconnected Universe: Conceptual Foundations of Transdisciplinary Unified Theory*, p.121.

Chapter Five: Physics

1. John Dobson, *Advaita Vedānta and Modern Science*, New York: Vivekananda Vedānta Society, 1979, p.35: "Mass and energy structure space and time (through particles), so also space and time structure mass and energy (through fields/forces)."
2. Ludwig Feuerbach, *The Essence of Christianity* (George Eliot, trans.), Amherst: Prometheus Books, 1989, appendix, p.286.
3. Peter Coveney and Roger Highfield, *The Arrow of Time*, London: W. H. Allen, 1990, p.78–79: "It is only when we have to deal with objects traveling at speeds approaching that of light itself that Newtonian mechanics breaks down. Such situations are usually quite alien to our everyday experience, except where light and electromagnetic effects are concerned.... Indeed, in the limit where the velocities of moving objects are small with respect to the speed of light...special relativity reduces to Newtonian or classical physics in these situations."

4. Ibid, p.76: "In 1905 Einstein put forward two fundamentally new postulates of physics. They appeared in his first scientific paper on relativity…'On the Electrodynamics of Moving Bodies.' These foundations of special relativity, which refer to observers moving at constant velocities, are: (1) A 'relativity principle': the laws of physics must be the same everywhere in the universe, no matter what the velocity of the observer. (2) The speed of light is constant and independent of the motion of the light source."
5. Brian Greene, *The Elegant Universe*, New York: W.W. Norton and Company, 1999, p.38–39: "Imagine that we are idly watching the passage of time by looking at a ticking light clock placed on a nearby table. Then…a second light clock slides by on the table, moving at constant velocity. The question we ask is whether the moving light clock will tick at the same rate as the stationary light clock?... The photon starts at the base of the sliding clock…and first travels to the upper mirror. Since…the clock is moving, the photon must travel at an angle. If the photon did not travel along this path, it would miss the upper mirror and fly off into space.... The photon bounces off the upper mirror and again travels a diagonal path to hit the lower mirror, and the sliding clock ticks.... The double diagonal path…is *longer* than the straight up-and-down path taken by the photon in the stationary clock… The constancy of the speed of light tells us that the sliding clock's photon travels at exactly the same speed as the stationary clock's photon. But since it must travel farther to achieve one tick it will tick *less frequently*. This simple argument establishes that the moving light clock, from our perspective, ticks more slowly than the stationary light clock. And since…the number of ticks directly reflects how much time has passed, we see that the passage of time has slowed down for the moving clock." By extrapolation, if the moving clock is traveling at a speed equivalent to the speed of a photon, the photon will never reach a mirror and there will be no ticks at all; time stops. Mathematically, time dilation is calculated as, $dt = \gamma dt_o$, where 'dt' is the change of time at velocity close to that of light in a vacuum, 'dt_o' is the change of time at velocity much less than that of light in vacuum, and $\gamma = 1/(1\text{-}v^2/c^2)^{1/2}$, the adjustment made in velocity, 'v,' as an object approaches the speed of light, 'c.'
6. Ervin Laszlo, *The Interconnected Universe: Conceptual Foundations of Transdisciplinary Unified Theory*, River Edge: World Scientific, 1995, p.28–29.
7. Brian Greene, *The Elegant Universe*, p.12.
8. David Bohm, *Wholeness and the Implicate Order*, Boston: Routledge and Kegan Paul, 1980, p.176.
9. Momentum is actually a product of mass and velocity rather than speed. The difference between velocity and speed is that the former refers to a specific coordinate direction for motion, while the latter indicates only the rate of motion.
10. Paul Janet, introduction to G. W. Leibniz, *Discourse on Metaphysics* and *Monadology* (George Montgomery, trans.), La Salle: Open Court Publishing Company, 1997, p.xx–xxi.
11. The Earth's magnetic field shifts its poles relative to the geographic poles every few million years. Currently, the geographic North Pole actually corresponds to the magnetic South Pole. Also, there is not an exact alignment between the types of poles; rather the magnetic poles are offset relative to the geographic poles by approximately five degrees.
12. Alpha particles are helium nuclei, two protons and two neutrons, stripped of electrons; beta particles are free electrons; positrons are anti-matter particles, the mirror image of electrons, but with a positive electrical charge in contrast to the typical negative charge on electrons. Designating the first three processes as "emissions" means these particles are spontaneously released by radioactive atoms; capture is the opposite of emission.

13. (a) Michio Kaku, *Hyperspace*, New York: Oxford University Press, 1994, p.114–115: "The weak force...is caused by the exchange of...a W particle (W stands for 'weak'). Similarly, the strong force holding the protons and neutrons together within the nucleus of the atom is caused by the exchange of subatomic particles called pi mesons. Both W bosons and pi mesons have been seen experimentally in the debris of atom smashers." (b) Brian Greene, *The Elegant Universe*, p.13: "The protons crammed together in atomic nuclei all repel one another electromagnetically; the strong force...overcomes this repulsion and tethers the protons tightly together. But a rather small change in the relative strengths of these two forces [the strong and weak forces] would easily disrupt the balance between them, and would cause most atomic nuclei to disintegrate."
14. Brian Greene, *The Elegant Universe*, p.12–13.
15. Ibid, p.121.
16. Ibid, p.122–123: "At high enough energy and temperature—such as occurred a mere fraction of a second after the big bang—electromagnetic and weak force fields *dissolve* into one another... When the temperature drops, as it has done steadily since the big bang, the electromagnetic and weak forces *crystallize* out...through a process known as *symmetry breaking*...and therefore appear to be distinct in the cold universe we currently inhabit."
17. Michio Kaku, *Hyperspace*, p. 25–26: "A *field* is a collection of numbers defined at every point in space that completely describes a force at that point.... In the 1860s, Scottish physicist James Clerk Maxwell wrote...the field equations for electricity and magnetism. In 1915, Einstein discovered the field equations for gravity.... The field equations for the subatomic forces were finally written down in the 1970s utilizing the earlier work of C. N. Yang and his student R. L. Mills.... The puzzle that has stumped physicists within this century is why the subatomic field equations look so...different from the field equations of Einstein—that is, why the nuclear forces seem so different from gravity.... Confined to three or four dimensions, the field equations of the subatomic world and gravitation are difficult to unify."
18. Ervin Laszlo, *The Interconnected Universe: Conceptual Foundations of Transdisciplinary Unified Theory*, p.15–16.
19. Brian Greene, *The Elegant Universe*, p.127–130. There are several parameters associated with Planck's constant, whose value is 6.63 x 10^{-34} joule-seconds: the Planck *length* is ~10^{-43} centimeters; the Planck *time* is ~10^{-33} seconds; the Planck *energy* is ~10^{19} billion electron volts.
20. Evan Harris Walker, *The Physics of Consciousness*, Cambridge: Perseus Publishing, 2000, p.81.
21. Ibid.
22. Peter Coveney and Roger Highfield, *The Arrow of Time*, p.128–129.
23. Michael Talbot, *The Holographic Universe*, New York: HarperCollins Publishers, 1991, p.121–122.
24. Ervin Laszlo, *The Interconnected Universe: Conceptual Foundations of Transdisciplinary Unified Theory*, p.57.
25. Amit Goswami, Maggie Goswami, and Richard E. Reed, *The Self-Aware Universe*, New York: Jeremy Tarcher Press, 1993, p.73–75.
26. Ibid, p.58–59: "Does the wave aspect of a single wavicle exist in a transcendental space, since it never manifests in ordinary space?... Between observations, the electron exists as a possibility form, like a Platonic archetype, in the transcendent domain of *potentia*."
27. Michael Talbot, *The Holographic Universe*, p.34.

28. (a) Ibid, p.47. (b) Ibid, p.41–42: "Imagine a fish swimming in an aquarium. Imagine also that…your only knowledge about them comes from two television cameras, one directed at the aquarium's front and the other at its side. When you look at the two television monitors you might mistakenly assume…the fish on the screens are separate entities…. But as you continue to watch you will eventually realize there is a relationship between the two fish."
29. Peter Coveney and Roger Highfield, *The Arrow of Time*, p. 120–121.
30. Sylvia Fraser, *The Quest for the Fourth Monkey: A Thinker's Guide to the Psychic and Spiritual Revolution*, Toronto: Key Porter Books, 1992, p.183: "Einstein…spent the last thirty years of his life in pursuit of a unified field theory that would bring all of science back under the same set of rules, thus restoring the Universe as objective and external."
31. Evan Harris Walker, *The Physics of Consciousness*, p.90–91.
32. Ibid, p.91–92.
33. John S. Bell, *Physics*, 1965, 1: 195.
34. (a) Sylvia Fraser, *The Quest for the Fourth Monkey: A Thinker's Guide to the Psychic and Spiritual Revolution*, p.180. (b) Ibid, p.182: "As Bell's theorem implied: either the observer created the effect through measurement, proving that the Universe is without objective reality, or the effect was created instantaneously through non-local causes—the equivalent of telepathy or synchronicity."
35. Evan Harris Walker, *The Physics of Consciousness*, p.124–125.
36. Ibid, p.131. A recitation of the various experiments was presented here along with a summary statement admitting some results conformed to Bell's theorem, while some supported quantum theoretical predictions.
37. A. Aspect, J. Dalibard, and G. Roger, *Physical Review Letters*, 1982, 49: 1804.
38. (a) Amit Goswami, Maggie Goswami, and Richard E. Reed, *The Self-Aware Universe*, p.120: "Alain Aspect used polarization-correlated photons that emerged in opposite directions from a source of calcium atoms. A detector was set up on the path of each beam of photons. The crucial feature of the experiment…was the inclusion of a switch that, in effect, changed the polarization setting of one of the detectors every ten-billionth of a second (shorter than the travel time of light or any other local signal between the two detector locations). Even so, the change of the polarization setting of the detector with the switch changed the outcome of the measurement in the other location—just as quantum mechanics says it should."
39. Evan Harris Walker, *The Physics of Consciousness*, p.134: "In deriving Bell's theorem, we assumed…any correlation had to be due to a plan…the photons already had. We excluded the possibility of any non-local, global, or faster-than-light influence."
40. Amit Goswami, Maggie Goswami, and Richard E. Reed, *The Self-Aware Universe*, p.127.
41. Ibid, p.133–134: "The only observer who could collapse a conventional wavefunction of the universe must be God."
42. Ervin Laszlo, *The Interconnected Universe: Conceptual Foundations of Transdisciplinary Unified Theory*, p.18: "The quantum vacuum is more than the source and sink of matter: recent evidence indicates that it can also be an active influence on the behavior of matter particles. For example, the zero-point energies of the vacuum interact with electrons around nuclei with effects that become observable when the electrons transit from one energy state to another."
43. Edwin A. Abbott, *Flatland*, New York: Dover Publications, 1992.
44. Brian Greene, *The Elegant Universe*, p.206: "Since tiny strings vibrate through all of the spatial dimensions, the precise way in which the extra dimensions are twisted up and curled back on each other strongly influences and tightly constrains the possible resonant

vibrational patterns.... *Extradimensional geometry determines fundamental physical attributes like particle masses and charges that we observe in the usual three large space dimensions of common experience.*"

45. Michio Kaku, *Hyperspace*, p.208.
46. Martin Rees, "The Thirteen Billion Year Bang," *New Scientist*, 1976 (2 December), 72: 512–515, from John Leslie (ed.), *Modern Cosmology and Philosophy*, Amherst: Prometheus Books, 1998, p.82: "But the consensus really swung towards acceptance of a 'big bang' cosmology after the discovery, in 1965, of the 2.7K microwave background radiation.... The fact...the radiation apparently has the same temperature in all directions with a precision of 1 in a 1000 provides direct evidence that the universe on a large scale is exceedingly homogeneous (it looks the same from all points) and isotropic (it looks the same in all directions from each point)."
47. Michio Kaku, *Hyperspace*, p.170.
48. Ibid, p.12.
49. (a) Ibid, p.38. (b) Ibid, p.39: "Riemann's idea was to introduce a collection of numbers at every point in space that would describe how much it was bent or curved.... The greater the value of the metric tensor, the greater the crumpling of the sheet ... The metric tensor gives us a simple means of measuring its curvature at any point." Note #7, chapter two, p.338, added: "The curvature of space at any given point can be measured by drawing a circle at that point and measuring the area inside that circle. In flat two-dimensional space, the area inside the circle is πr^2. However, if the curvature is positive, as in a sphere, the area is $< \pi r^2$. If the curvature is negative, as in a saddle, the area is $> \pi r^2$." (c) Ibid, p.43: "Riemann was unable to complete his work on force fields because he lacked the field equations that electricity and magnetism and gravity obey. In other words, he did not know precisely how the universe would be crumpled in order to yield the force of gravity.... These crucial developments would be left to Maxwell and Einstein."
50. Ibid, p.100–103.
51. Ibid, p.105–107.
52. Ervin Laszlo, *The Interconnected Universe*: *Conceptual Foundations of Transdisciplinary Unified Theory*, p.17–18.
53. Evan Harris Walker, *The Physics of Consciousness*, p.309.

Chapter Six: Self-Consciousness

1. (a) Alan M. Laibelman, *Going Against the Flow: An Exercise in Ethical Syncretism*, New York: Peter Lang Publishers, 2004. (b) Alan M. Laibelman, "Pure Consciousness as Ultimate Reality," *Ultimate Reality and Meaning*, Toronto: University of Toronto Press, 2003, 26(1): 49–73. (c) Alan M. Laibelman, *The Other Perennial Philosophy: A Metaphysical Dialectic*, Lanham: University Press of America, 2000. (d) Alan M. Laibelman, "Ultimate Reality and Meaning According to the Perennial Philosophy: Evidence from the Mathematical and Physical Sciences," *Ultimate Reality and Meaning*, Toronto: University of Toronto Press, 1992, 15(3): 216–236.
2. Owen Flanagan, "Conscious Inessentialism and the Epiphenomenalist Suspicion," from Ned Block, Owen Flanagan, and Güven Güzeldere (ed.), *The Nature of Consciousness*: *Philosophical Debates*, Cambridge: MIT Press, 1997, p.357.
3. Jean-Paul Sartre, *Being and Nothingness* (Hazel E. Barnes, trans.), New York: Washington Square Press, 1992, p.16: "This self-determination of consciousness must not be conceived as a genesis, as a becoming, for that would force us to suppose that consciousness is prior to its own existence.... Consciousness is a plenum of existence, and this determination of

itself by itself is an essential characteristic.... The paradox is not that there are self-activated existences but that there is no other kind."

4. (a) Hugh Lawson-Tancred, introduction to Aristotle, *De Anima* [*On the Soul*] (Hugh Lawson-Tancred, trans.) Middlesex: Penguin Books, 1986, p.27–28: "According to Aristotle, an account of the soul had for the speculations of the fourth and fifth centuries...to meet two requirements: it had (i) to explain how the bodies of two kinds of things, plants and animals including man, come to be able to originate physical processes...and it had (ii) to account for the fact...some of these existing things were...capable of having something that can loosely be labeled cognition of the surrounding world." (b) Samuel Alexander, *Space, Time, and Deity*, two volumes, London: Macmillan and Company, 1966, volume II, p.5: "That...experienced from the inside...is a conscious process, is as experienced from the outside...a neural one." (c) David J. Chalmers, *The Conscious Mind*, New York: Oxford University Press, 1996, p.275: "In a certain sense, we can say that not only does consciousness supervene on the physical, but it supervenes on the organizational."
5. John R. Searle, "Reductionism and the Irreducibility of Consciousness," from Ned Block, Owen Flanagan, and Güven Güzeldere, *The Nature of Consciousness: Philosophical Debates*, p.451.
6. Dermot Moran, *Introduction to Phenomenology*, London: Routledge, 2000, p.39. Moran's reference to 1874 pertained to the date of publication of Brentano's *Psychology from an Empirical Standpoint*, the work most representative of his philosophy.
7. David J. Chalmers, *The Conscious Mind*, p.171.
8. Andrew B. Newberg, Eugene D'Aquili and Vince Rause, *Why God Won't Go Away*, New York: Ballantine Books, 2001, p.19–20.
9. Jeffrey Mishlove, *The Roots of Consciousness*, New York: Random House, 1975, p.210.
10. Andrew B. Newberg, Eugene D'Aquili and Vince Rause, *Why God Won't Go Away*, p.43–44.
11. Ibid, chapter 3, note #13, p.194.
12. Ibid, p.44–45.
13. Rhawn Joseph, "After Death, Astral Projection, Judgment Day, and the Second Death," from Rhawn Joseph (ed.), *Neurotheology*, San Jose: University Press, 2002, p.363.
14. Ibid, p.412.
15. Andrew B. Newberg, Eugene D'Aquili and Vince Rause, *Why God Won't Go Away*, p.45–46.
16. Rhawn Joseph, "After Death, Astral Projection, Judgment Day, and the Second Death," from Rhawn Joseph, *Neurotheology*, p.361.
17. Andrew B. Newberg and Jeremy Iversen, "On the 'Neuro' in Neurotheology," from Rhawn Joseph, *Neurotheology*, p.263–264.
18. Carol Raush Albright, "Religious Experience, Complexification, and the Image of God," from Rhawn Joseph, *Neurotheology*, p.171.
19. Erich Jantsch, *The Self-Organizing Universe*, Oxford: Pergamon Press, 1980, p.165–166: "The *reptilian brain* originated approximately 250 to 280 million years ago. It serves the coordination of an already very rich spectrum of behavioral patterns...it does not have good learning capability.... It provides a possibility for...development of a...repertoire of self-expression.... Self-expression must not be understood in a strictly functional way as the securing of survival, but as true symbolic self-representation." *Homo habilis* arrived on the world stage ~2.2–1.6 million years ago; *Homo erectus*, ~2.0–0.4 million years ago; *Homo sapiens archaic* ~400,000–200,000 years ago; *Homo sapiens neanderthalensis* ~200,000–30,000 years ago; *Homo sapiens sapiens* ~200,000 years ago.

20. Antonio Damasio, *The Feeling of What Happens: Body and Emotion in the Making of Consciousness*, San Diego: Harcourt, 1999, p.275–276.
21. Jean-Claude Marquet and Michel Lorblanchet, *Antiquity*, 2003 referenced by a news account reporting the finding published by the British Broadcasting Company on 2 December 2003. The detailed information quoted in the chapter was recovered from the corresponding Internet story: http://news.bbc.co.uk, entitled "Neanderthal 'Face' found in Loire."
22. H. S. Micklen, "Immunology," from Ronald Duncan and Miranda Weston-Smith (ed.), *The Encyclopedia of Ignorance*, New York: Pergamon Press, 1977, page unstated.
23. Michael Talbot, *The Holographic Universe*, New York: HarperCollins Publishers, 1991, p.13.
24. Carol Raush Albright, "Religious Experience, Complexification, and the Image of God," from Rhawn Joseph, *Neurotheology*, p.171 and p.175.
25. Andrew B. Newberg, Eugene D'Aquili and Vince Rause, *Why God Won't Go Away*, p.28.
26. Ibid, p.29–30.
27. Ibid, p.31.
28. Roger Crisp, "Evolution and Psychological Unity," from Marc Bekoff and Dale Jamieson (ed.), *Readings in Animal Cognition*, Cambridge: MIT Press, 1999, p.312.
29. Damage to the corpus callosum connecting the two hemispheres of the brain is frequently observed to lead to greater integration of the hemispheres functionally. It is the physical condition said to underlie the performance capabilities of autistic individuals, who used to be called "idiot savants."
30. David J. Chalmers, *The Conscious Mind*, p.5.
31. Ibid, p.288–289.
32. Max Velmans, *Understanding Consciousness*, London: Routledge, 2000, p.275.
33. Ibid, p.276.
34. (a) Rom Harré, *One Thousand Years of Philosophy*, Malden: Blackwell Publishers, 2000, p.274: "The mental and material aspects of a person will always *seem* to be radically different. At the very best empirical research can show only that there are systematic and exceptionless correlations between material and mental states." (b) Nicholas Rescher, *Process Metaphysics*, Albany: State University of New York Press, 1996, p.114: "Mental and physical processes are not *reducible*, the one to the other; they are *coordinated* as different aspects of one unified whole.... Mental and material operations can...be seen as two modes of natural process at large, representing a difference in sort but not in kind."
35. Leonard Angel, *Enlightenment East and West*, Albany: State University of New York Press, 1994, p.261.
36. Ibid, p.262.
37. Dermot Moran, *Introduction to Phenomenology*, p.251: "Scientific understanding has always assumed the possibility of...timeless knowledge...whereas...all understanding, including scientific explanation, is historically conditioned, is partial, and always comes from a point of view."
38. (a) Max Velmans, *Understanding Consciousness*, p.17. (b) W. D. Hart, "Dualism," from S. Guttenplan (ed.), *A Companion to the Philosophy of Mind*, Oxford: Blackwell, 1994.
39. Lenn E. Goodman, *Avicenna*, London: Routledge, 1992, p.161–162.
40. William James, "Does Consciousness Exist?" from William James, *Writings 1902–1910*, New York: Library of America, 1987, p.1145.
41. Edmund Husserl, *Ideas Pertaining to a Pure Phenomenology and to a Phenomenological Philosophy*, part I, section 57, p.132, from Dermot Moran, *Introduction to Phenomenology*,

p.171: "This pure ego...is not a part of any mental process, but is a necessary condition of these processes."

42. Sydney Shoemaker, "The First-Person Perspective," from Ned Block, Owen Flanagan, and Güven Güzeldere, *The Nature of Consciousness: Philosophical Debates*, p.514.
43. Max Velmans, *Understanding Consciousness*, p.218–219.
44. David J. Chalmers, *The Conscious Mind*, p.16: "At a first approximation, phenomenal concepts deal with the first-person aspects of mind, and psychological concepts deal with the third-person aspects."
45. John Locke, *An Essay Concerning Human Understanding* (Peter H Nidditch, ed.), New York: Oxford University Press, 1975, book II, chapter I, section 19, p.115.
46. David J. Chalmers, *The Conscious Mind*, p.243.
47. Johann Gottlieb Fichte, *Foundations of Transcendental Philosophy* (Daniel Breazeale, ed. and trans.), Ithaca: Cornell University Press, 1992, section 1, p.113.
48. (a) G. W. F. Hegel, *Phenomenology of Spirit* (A. V. Miller, trans.), Oxford: Oxford University Press, 1977, section 197, p.120: "For *to think* does not mean to be an *abstract* 'I,' but an 'I' which has at the same time the significance of *intrinsic* being, of having itself for object, or of relating itself to objective being in such a way that its significance is the *being-for-self* of the consciousness for which it is an object." (b) Julius Lipner, *The Face of Truth*, Albany: State University of New York Press, 1986, p.55–56: "A transitive act of awareness is an act in which the explicit object of attention is something external to the knowing subject. Most...cognitive acts are of this kind, ranging from everyday perception to contemplation of the deity.... A reflexive act of awareness is an act in which the chief object attended to is the knowing subject... Rāmānuja is affirming...every transitive conscious act is stretched between, so to speak, a subjective pole, the knower, and an objective pole, the not-I.... He says...in such acts the self is aware not only of the outside object, the not-I, made manifest to it by consciousness, but that it is also aware, *at the same time*, at least implicitly, of *itself*." (c) F. W. J. Schelling, *System of Transcendental Idealism* (Peter Heath, trans.), Charlottesville: University Press of Virginia, 1978, p.24–25: "Self-consciousness is the act whereby the thinker immediately becomes an object to himself, and, conversely, this act and no other is self-consciousness.... The concept of the self arises through the act of self-consciousness, and thus *apart* from this act the self is nothing; its whole reality depends solely on this act, and *it is itself nothing other than this act*."
49. (a) Antonio Damasio, *The Feeling of What Happens: Body and Emotion in the Making of Consciousness*, p.18: "A theory of consciousness should *not* be just a theory of how the brain attends to the image of an object. As I see it, natural low-level attention precedes consciousness, while focused attention follows the unfolding of consciousness. Attention is as necessary to the function of consciousness as having images. But attention is not sufficient for consciousness and is not the same as consciousness." (b) Ibid, p.93: "For my purposes, the term *alertness* should signify that the subject is not just awake, but apparently disposed to perceive and act. The proper meaning of *alert* is somewhere between 'awake' and 'attentive'."
50. Evan Harris Walker, *The Physics of Consciousness*, Cambridge: Perseus Publishing, 2000, p.203–204.
51. Thomas Nagel, *The View from Nowhere*, Oxford: Oxford University Press, 1986, p.7–8.
52. (a) Evan Harris Walker, *The Physics of Consciousness*, p.151. (b) Tenzin Gyatso (14th Dalai Lama), *The World of Tibetan Buddhism: An Overview of its Philosophy and Practice* (Geshe Thupten Jinpa, trans.), Somerville: Wisdom Publications, 1995, page unstated, from Jean Smith (ed.), *Radiant Mind*, New York: Riverhead Books, 1999, p.167: "What

remains between the vanishing of one thought, and the arising of another, is simply awareness, empty and without obstruction, like space." I would call this alertness, rather than awareness for the reason mentioned in the text: it exists as a subject without an object. (c) Jiddu Krishnamurti, *The Awakening of Intelligence*, San Francisco: Harper Publishers, 1973, p.448: "When there is this absolute quietness of the mind, of the whole organism, including the brain, measurement as thought ceases. Then one can enquire if there is such a thing as the immeasurable.... So when thought is absolutely quiet, then there is a state, or a dimension, in which the immeasurable has its own movement."

53. (a) Jean-Paul Sartre, *The Transcendence of the Ego* (Forrest Williams and Robert Kirkpatrick, trans.), New York: Farrar, Straus, and Giroux, 1991, p.38–39: "The individuality of consciousness...stems from the nature of consciousness. Consciousness...can be limited only by itself.... It is consciousness...which makes possible the unity and the personality of my I.... The existence of consciousness is an absolute because consciousness is consciousness of itself." (b) Frithjof Schuon, *Survey of Metaphysics and Esoterism*, Bloomington: World Wisdom Books, 1986, p.9–10: "Subjectivity as such participates in necessary Being because the Absolute is pure Consciousness... This is to say...a particular subjectivity (i.e., ego) is a possibility; its principle pertains to the Absolute; and its particularity pertains to relativity or contingency."
54. Dermot Moran, *Introduction to Phenomenology*, p.161.
55. Donald Rothberg, "Contemporary Epistemology and the Study of Mysticism," from Robert K. C. Forman (ed.), *The Problem of Pure Consciousness: Mysticism and Philosophy*, New York: Oxford University Press, 1990, p.173: "Psychotherapeutic investigators... have suggested the presence...of unconscious mediations: libidinal drives, unresolved conflicts and traumas, deeply rooted potentialities toward creativity and self-actualization."
56. (a) John Locke, *An Essay Concerning Human Understanding* (Peter H Nidditch, ed.), book I, chapter II, section 5, p.49–50. (b) Franz Brentano, *Psychology from an Empirical Standpoint* (A. C. Rancurello, D. B. Terrell, and L. L. McAllister, trans.), London: Routledge, 1995, p.153, from Dermot Moran, *Introduction to Phenomenology*, p.43: "All consciousness of an object is accompanied by a consciousness of itself as act... There is no perceiving without the possibility of apperception...there can be no unconscious mental acts."
57. Güven Güzeldere, introduction to Ned Block, Owen Flanagan, and Güven Güzeldere, *The Nature of Consciousness: Philosophical Debates*, p.20.
58. F. W. J. Schelling, *System of Transcendental Idealism*, p.60: "The self has sensation, in that it intuits itself as originally limited. This intuition is an activity, but the self cannot at once both intuit, and intuit itself as intuiting. In this act, therefore, it is not aware of any activity at all; so in sensation the concept of an action is nowhere entertained, but only that of a passivity. In the present moment, the self is *for itself* merely the *sensed*."
59. Franklin Merrill-Wolff, *Consciousness and Dualism*, unpublished manuscript.
60. (a) Douglas Dean, "The Plethysomograph as an Indicator of ESP," *Journal of the Society for Psychical Research*, 1962, 41: 351–353. (b) Jean Barry, "Telepathy and Plethysomography," *Revue Metapsychique*, 1967, 6: 56–74. (c) Erlendur Haraldsson, "Vasomotor Reactions as Indicators of Extrasensory Perception," Ph. D. Dissertation, University of Freiburg, Germany.
61. Sylvia Fraser, *The Quest for the Fourth Monkey: A Thinker's Guide to the Psychic and Spiritual Revolution*, Toronto: Key Porter Books, 1992, p. 87.
62. (a) Dean I. Radin, *The Conscious Universe*, p.153–154. A receiver was wired up to a monitor that continuously measured the electrical conductivity of his or her skin, which allows for the detection of changes in one's emotional state. At random intervals of time, a sender in another room is told to attempt to either agitate or calm the subject merely by

thoughtful concentration on him or her. As a control, during other randomly selected times, the sender is told to think about anything other than the receiver and his/her mental or physical condition. Results were determined by observing the degree of change in the desired direction (agitated or calmed) relative to the total change in skin electrical conductivity during the control periods. If there were no net effect on the receiver by the remote sender, this result would be calibrated at the fifty percent level. (b) W. G. Braud and M. J. Schlitz, "Psychokinetic Influence on Electrodermal Activity," *Journal of Parapsychology*, 1983, 47: 95–119. (c) W. G. Braud and M. J. Schlitz, "Consciousness Interactions with Remote Biological Systems: Anomalous Intentionality Effects," *Subtle Energies*, 1991, 2(1): 1–46.

63. (a) G. W. Leibniz, *Discourse on Metaphysics*, section XXIV, from G. W. Leibniz, *Discourse on Metaphysics* and *Monadology* (George Montgomery, trans.), La Salle: Open Court Publishing Company, 1997, p.41. (b) René Descartes, *Rules for the Direction of the Mind*, from René Descartes, *Philosophical Essays* (Laurence J. LaFleur, trans.), New York: Macmillan Publishing Company, 1964, p.152: "Concerning the subjects proposed for investigation, we should seek to determine, not what others have thought, nor what we ourselves conjecture, but what we can clearly and evidently intuit, or deduce with certainty; for in no other way is knowledge obtained."
64. George Berkeley, *Three Dialogues between Hylas and Philonius* (Robert Merrihew Adams, ed.), Indianapolis: Hackett Publishing Company, 1979, first dialogue, p.38: "To speak the truth, Philonius, I think there are two kinds of objects, the one perceived immediately, which are likewise called *ideas*; the other are real things or external objects perceived by the mediation of ideas, which are their images or representations. Now I own, ideas do not exist without the mind; but the latter sort of objects do."
65. Johann Gottlieb Fichte, *Foundations of Transcendental Philosophy*, section 9, p.227.
66. (a) Arthur Schopenhauer, "Free Will and Fatalism," from Arthur Schopenhauer, *On Human Nature* and *Essays*, London: Allen and Unwin, 1951, p.18: "Self-consciousness contains only the willing, but not the grounds which determine the willing. The grounds are found in the consciousness of other things...and lie outside self-consciousness, in the world of objects, to which belong both the motive and the man as an object." (b) Johann Gottlieb Fichte, *Foundations of Transcendental Philosophy* (Daniel Breazeale, ed. and trans.), section 8, p.223: "My ability to be conscious of anything whatsoever has its foundation within me and not within things. I am conscious of *something*; but I myself am the sole thing of which I am immediately conscious.... Through self-consciousness, I become conscious of the world." Schopenhauer and Fichte describe the relationship between the world and self-consciousness in diametrically *opposite* ways: Schopenhauer focuses on the object as the origin of the connection, while Fichte concentrates on the subject as the origin of the connection. They are both cited here despite this contrast because they *agree* on the point of concern in the text: self-consciousness determines *our* involvement in the world.
67. (a) Immanuel Kant, *Critique of Pure Reason* (F. Max Muller, trans.), Garden City: Doubleday and Company, 1966, p.436: "The understanding forms an object for reason in the same manner as sensibility for the understanding. It is the proper business of reason to render the unity of all possible empirical acts of the understanding systematical, in the same manner as the understanding connects the manifold of phenomena by concepts, and brings it under empirical laws." (b) Ibid, p.279: "One may therefore say of the thinking I (the soul)...that it *does not know itself through the categories*, but knows the *categories* only, and through them all objects, in the absolute unity of apperception, that is, *through itself*."
68. (a) Noel E. Boulting, "Charles S. Peirce's Idea of Ultimate Reality and Meaning Related to Humanity's Ultimate Future as seen through Scientific Inquiry," *Ultimate Reality and Meaning*, Toronto: University of Toronto Press, 1993, 16(1): 9–26, p.24: "What we have

in Peirce is an account of human beings as an enquiring being and this is not to be understood psychologically but logically. The ultimate reality of the human being is to be understood in terms of four basic dispositions: to be able to experience, to reflect on that experience, to logically deduce consequences from those reflections, and then to test those deductions." (b) Chapter one, note #32, p.294–295.

69. (a) Forrest Williams and Robert Kirkpatrick, introduction to Jean-Paul Sartre, *The Transcendence of the Ego*, p.21. (b) Dermot Moran, *Introduction to Phenomenology*, p.377: "Sartre criticizes Husserl for positing a pure ego behind consciousness, whereas in fact the self is a projection from consciousness and lies outside it like an object.... Sartre proposes a distinction between the empirical ego...and what he calls the 'transcendental field,' which is pre-personal, a non-I, a 'nothing.' This field of consciousness is a nothing precisely because it is the consciousness of all things."
70. Plato, *Parmenides*, from Benjamin Jowett (trans.), *The Dialogues of Plato*, five volumes, London: Oxford University Press, 1892, paragraph 133: "When ideas are what they are in relation to one another, their essence is determined by a relation among themselves, and has nothing to do with the resemblances, or whatever they are to be termed, which are in our sphere, and from which we receive this or that name when we partake of them."
71. (a) Jack J. Cohen, *Major Philosophers of Jewish Prayer in the Twentieth Century*, New York: Fordham University Press, 2000, p.117–118. (b) Hannah Arendt, *The Life of the Mind*, one volume edition, New York: Harcourt Brace and Company, 1971, volume I: Thinking, p.6: "Thinking aims at and ends in contemplation, and contemplation is not an activity but a passivity; it is the point where mental activity comes to rest."
72. Charles W. Leadbeater, *The Inner Life*, Wheaton: The Theosophical Publishing House, 1978, p.334: "As regards the size of the aura, that of an ordinary man extends about eighteen inches on each side of the body. If he puts his elbow against his side and stretches out his arm and hand, his fingertips will be near the circumference."
73. (a) Valerie V. Hunt, *Infinite Mind: Science of the Human Vibrations of Consciousness*, Malibu: Malibu Publishing Company, 1989, p.19–20. (b) Ibid, p.21: "On every recording taken from the body where there are muscles, there is a resting baseline of electrical activity between muscle contractions. I believe...this is produced by all living tissue and can be described as life activity.... When we filtered the baseline data to remove the brain, heart, and muscle frequencies, we discovered...there is a void (no electrical activity) between 250 cycles to 450 or 500 cycles per second. But from 500 cycles per second up to 20,000 cycles per second, the frequency information was continuous.... This electromagnetic energy is about one-half to one-third as strong as the millivoltage of a resting muscle."
74. (a) Manly Palmer Hall, *An Encyclopedic Outline of Masonic, Hermetic, Qabbalistic, and Rosicrucian Philosophy*, Los Angeles: Philosophical Research Society, 1928, Golden Anniversary edition, p.117. (b) Thelma Moss, *The Probability of the Impossible*, New York: New American Library, 1974. (c) Robert E. Massy, *Christopher Hills' Theory of Consciousness: An Interpretation of Nuclear Evolution*, Boulder Creek: University of the Trees Press, 1976.
75. Ervin Laszlo, *The Interconnected Universe: Conceptual Foundations of Transdisciplinary Unified Theory*, River Edge: World Scientific, 1995, p.100: "The variety of field theory espoused here...claims that the events, images, and other items recalled in long-term memory are not stored *in* the brain; they are only *accessed* by the brain from an ambient field."
76. Brian Greene, *The Elegant Universe*, New York: W.W. Norton and Company, 1999, p.61: "Einstein called the indistinguishability between accelerated motion and gravity the *equivalence principle*.... Through the inclusion of gravity, general relativity ensures that all possible observational vantage points are on equal footing."

77. John Dobson, *Advaita Vedānta and Modern Science*, New York: Vivekananda Vedānta Society, 1979, p.14: "Gravity is the undividedness seen in the divided. Electricity is the infinitude seen in the finite."
78. Hannah Arendt, *The Life of the Mind*, volume I: Thinking, p.85–86.
79. Valerie V. Hunt, *Infinite Mind: Science of the Human Vibrations of Consciousness*, p.139: "One's thoughts, especially when amplified by intent or emotion, leave an imprint on matter; that is, will directs all energy."
80. Arthur Schopenhauer, *The World as Will and Representation* (E. F. J. Payne, trans.), New York: Dover Publications, 1969, p.128.
81. Hans Jonas, *The Gnostic Religion*, Boston: Beacon Press, 1970.
82. Hannah Arendt, *The Life of the Mind*, volume II: Willing, p.100.
83. (a) Alan M. Laibelman, *Going Against the Flow: An Exercise in Ethical Syncretism*, p.296–297. (b) Alan M. Laibelman, "Pure Consciousness as Ultimate Reality," *Ultimate Reality and Meaning*, p.66–68.
84. Hannah Arendt, *The Life of the Mind*, volume II: Willing, p.130.
85. Mary Beth Ingham and Mechtild Dreyer, *The Philosophical Vision of John Duns Scotus*, Washington: The Catholic University of America Press, 2004, p.190.
86. Augustine, *On Free Choice of the Will* (Thomas Williams, trans.), Indianapolis: Hackett Publishing Company, 1993, book 2, chapter 1, p.30.
87. (a) Patricia Smith Churchland has recorded a strong current that persists in the brain even during the various stages of sleep. It is suspected this baseline brain activity is a physical manifestation of the continued activity of the self-conscious mind in its state of alertness. (b) Patricia Smith Churchland, "Can Neurobiology Teach Us Anything About Consciousness?" from Ned Block, Owen Flanagan, and Güven Güzeldere, *The Nature of Consciousness: Philosophical Debates*, p.136.
88. Dean I. Radin, *The Conscious Universe*, New York: HarperCollins, 1997, p.255–256.

Chapter Seven: Mysticism

1. Bernard McGinn, *The Flowering of Mysticism*, New York: Crossroad Publishing Company, 1998, p.26.
2. (a) Renford Bambrough, introduction to J. L. Creed and A. E. Wardman (trans.), *The Philosophy of Aristotle*, New York: New American Library, 1963, p.12–13: "It was part of Aristotle's...achievement to help forward the process of division of labor among the arts and sciences, to distinguish the types and branches of knowledge, with the result that much of what he included in his encyclopedic scope is now in the hands of specialized experts—physicists, chemists, botanists, zoologists, logicians—whose fields are sharply distinguishable from philosophy. Much of his work therefore joins that of Aristarchus and Darwin on the history-of-science shelves." (b) Sylvia Fraser, *The Quest for the Fourth Monkey: A Thinker's Guide to the Psychic and Spiritual Revolution*, Toronto: Key Porter Books, 1992, p.183: "Newton also spent most of his like looking for a unified theory, but for an opposite reason: judging his own materialistic physics incomplete, he yearned for a...truth that would embrace metaphysical values. Whereas Einstein looked to mathematical equations, Newton looked to alchemical ones."
3. Edward H. Madden, "Cosmic Doing and Undoing Without End: Chauncey Wright's Idea of Ultimate Reality," *Ultimate Reality and Meaning*, Toronto: University of Toronto Press, 1993, 16(2): 27–44, p.32.
4. Peter A. Bertocci, "The Personalism of Edgar S. Brightman and Ultimate Reality," *Ultimate Reality and Meaning*, Toronto: University of Toronto Press, 1983, 6(1): 32–50, p.41: "As he argues with regard to sensory data, so he argues with regard to the data of

religious experiences—the cognitive claims presumably intrinsic to them cannot be accepted at face value independently of experience as a whole. Religious claims cannot be ruled out of court; they cannot be evaluated arbitrarily by norms drawn from other realms of experience.... Neither can they be granted autonomy, independence, or priority over other aspects of experience."

5. William James, "A Suggestion about Mysticism," from William James, *Writings 1902–1910*, New York: Library of America, 1987, p.1272.
6. Dwight Goddard (comp.), *The Laṅkāvatāra Sūtra* (D. T. Suzuki, trans.; Georg Feuerstein, ed.), San Rafael: Dawn Horse Press, 1983, p.86.
7. (a) Saint John of the Cross, *Ascent of Mount Carmel* (E. Allison Peers, trans. and ed.), Garden City: Image Books, 1958, book 2, chapter 17, p161. (b) Thomas Merton, *The Ascent to Truth*, London: Burns and Oates, 1976, p.140–142: "When the will begins to be absorbed in the mysterious and delightful love which God produces passively in it...this faculty becomes momentarily independent of reason, at least in the sense...it now begins to be guided directly by God in the intimacy of an experience of values that are not made known to it through the intelligence."
8. Thomas Merton, *Contemplative Prayer*, New York: Doubleday, 1996, p.34–38. This work was originally published as: *The Climate of Monastic Prayer*, Kalamazoo: Cistercian Publications, 1971.
9. Aryeh Kaplan, *Jewish Meditation: A Practical Guide*, New York: Schocken Books, 1985, p.61.
10. (a) Thomas Merton, *The Ascent to Truth*, p.153. (b) Thomas Merton, *The Monastic Journey*, Garden City: Doubleday, 1977, p.70: "The Fathers of the Church used the term 'active life' for the life of ascetic purification and the practice of virtue which leads to the pacification of our passions and brings them under the control of the Spirit.... The whole function of the active life...is to liberate a man from attachment to...things of the world, to deliver him from a life subject to sin, and to place him in a state of peace and interior purity which disposes his soul for contemplation."
11. William James, *The Varieties of Religious Experience*, from William James, *Writings 1902–1910*, p.342–343.
12. William Braud, "Brains, Science, Nonordinary and Transcendent Experiences," from Rhawn Joseph (ed.), *Neurotheology*, San Jose: University Press, 2002, p.129–130.
13. (a) Arthur Osborne, *Ramana Maharshi and the Path of Self-Knowledge*, New York: Sophia Perennis, 2006. (b) C. Bernard Ruffin, *Padre Pio: The True Story*, Huntington: Our Sunday Visitor, 1982.
14. Annemarie Schimmel, *Mystical Dimensions of Islam*, Chapel Hill: University of North Carolina Press, 1975, p.99–100: "Once the mystics had identified these three main parts of religious life, they began to analyze the different stages and stations that the wayfarer has to pass on his way. They distinguished between *maqam*, station, and *hal*, state.... The *maqam* is a lasting stage, which man reaches, to a certain extent, by his own striving. It belongs to the category of acts, whereas the states are the gifts of grace."
15. (a) Thomas Aquinas, *Summa Theologiae* (Thomas Gilby, general editor), Garden City: Doubleday and Company, 1969, question 12, p.190: "God can work miracles with minds as well as bodies...raising them beyond the normal order of things to a supernatural level.... He may raise up certain minds to see His essence in this life but not by making use of their bodily senses." (b) Robert Gimello, "Mysticism and Meditation," from Steven Katz (ed.), *Mysticism and Philosophical Analysis*, New York: Oxford University Press, 1978, p.172: "A numinous experience...is one of an encounter with a being wholly other

than oneself and altogether different from anything else. Such an encounter is...said to be gratuitous...those subject to it are not themselves responsible for its occurrence."

16. (a) P. D. Ouspensky, *Tertium Organum*, revised translation, Garden City: Vintage Books, 1982, p.204: "In the vision of God, says Plotinus, what sees is not our reason, but something prior and superior to our reason... He who thus sees does not properly see, does not distinguish or imagine two things (the seer and the seen). He changes, he ceases to be himself, preserves nothing of himself. Absorbed in God, he makes but one with Him, like a center of a circle coinciding with another circle." (b) Majid Fakhry, *A History of Islamic Philosophy*, New York: Columbia University Press, 1983, p.250: "The most privileged are those who, having attained to this stage, are completely annihilated or absorbed in the Supreme One and consequently see neither themselves nor any other being but only God's face."
17. Reynold A. Nicholson, *The Mystics of Islam*, New York: World Wisdom Books, 2002, p.60–61.
18. (a) Plotinus, *Enneads* (Stephen Mackenna, trans.), London: Penguin Books, 1991, IV.4, p.287–288: "In this connection it should be borne in mind that, in contemplative vision, especially when it is vivid, we are not at the time aware of our own personality; we are in possession of ourselves, but the activity is towards the object of vision with which the thinker becomes identified; he has made himself over as matter to be shaped; he takes ideal form under the action of the vision while remaining, potentially, himself."
19. (a) John W. Harvey, preface to Rudolf Otto, *The Idea of the Holy* (John W. Harvey, trans.), London: Oxford University Press, 1958, pg.xvi: "The word 'numinous' has been widely received as a happy contribution to the theological vocabulary, as standing for that aspect of deity which transcends or eludes comprehension in rational or ethical terms." (b) Alice A. Bailey, *From Intellect to Intuition*, New York: Lucis Press, 1978, p.128: "The word 'numinous,' we are told, comes from the Latin *numen*, meaning 'supernatural divine power.' It stands for the specific nonrational religious apprehension and its object, at all its levels, from the first dim stirrings where religion can hardly yet be said to exist to the most exalted forms of spiritual experience."
20. Peter Moore, "*Mystical Experience, Mystical Doctrine, Mystical Technique*," from Steven Katz, *Mysticism and Philosophical Analysis*, p.107.
21. Steven Katz, "Language, Epistemology, and Mysticism," from Steven Katz, *Mysticism and Philosophical Analysis*, p.55–56.
22. Evan Harris Walker, *The Physics of Consciousness*, Cambridge: Perseus Publishing, 2000, p.190.
23. Samuel Alexander, *Space, Time, and Deity*, two volumes, London: Macmillan and Company, 1966, volume II, p.394.
24. Ann and Barry Ulanov, *Primary Speech: A Psychology of Prayer*, Atlanta: Westminster John Knox Press, 1982, p.90.
25. Donald Rothberg, "Contemporary Epistemology and the Study of Mysticism," from Robert K. C. Forman (ed.), *The Problem of Pure Consciousness: Mysticism and Philosophy*, New York: Oxford University Press, 1990, p.191.
26. Dermot Moran, *Introduction to Phenomenology*, London: Routledge, 2000, p.124: "Whereas, in 1901, he had conceived of phenomenology...in relation to problems of logic and epistemology, gradually he came to conceive of the field of phenomenology as encompassing all conscious experiences, their correlates, and their essential structures, as a science of all essential possibilities."
27. Ibid, p.149: "We need to *bracket* certain fundamental structures in order to allow more basic objectifying acts of consciousness to become visible in themselves.... It is like putting

brackets round an expression in an equation [e.g., 2+2 = (8/4)+2] which allows one to employ an expression without subjecting what is inside...to the operations going on outside the brackets."

28. (a) Bernard McGinn, *The Foundations of Mysticism*, New York: Crossroad Publishing Company, 1994, note #145, p.433. (b) Walter T. Stace, *Mysticism and Philosophy*, New York: Macmillan, 1960, p.131–133. (c) Robert K. C. Forman, introduction to Robert K. C. Forman, *The Problem of Pure Consciousness: Mysticism and Philosophy*, p.8: "Introvertive mysticism denotes a transient state, whereas extrovertive mysticism denotes a more permanent state, one that lasts even while one is engaged in activity."
29. Richard H. Jones, *Mysticism Examined: Philosophical Inquiries into Mysticism*, Albany: State University of New York Press, 1993, p.1–2.
30. Mary Beth Ingham and Mechtild Dreyer, *The Philosophical Vision of John Duns Scotus*, Washington: The Catholic University of America Press, 2004, p.205: "It is the beatific vision, along with its natural conditions, that inform the development of the double cognitive capacity of abstractive and intuitive intellection. Human nature, rather than the human condition, serves as the point of departure for a reflection that argues for the possibility of certain knowledge, the immediate experience of extra-mental reality, and the subjective awareness of internal states."
31. Walter T. Stace, *Mysticism and Philosophy*, p.85–86.
32. Steven Katz, "Language, Epistemology, and Mysticism," from Steven Katz, *Mysticism and Philosophical Analysis*, p.26.
33. Ibid, p.60: "Even in mystical experience there seems to be epistemological activity of the sort we know as discrimination and integration and, in certain cases at least, of further mental activities such as relating the present experience to past and future experience, as well as to traditional theological claims and metaphysics."
34. Walter T. Stace, *Mysticism and Philosophy*, p.31: "It is probably impossible...to isolate 'pure' experience.... Although we may never be able to find sense experience completely free of any interpretation, it can hardly be doubted that a sensation is one thing and its conceptual interpretation is another thing. That is to say, they are distinguishable though not completely separable."
35. Paul J. Griffiths, "Pure Consciousness and Indian Buddhism," from Robert K. C. Forman, *The Problem of Pure Consciousness: Mysticism and Philosophy*, p.76–77: "Katz suggests ...the occurrence of any (mystical) experience to any subject at any time requires the possession by that subject at that time of a specifiable conceptual scheme.... You just cannot have certain experiences unless you have the appropriate equipment (conceptual, religious, sociocultural, and the like) to make them possible. This thesis seems, though not obviously universally true, at least intuitively plausible. But it is often combined with a second thesis...which has rather less initial plausibility. This second thesis states...the phenomenological attributes and content of any (mystical) experience occurring to any subject at any time are constituted by...the content of the conceptual scheme possessed by that subject... The adherent of the unmediated consciousness thesis will, of course, reject this second thesis, even if he or she accepts the former."
36. Gershom Scholem, *Kabbalah*, New York: New American Library—Meridian Press, 1978, p.176: "Devekuth results in a sense of beatitude and intimate union, yet it does not entirely eliminate the distance between the creature and its Creator, a distinction that most kabbalists, like most Hasidim, were careful not to obscure by claiming that there could be a complete unification of the soul and God." *Devekuth* means "cleaving" unto the Lord.

37. Donald Rothberg, "Contemporary Epistemology and the Study of Mysticism," from Robert Forman, *The Problem of Pure Consciousness: Mysticism and Philosophy*, p.166.
38. (a) Dermot Moran, *Introduction to Phenomenology*, p.121–122. (b) Ibid, 133–134: "Husserl wants to affirm both that phenomenology is attention to the phenomena and...its insights yield truths of a purely *a priori* character.... Husserl believed...it was possible to have an insight into the essential nature of things, that these could be 'seen' in a manner analogous to perceptual seeing of a physical object."
39. Richard Popkin and Avrum Stroll, *Skeptical Philosophy for Everyone*, Amherst: Prometheus Books, 2002, p.108–109: "Direct realism is the view that at least some of the time we apprehend physical objects...as they are.... Representative realism, in contradistinction, holds that even in normal cases of perception one's apprehension of the object is always mediated by the sensations one has.... The brain 'interprets' these data and 'reproduces' an image of the object. It is not the object but the image that is directly seen. The nature of the object is *inferred from* the image. On such a view, it is always possible that our perceptions may misrepresent the way the world is."
40. Johann Gottlieb Fichte, *Foundations of Transcendental Philosophy* (Daniel Breazeale, ed. and trans.), Ithaca: Cornell University Press, 1992, second introduction, p.95: "The idealist does not presuppose anything outside of his own consciousness; he merely postulates... this free activity of his I is that principle that cannot be derived from anything else. Idealism is *transcendental*; for though it remains within consciousness, it shows how it is possible to go beyond consciousness.... It shows how we come to assume...there are things outside of ourselves which correspond to our representations." Fichte's statement would be more accurate if rearranged: "Idealism is *transcendental*; for though it remains within Consciousness, it shows how it is possible to go beyond self-consciousness."
41. Dermot Moran, *Introduction to Phenomenology*, p.458–459: "Husserl had assumed that in the interiority of self-consciousness no indicative signs would be needed...since we do not have to signal our inner mental states to ourselves.... Derrida attacks this claim... There is...direct experience of one's mental life, which nevertheless may be represented as a kind of inner monologue."
42. Stephen Bernhardt, "Are Pure Consciousness Events Unmediated?" from Robert K. C. Forman, *The Problem of Pure Consciousness: Mysticism and Philosophy*, p.232.
43. Livia Kohn, *Early Chinese Mysticism*, Princeton: Princeton University Press, 1992, p.25–26: "First, there is the trigger. This refers to the situational setup in which the experience takes place, and the outside or psychological stimulus that sets it off.... Second, there is the moment of the experience proper.... Third, there is the afterglow.... The intensity of the experience recedes, but a sense of heightened consciousness remains.... The experience is embedded in a framework of associations, taken from the background influence that gave rise to the trigger... As a rule, the experience is then interpreted meaningfully—a process that can be divided into several stages. First, the aforementioned afterglow associations take place. Next, relevant supporting information is collected outside the experience, from friends and co-seekers, from books and records.... Only in a third step is the more universal relevance of the experience asserted. It is actively related to established theories and mystical practices and gradually shaped into an organized mystical doctrine."
44. Alan M. Laibelman, *The Other Perennial Philosophy: A Metaphysical Dialectic*, Lanham: University Press of America, 2000.
45. Richard H. Jones, *Mysticism Examined: Philosophical Inquiries into Mysticism*, p.22.
46. Norman Prigge and Gary Kessler, "Is Mystical Experience Everywhere the Same?" from Robert K. C. Forman, *The Problem of Pure Consciousness: Mysticism and Philosophy*,

p.281–283: "The qualitative sameness of two...distinct experiences...can be defined either affirmatively, in terms of the presence of similarities, or negatively, in terms of the absence of differences.... To say...two numerically distinct experiences are qualitatively the same is to say that they have qualities or aspects which are the same. It is not required that they have all the same qualities or aspects. Nor need the qualities or aspects which are the same be identical... Two...experiences...are different is to say that they have qualities or aspects which are different. It is not required, nor is it possible, that they have all different qualities or aspects. Nor need the qualities or aspects which are different be radically dissimilar.... The only thing intrinsic to an experience that can have qualities and aspects is a content of consciousness.... Since unitive mystical experiences cannot... have a content of consciousness, there cannot obtain any similarity between them *qua* contents.... Since unitive mystical experiences cannot...have a content of consciousness, there cannot obtain any difference between them *qua* contents—and this, not because their contents are similar, but because they have none whatsoever. And since two such experiences cannot be different, then, given the negative definition of sameness, they must be the same." This argument is completely fallacious. *If* the issue of sameness or differentness of experiences is, by this definition, dependent upon the sameness or differentness of aspects or qualities of an experience; *if* "the only thing intrinsic to an experience that can have a quality or aspect is a content of consciousness"; *if* "unitive mystical experiences cannot...have a content of consciousness"; *then*, the only logical conclusion is that these definitions of sameness and differentness *do not apply* to unitive mystical experiences, since the qualities or aspects upon which the assessment is made do not exist, thus rendering the assessment impossible. Hence, under these guidelines, the best that can be said about two unitive mystical experiences is that they are neither the same nor different.

47. Joseph Dan (ed.), introduction to *The Heart and the Fountain: An Anthology of Jewish Mystical Experiences*, New York: Oxford University Press, 2002, p.6.
48. William James, *The Varieties of Religious Experience*, from William James, *Writings 1902–1910*, p.381.
49. Jan van Ruysbroeck, *The Sparkling Stone*, from Ray C. Petry (ed.), *Late Medieval Mysticism*, Philadelphia: Westminster Press, 1957, p.312.
50. (a) Angela of Foligno, *The Book of Blessed Angela*, from Bernard McGinn, *The Flowering of Mysticism*, p. 144. (b) Evelyn Underhill, *Mysticism*, New York: E. P. Dutton, 1961. (c) Ray C. Petry, *Late Medieval Mysticism*, p.21: "Underhill's classic work, as indeed almost every book on the contemplative life, is developed around the triple mystic route of purgation, illumination, and union."
51. M. O. Walshe (trans.), *Meister Eckhart: German Sermons and Treatises*, two volumes, London: Watkins, 1979, volume 2, p.272–273.
52. Abū Hāmid Muhammad al-Ghazālī, *The Niche of Lights* (David Buchman, trans.), Provo: Brigham Young University Press, 1998, p.36–37.
53. (a) Ludwig Feuerbach, *The Essence of Christianity* (George Eliot, trans.), Amherst: Prometheus Books, 1989, p.12: "The object of any subject is nothing else than the subject's own nature taken objectively. Such as are a man's thoughts and dispositions, such is his God; so much worth as a man has, so much and no more has his God. Consciousness of God is self-consciousness; knowledge of God is self-knowledge." (b) Hans Küng, *Does God Exist?* (Edward Quinn, trans.), Garden City: Vintage Books, 1981, p.133: "Spinoza's God does not live apart from the universe: God is in the world and the world is in God. Nature is a particular way in which God himself exists; human consciousness is a

particular way in which God himself thinks. The individual self and all finite things are not autonomous substances but only modifications of the *one and only divine substance*."

54. (a) William C. Chittick, *Imaginal Worlds*, Albany: State University of New York Press, 1994, p.146: "Perfection is achieved when the latent divine form becomes the soul's actuality.... Perfection...does not imply some sort of static finality. It represents fully conscious participation in the dynamic and never-ending self-disclosure of *wujud.* Perfect human beings participate with total presence and undistracted awareness in the process whereby the cosmos is reborn at each instant." (b) Annemarie Schimmel, *Mystical Dimensions of Islam*, p.267: "Ibn 'Arabi's entire system is generally designated by the term *wahdat al-wujud*, 'unity of being.'... The term *wujud*, which is usually translated as 'being,' 'existence,' means basically, 'finding,' 'to be found,' and is thus more dynamic than...'existence.' At the end of the Path only God is present; is 'found'." (c) Rajmani Tigunait, *Seven Systems of Indian Philosophy*, Honesdale: Himalayan Institute of Yoga, 1983, p.26: "Each system of Indian philosophy proclaims that there is an eternal center of consciousness in man, and that the realization of this center of consciousness, or Self, is the goal of life. The Self is considered to be omniscient, omnipotent, and omnipresent." (d) Abraham Joshua Heschel, *God in Search of Man*, New York: Farrar, Straus, and Giroux, 1955, p.33: "*The grand premise* of religion is that *man is able to surpass himself*; that man who is part of this world may enter into a relationship with Him who is greater than the world; that man may lift up his mind and be attached to the absolute."
55. Jeffrey Mishlove, *The Roots of Consciousness*, New York: Random House, 1975, p.286: "The idealist physicist points to the fact that even a...slight change in the fine structure constant, $e^2/hc = 1/137$, that determines the strength of coupling of light to matter and the size of atoms (10^{-8} cm), or a slight change in the age (10^{10} years) or mass (10^{56} g) or size (10^{28} cm) of the visible universe would make life and consciousness impossible!"
56. Meister Eckhart, *Sermon 6* from Edmond Colledge and Bernard McGinn (trans.), *Meister Eckhart*, Ramsey: Paulist Press, 1981, p.187.
57. (a) G. W. F. Hegel, *Phenomenology of Spirit* (A. V. Miller, trans.), Oxford: Oxford University Press, 1977, section 165, p.103. (b) Johann Gottlieb Fichte, *Foundations of Transcendental Philosophy*, second introduction, p.96–97: "Our I is nothing other than consciousness itself.... It is not correct to think...the I becomes conscious by means of something else." (c) Dwight Goddard, *The Laṅkāvatāra Sūtra*, p.83: "Between Universal Mind and... individual discriminating mind is the intuitive mind which is dependent upon Universal Mind for its cause and support and enters into relations with both. It partakes of the universality of Universal Mind, shares its purity, and like it, is above form and momentariness.... The intuitive mind enters into relations with the lower mind-system, shares its experiences, and reflects upon its activities." (d) Jean-Paul Sartre, *The Transcendence of the Ego* (Forrest Williams and Robert Kirkpatrick, trans.), New York: Farrar, Straus, and Giroux, 1991, p.35–36: "It is...transcendental consciousness which constitutes our empirical consciousness...with its psychic and psychophysical *me*. For our part, we readily acknowledge the existence of a constituting consciousness.... Is not this psychic and psychophysical *me* enough? Need one double it with a transcendental I, a structure of absolute consciousness?... If the reply is negative, the consequences are: First, the transcendental field becomes impersonal...*without an I.* Second, the I appears only at the level of humanity and is only one aspect of the *me*, the active aspect."
58. (a) Pierre Teilhard de Chardin, *The Phenomenon of Man*, New York: Harper and Row, 1965, p.60. (b) Elisabeth Haich, *Initiation*, Garberville: Seed Center Press, 1974, p.17: "The true divine self is perfection itself, so a development in it is not possible. *Only the body must develop in order to be able to manifest higher and higher vibrations and higher*

and higher frequencies of the self." (c) Valerie V. Hunt, *Infinite Mind: Science of the Human Vibrations of Consciousness*, Malibu: Malibu Publishing Company, 1989, p.286–287: "I concur with the statement of contemporary theologians, 'God is indeed within man.' But I do not believe that man can perceive his 'God-like' qualities until his field reaches higher vibrations and attains a greater degree of coherency."

59. John Horgan, *Rational Mysticism*, New York: Houghton Mifflin Company, 2003, p. 80–81: "A more generic limitation of his [Andrew Newberg's] experiments is that he must rely on his subject's assessments of their own states of mind; one subject's self-described peak experience might be a trivial event for someone else... He conceded his study might be capturing brain states related...to meditation and prayer...rather than to deep mystical states *per se*."
60. Ibid, p.75.
61. (a) Roland Fischer, "A Cartography of Understanding Mysticism," *Science*, 26 Nov 1971, 174: 897–904. (b) Aryeh Kaplan, *Jewish Meditation: A Practical Guide*, p.30: "It seems that the mind has two modes in which it possesses abnormal ability to solve problems.... In 'hot' concentration, the entire body is brought into play and, as it were, the adrenaline is made to flow. In 'cool' concentration, body and mind are quieted down as much as possible, so that the mind is able to focus on the problem like a laser beam." (c) Andrew B. Newberg, Eugene D'Aquili and Vince Rause, *Why God Won't Go Away*, New York: Ballantine Books, 2001, p.116–117: "Meditative techniques assume different forms and functions. Some mystics meditate to bring the mind to a laser-sharp focus; others meditate to undo the mind's focus and sweep away all thoughts. Some engage in reflective prayer... Others pray more passively by simply opening their minds to the possibility of God." (d) Robert K. C. Forman, introduction to Robert K. C. Forman, *The Problem of Pure Consciousness: Mysticism and Philosophy*, p.7: "Physiological parameters ...heart rate, skin temperature, spontaneous galvanic skin response, and electroencephalogram patterns all differ markedly. Given the differences, these two sets of states are unlikely to have identical psychological characteristics, causes, and etiologies."
62. Paul Weiss, "God and the World," *Science, Philosophy and Religion: A Symposium*, New York: Conference on Science, Philosophy, and Religion in their Relation to the Democratic Way of Life, 1941, p.429–430, from Charles Hartshorne and William L. Reese, *Philosophers Speak of God*, Amherst: Prometheus Press, 2000, p.320.
63. Michael A. Persinger, "Experimental Simulation of the God Experience," from Rhawn Joseph, *Neurotheology*, p.280.
64. Ibid, p.286: "Several experiments, involving about one hundred subjects, showed that stimulation over the right hemisphere first by a continuous frequency modulated field for about fifteen minutes followed by a bilateral stimulation over the temporal-parietal lobes with a burst-firing field once every three or four seconds markedly facilitates the sensed presence. If the presence occurs during the right hemispheric stimulation, the experience appears to be along the left side and to be more emotional (or fearful). If the experience does not occur until the bilateral stimulation, the location of the "being" is perceived to be more to the right or on both sides, and to be pleasant, even spiritual. About eighty percent of volunteers exposed to this sequence report a presence. The reverse procedure (burst-firing followed by frequency modulation) is less effective." Three methods were used to determine the onset of "sensed presence" by Persinger's subjects: (a) they were asked to articulate feelings as the tests proceeded; (b) they were questioned about their experience after the experiments were concluded; (c) they were instructed "to press one of two buttons to indicate when a presence occurred and on which side it was experienced."

65. One tesla is equal to one newton-second/coulomb-meter; since one newton equals one kilogram-meter/second^2, then, one tesla is also equal to one kilogram/coulomb-second. In English, magnetic field strength depends upon the force (in newtons) developed by a moving charge (in coulombs) covering a given distance (in meters) over a specified time (in seconds).
66. (a) Andrew B. Newberg, Eugene D'Aquili and Vince Rause, *Why God Won't Go Away*, p.4–5. The book's title exemplifies the supposition of its authors, in complete alignment with the approach pursued by Persinger, that God is, to use the vernacular, "hard-wired into the brain." It must be emphasized that the ability to analyze some phenomenon using techniques imported from physical science does not imply the topic in question is entirely within the purview of scientific mechanisms of action: correlations can be identical, causative, or associative. (b) Max Velmans, *Understanding Consciousness*, London: Routledge, 2000, p.36: "Ontological identity is *symmetrical*; that is, if A is identical to B, then B is identical to A. Ontological identity also *obeys Leibniz's Law*: if A is identical to B, all the properties of A are also properties of B, and vice versa.... Correlation is also *symmetrical*: if A correlates with B, then B correlates with A. But correlation *does not obey Leibniz's Law*: if A correlates with B, it does not follow that all the properties of A and B are the same.... Causation, by contrast, is *asymmetrical*: if A causes B, it does not follow that B causes A.... Causation *does not obey Leibniz's Law*.... Under appropriate conditions, brain states may be shown to cause, or correlate with, conscious experiences, but it does not follow...conscious experiences are nothing more than states of the brain."
67. (a) Andrew B. Newberg and Jeremy Iversen, "On the 'Neuro' in Neurotheology," from Rhawn Joseph, *Neurotheology*, p.254–255. (b) A. N. Elias and A. F. Wilson, "Serum Hormonal Concentrations Following Transcendental Meditation—Potential Role of γ-Aminobutyric Acid (GABA)," *Medical Hypotheses*, 1995, 44: 287–291. (c) M. Bujatti and P. Riederer, "Serotonin, Noradrenaline, and Dopamine Metabolites in Transcendental Meditation," *Journal of Neural Transmission*, 1976, 39(3): 257–267. (d) G. A. Tooley, S. M. Armstrong, T. R. Norman, and A. Sali, "Acute Increases in Night-Time Plasma Melatonin Levels Following a Period of Meditation," *Biological Psychiatry*, 2000, 53(1): 69–78. (e) R. Sudsuang, V. Chentanez, and K. Veluvan, "Effects of Buddhist Meditation on Serum Cortisol and Total Protein Levels, Blood Pressure, Pulse Rate, Lung Volume, and Reaction Time," *Physiology of Behavior*, 1991, 50: 543–548. (f) J. R. Infante, M. Torres-Aviabol, and P. Pinel, "Catecholamine Levels in Practitioners of the Transcendental Meditation Technique," *Physiology and Behavior*, 2001, 72(1–2): 141–146.
68. Andrew B. Newberg, Eugene D'Aquili and Vince Rause, *Why God Won't Go Away*, p.86.
69. Plotinus, *Enneads* (Stephen Mackenna, trans.), London: Penguin Books, 1991, IV.8, p.334.
70. (a) Saint John of the Cross, *Dark Night of the Soul* (E. Allison Peers, trans.), New York: Doubleday and Company, 1990, p.68. (b) Ibid, p.155–156: "Apart from the fact...the soul has no desire to speak of it, it can find no suitable way or manner or similitude by which it may be able to describe such lofty understanding and such delicate spiritual feeling.... For as that inward wisdom is so simple, so general, and so spiritual that it has not entered into the understanding enwrapped or cloaked in any form or image subject to sense, it follows that sense and imagination cannot account for it or imagine it, so as to say anything concerning it, although the soul be clearly aware that it is experiencing and partaking of that rare and delectable wisdom. It is like one who sees something never seen before, whereof he has not even seen the like; although he might understand its nature and have experience of it, he would be unable to give it a name, or say what it is, however much he tried to do so, and this in spite of its being a thing which he had

perceived with the senses. How much less, then, could he describe a thing that has not entered through the senses!"

71. Richard H. Jones, *Mysticism Examined: Philosophical Inquiries into Mysticism*, p.20: "The depth-mystical experiences result...from a radical 're-collecting' of the senses and a 'purging' of the mind of all dispositional and cognitive content—especially any sense of 'I.' The resulting one-pointedness produces a stillness of mind where all sensory-conceptual apparatuses are in total abeyance."
72. William James, *The Varieties of Religious Experience*, p.228: "The characteristics of the affective experience...can be easily enumerated, though it is probably difficult to realize their intensity, unless one has been through the experience oneself. The central one is the loss of all the worry, the sense...all is ultimately well with one, the peace, the harmony, the *willingness to be*, even though the outer conditions should remain the same."
73. Franklin Merrill-Wolff, *The Philosophy of Consciousness Without an Object*, New York: Julian Press, 1973, p.67–68.
74. Charles N. Alexander, Ken Chandler, and Robert W. Boyer, "Experience and Understanding of Pure Consciousness in the Vedic Science of Maharishi Mahesh Yogi," unpublished paper, p.6, from Robert K. C. Forman, *The Problem of Pure Consciousness: Mysticism and Philosophy*, p.27–28.
75. Leonard Angel, *Enlightenment East and West*, Albany: State University of New York Press, 1994, p.303: "Mystics the world over have reported the awakening of subtle energy circulating through the body.... It is often described as coalescing in bundles in various parts of the body, or as being particularly strong in various parts of the body. Some people report...it awakens in stages in the various parts of the body, others that it awakens all at once. People...report a strong link between the awakening of this subtle energy and the experience of bliss."
76. Alice A. Bailey, *From Intellect to Intuition*, p.67: "Four degrees of prayer might be recognized: 1. Prayer for material benefits, and for help; 2. Prayer for virtues and for graces of character; 3. Prayer for others, intercessory prayer; 4. Prayer for illumination and for divine realization." Petitionary prayer might fall under any of these categories: the first subset if the prayer is meant to aid oneself, or the third class if meant to aid others; if specifically intended for mystical unfolding, it would clearly be a subset of both the second and fourth groupings. Prayers of thanksgiving would most likely be a subset of the first two classes, while prayers of glorification would also be subsumed in sets two and four.
77. Jerome Gellman, "Judaic Perspectives on Petitionary Prayer," from Paul Helm (ed.), *Referring to God: Jewish and Christian Philosophical and Theological Perspectives*, New York: St. Martin's Press, 2000, p.129–130.
78. Ibid, p.147.
79. Ann and Barry Ulanov, *Primary Speech: A Psychology of Prayer*, p.67.
80. Simone Weil, *Waiting for God* (Emma Craufurd, trans.), New York: Capricorn, 1951, p.110–111, from Ann and Barry Ulanov, *Primary Speech: A Psychology of Prayer*, p.99–100.
81. Ann and Barry Ulanov, *Primary Speech: A Psychology of Prayer*, p.13: "Desire is the motivating force that leads us toward that meeting and prayer the language in which that movement explains itself."
82. Thomas Merton, *The Ascent to Truth*, p.142–144.
83. Leonard Angel, *Enlightenment East and West*, p.143.
84. Edward Conze, *Buddhism: Its Essence and Development*, New York: Harper and Row, 1975, p.50: "During the course of one *kalpa*, a world system completes its evolution from its initial condensation to the final conflagration. One world system follows the other, without

beginning and end, quite interminably." By tradition, one *kalpa* is equal to 4.32 billion earth years.

Chapter Eight: Parapsychology

1. (a) Patanjali, *Aphorisms of Yôga* (Purohit Swāmi, trans.), London: Faber and Faber, 1979, part III, aphorism 19: "Concentrate on another's mind; know that mind"; aphorism 22: "Concentrate on immediate or future karma, know the time and cause of death"; aphorism 37: "These powers of knowledge are obstacles to illumination"; aphorism 50: "Finally, by renouncing even these powers, the seed of bondage being destroyed, the yôgi attains liberation"; aphorism 51: "The yôgi should not be allured or wonderstruck by the courtship of celestial powers, lest he fall into the undesirable world again." (b) Evelyn Underhill, *Mysticism*, New York: E.P. Dutton, 1961, p.149: "In every period of true mystical activity we find an outbreak of occultism, illuminism, or other perverted spirituality and...a borderland region where the mystical and psychical meet."
2. Dean I. Radin, *The Conscious Universe*, New York: HarperCollins, 1997, p.40: "Psi effects do not fall into the class of *easily* replicated effects. There are eight typical reasons why replication is difficult to achieve: (1) the phenomenon may not be replicable; (2) the written experimental procedures may be incomplete, or the skills needed to perform the replication may not be well understood; (3) the effect under study may change over time or react to the experimental procedure; (4) investigators may inadvertently affect the results of their experiments; (5) experiments sometimes fail for sociological reasons; (6) there are psychological reasons that prevent replication from being easy to conduct; (7) the statistical aspects of replication are much more confusing than most people think; and (8) complications in experimental design affect some replications."
3. C. E. M. Hansel, *The Search for Psychic Power*, Buffalo: Prometheus Books, 1989, p.19.
4. William Braud, "Brains, Science, Nonordinary and Transcendent Experiences," from Rhawn Joseph (ed.), *Neurotheology*, San Jose: University Press, 2002, p.126.
5. C. E. M. Hansel, *The Search for Psychic Power*. Experiments included: (a) J. B. Rhine's studies with Zener cards; (b) Montague Ullman's study of telepathic dreams in matching dream content with targeted pictures; (c) Harold Putoff and Russell Targ's studies of remote viewing, (d) Helmut Schmidt's investigations on telepathic influences on random number generation by machine.
6. Paul Weiss, *Being and Other Realities*, Peru: Open Court Publishing Company, 1995, p.293.
7. Dean I. Radin, *The Conscious Universe*, p.48.
8. Seymour Epstein, "The Stability of Behavior. II. Implications for Psychological Research," *American Psychologist*, 1980, 35: 790–806, p.790.
9. J. D. Bozarth and R. R. Roberts, "Signifying Significant Significance," *American Psychologist*, 1972, 27: 774–775.
10. (a) Dean I. Radin, *The Conscious Universe*, p.49–50. (b) V. E. Barns, *et al.*, "Confirmation of the Existence of the Omega Minus Hyperion," *Physics Letters*, 1964, 12: 134–136.
11. (a) Dean I. Radin, *The Conscious Universe*, p.97. The accumulated data derive from "186 publications by dozens of investigators around the world." Odds against chance are 1 in 5 for a standard deck of twenty-five cards containing five each of five different symbols. (b) J. G. Pratt, J. B. Rhine, B. M. Smith, C. E. Stuart, and J. A. Greenwood, *Extrasensory Perception After Sixty Years*, Boston: Bruce Humphries, 1966, p.42.
12. Dean I. Radin, *The Conscious Universe*, p.70–73. Test subjects were monitored by electroencephalograph to determine when he or she entered a rapid eye movement dream cycle. Then, the sender viewed photographs of targets and tried to telepathically convey

the contents of that picture to the dreaming receiver. When awakened, subjects were asked to verbally record what they remembered of their dreams. Independent judges not otherwise involved in the conduct of the experiment were asked to score the degree of match between recorded dreams and eight possible picture targets on a scale of 1 to 8, with a score of 1 being a perfect match and a score of 8 being a complete miss. Any judged score between 1 and 4 was considered a "hit," which meant there was statistically a 50 percent chance of scoring a hit. Clearly, a significant factor is the liberality of each judge's determination of a match between a recorded dream and any particular potential picture target. In addition, as Radin remarked: "Left out of this analysis were other important factors...assessing how many studies might have been conducted that were not published, evaluating the quality of individual studies, and assessing the degree of replication across different experiments and experimenters."

13. (a) Dean I. Radin, *The Conscious Universe*, p.101. The standard procedure for remote-viewing experiments consists of an investigator traveling to some location geographically distant from the subject, but near some uniquely identifiable structure such as a bridge or a factory smokestack. Distances studied have varied from a few miles to several thousand miles—the distance between the East and West Coast of the United States. The traveler is directed to focus on the unique structure at a time predetermined in conjunction with his co-researcher back in the lab observing the subject. It is important that this co-researcher have no advance knowledge of the location or structure serving as the target. The subject is then requested to focus his or her attention on the distant target and draw a picture of whatever occurs to their perception. A panel of independent investigators, typically three to five in number, who are unfamiliar with either the subject's drawing, the location or the structure at the target site, are given a set of four to eight professional photographs depicting various possible targets, one of which is the correct choice. They are asked to separately compare these possible targets with the subject's drawing, and rank the pictures in order of closest to furthest approximation to the drawing. If the actual target is among the top half of the ranked pictures, this is considered a success; if the actual target is among the bottom half of ranked pictures, this is recorded as a failure. Odds of success are therefore 2:1 with each target for each judge. (b) E. C. May, J. M. Utts, V. V. Trask, W. W. Luke, T. J. Frivold, and B. S. Humphrey, "Review of the Psychoenergetic Research Conducted at SRI International (1973–1988)," SRI *Technical Report*, March 1988.
14. (a) Dean I. Radin, *The Conscious Universe*, p.113–116. In a forced-choice precognition study, a person is asked to guess which one of a fixed number of targets will be selected *later*, with the choice of targets taken from a set of colored lamps, or standard ESP card symbols, or a standard die face. One target is randomly selected and if the person's guess matches the selected symbol, this is counted a "hit." Radin reported: "Charles Honorton and Diane Ferrari...found 309 studies, reported in 113 articles...contributed by sixty-two different investigators," and further remarked, "to be sure the same results have been replicated, a homogeneous set of effects is commonly created by trimming from the full set ten percent of the studies producing the largest effects and ten percent of the studies producing the smallest effects. After performing this trimming, Honorton and Ferrari were left with 248 studies, and the total number of investigators was reduced from sixty-two to fifty-seven. But the combined effect of the remaining 80 percent of the data still produced odds against chance of a billion to one." (b) C. Honorton and D. C. Ferrari, "Fortune Telling: A Meta-analysis of Forced-Choice Precognition Experiments, 1935–1987," *Journal of Parapsychology*, 1989, 53: 281–308.
15. (a) Dean I. Radin, *The Conscious Universe*, p.88. The *ganzfeld* experimental setup consisted of a receiver who sat in a room electromagnetically shielded and soundproofed

against external noise. His or her eyes were covered and he or she wore headphones, thus depriving the subject of external sensory stimulation. A sender, sitting in another room, picks a target at random out of a set of four uniquely different targets—photos, paintings, or video clips from TV show scenes or movie films—with the selection being completely unknown to the experimenter. The sender attempted to send the information contained in the target telepathically for twenty minutes. Afterwards, the receiver was given a duplicate set of the four targets, and asked to rank them from 1 to 4 based on his or her mental imagery during the twenty-minute session. A direct hit is recorded when the sender ranks the proper target as "1." There is a 25 percent chance expectation of being right. (b) C. Honorton, "Meta-Analysis of Psi Ganzfeld Research," *Journal of Parapsychology*, 1985, 49: 51–91. (c) C. Honorton, R. E. Berger, M. P. Varvoglis, M. Quant, P. Derr, E. I. Schechter, and D. C. Ferrari, "Psi Communication in the Ganzfeld: Experiments with an Automated Testing System and a Comparison with a Meta-analysis of Earlier Studies," *Journal of Parapsychology*, 1990, 54: 99–139.

16. Dean I. Radin, *The Conscious Universe*, p.140–142. In random event precognition experiments, success by random chance is fifty percent; that is, half the time the guesser will be correct, and half the time he or she will be wrong. Thus, overall success through multiple experiments is determined by the amount of successful guesses exceeding fifty percent. (b) R. D. Nelson and D. I. Radin, "Effects of Human Intention on Random Event Generators: A Meta-analysis," *Proceedings of Abstracts of the Eastern Psychological Association Conference*, Philadelphia, March 29–April 1, 1990.
17. Robert Morris, "Biology and Psychical Research," from Gertrude R. Schmeidler (ed.), *Parapsychology: Its Relation to Physics, Biology, Psychology, and Psychiatry*, Metuchen: The Scarecrow Press, 1976, p.50: "The reader should bear in mind the following problems with anecdotal evidence: deliberate fraud; distortion in perception of the original events; distortion in retention of the original events; impossibility of exactly recreating the original events to look for additional factors not known to the reporters; distortion and/or selective reporting by the investigator; and difficulty in assessing how unusual the event is, i.e., how often, if at all, it would occur by chance."
18. Dean I. Radin, *The Conscious Universe*, p.97.
19. Ibid, p.116.
20. Rhawn Joseph, "Creation Science and the Myth of the Organic Soup," from Rhawn Joseph, *Neurotheology*, p.48.
21. Sylvia Fraser, *The Quest for the Fourth Monkey: A Thinker's Guide to the Psychic and Spiritual Revolution*, Toronto: Key Porter Books, 1992, p.34–35.
22. Stanley Krippner, *Song of the Siren: A Parapsychological Odyssey*, New York: Harper and Row, 1975, p.247.
23. Ibid, p.281.
24. R. B. Roberts, "A Theory for Psi," from Gertrude R. Schmeidler, *Parapsychology: Its Relation to Physics, Biology, Psychology, and Psychiatry*, p.41.
25. Max Velmans, *Understanding Consciousness*, London: Routledge, 2000, p.17.
26. The gravitational constant, G, has a value of 6.67×10^{-11} newton-meter^2/kg^2. The electromagnetic constant, k, has a value of 8.99×10^{9} newton-meter^2/coulomb^2. It should be understood that: 1 newton = 1 kg-meter/sec^2; 1 coulomb = 1.6×10^{-19} joules of energy.
27. Russell Targ and Jane Katra, *Miracles of Mind*, Novato: New World Library, 1998, p.274–275.
28. William Braud, "Brains, Science, Nonordinary and Transcendent Experiences," from Rhawn Joseph, *Neurotheology*, p.129.

29. S. David Kahn, "Myers' Problem Revisited," from Gertrude R. Schmeidler, *Parapsychology: Its Relation to Physics, Biology, Psychology, and Psychiatry*, p.219.
30. (a) F. W. H. Myers, *Human Personality and its Survival of Bodily Death*, London: Longmans and Green, two volumes, 1903. (b) Dean I. Radin, *The Conscious Universe*, p.281: "In ordinary states, the mind is more particle-like and is firmly localized in space and time. This is supported by the ordinary subjective experience of being an isolated, independent creature. But in unusual, nonordinary states of awareness, our minds may be more wavelike, and no longer localized in space or time. This is supported by subjective experiences of timelessness, mystical unity, and psi." It is not a great stretch to correlate a "particle-like mind" with self-consciousness and a "wavelike mind" with Myers' group mind or global consciousness. (c) Robert G. Jahn and Brenda J. Dunne, *Margins of Reality*, New York: Harcourt Brace Jovanovich, 1987.
31. Sylvia Fraser, *The Quest for the Fourth Monkey: A Thinker's Guide to the Psychic and Spiritual Revolution*, p.42–43.
32. (a) Ibid, p.55–56. (b) Valerie V. Hunt, *Infinite Mind: Science of the Human Vibrations of Consciousness*, Malibu: Malibu Publishing Company, 1989, p.133: "Thought is an organized field of energy composed of complex patterns of vibrations which consolidate information."
33. Sylvia Fraser, *The Quest for the Fourth Monkey: A Thinker's Guide to the Psychic and Spiritual Revolution*, p.110.
34. Itzhak Bentov, *Stalking the Wild Pendulum*, New York: Bantam Books, 1981, p.53.
35. Valerie V. Hunt, *Infinite Mind: Science of the Human Vibrations of Consciousness*, p.63–64.
36. E. Gellhorn and W. F. Kiely, "Mystical States of Consciousness: Neurophysiological and Clinical Aspects," *Journal of Nervous and Mental Disease*, 1972, 154: 399–405, p.400–401: "The physiologic change which accompanies the mental state of meditation is a shift ...to the trophotropic side. The electroencephalograph shows four stages as the subject develops proficiency in meditation as judged by the Zen master. The presence of alpha potentials...characterizes stage I. These potentials increase in amplitude in stage II and decline in frequency in stage III, whereas in stage IV theta potentials appear. The decrease in heart and respiratory rate and in oxygen consumption—the latter due to the relaxation of skeletal muscles—confirms the interpretation that...dominance of the trophotropic system is involved."
37. K. Badawi, R. K. Wallace, D. W. Orme-Johnson, A. M. Rouzere, "Electrophysiological Characteristics of Respiratory Suspension Periods Occurring During the Practice of the Transcendental Meditation Program," *Psychosomatic Medicine*, 1984, 46: 267–276, p.273.
38. Sylvia Fraser, *The Quest for the Fourth Monkey: A Thinker's Guide to the Psychic and Spiritual Revolution*, p.36–37.
39. Ibid, p.117.
40. (a) Russell Targ and Jane Katra, *Miracles of Mind*, p.75. (b) James Spottiswoode, "Geomagnetic Fluctuations and Free-Response Anomalous Cognition: A New Understanding," *Journal of Parapsychology*, March 1997, 61(1): pages unstated.
41. John Horgan, *Rational Mysticism*, New York: Houghton Mifflin Company, 2003, p.96.
42. Ervin Laszlo, *The Interconnected Universe: Conceptual Foundations of Transdisciplinary Unified Theory*, River Edge: World Scientific, 1995, p.30–31.
43. Ibid, p.32: "Scalars are longitudinal waves, and it is...recognized that their propagation is proportional to the medium's mass density.... It is a variable quantity, higher in regions of dense mass, in or near stars and planets, and lower in deep space.... Scalars travel faster through matter-dense regions of the vacuum than in deep space."

44. Ibid, p.33.
45. Boethius, *The Consolation of Philosophy* (P. G. Walsh, trans.), Oxford: Oxford University Press, 1999, book V, chapter 6, p.111–112.
46. (a) Robert Bunge, "Concept of Ultimate Reality and Meaning of the Teton Sioux," *Ultimate Reality and Meaning*, Toronto: University of Toronto Press, 1987, 10(2): 83–100, p.92: "The Wakantanka (Great Mystery) is the ultimate reality on the cosmic level. The Great Mystery is a genderless...concept of power." (b) Thomas Aquinas, *Summa Theologica* (Fathers of the English Dominican Province, trans,), London: Burns, Oates, and Washbourne, 1920, part I, questions 1–26, p.209, from Charles Hartshorne and William L. Reese, *Philosophers Speak of God*, Amherst: Prometheus Press, 2000, p.126–127: "Although contingent things become actual successively, nevertheless God knows contingent things not successively...but simultaneously. The reason is...His knowledge is measured by eternity...and eternity being simultaneously whole comprises all time."
47. (a) Ervin Laszlo, *The Interconnected Universe*: *Conceptual Foundations of Transdisciplinary Unified Theory*, p.96. (b) Glen Rein, "Biological Interactions with Scalar Energy: Cellular Mechanisms of Action," *Proceedings of the Seventh International Association of Psychotronics Research*, Georgia, December 1988.
48. (a) Daniel C. Matt, "Ayin: The Concept of Nothingness in Jewish Mysticism," from Robert K. C. Forman (ed.), *The Problem of Pure Consciousness: Mysticism and Philosophy*, New York: Oxford University Press, 1990, p.142. (b) Uziel Meizlish, *Tif'eret 'Uzzi'el*, Tel Aviv: unpublished, 1962, p.53b.
49. (a) Valerie Hunt, *Infinite Mind: Science of the Human Vibrations of Consciousness*, p.289. (b) Jack J. Cohen, *Major Philosophers of Jewish Prayer in the Twentieth Century*, New York: Fordham University Press, 2000, p.44: "In [Abraham Isaac] Kook's eyes, prayer is part of a cosmic process struggling toward unity.... On its part, the human soul engages in ceaseless prayer in order to establish its place in the cosmos and to elevate and unify all of reality with the Source of blessing and life."
50. Larry Dossey, *Prayer is Good Medicine*, New York: HarperCollins Publishers, 1996, p.130.
51. Russell Targ and Jane Katra, *Miracles of Mind*, p.76.
52. Sylvia Fraser, *The Quest for the Fourth Monkey: A Thinker's Guide to the Psychic and Spiritual Revolution*, p.46–47.
53. Ibid, p.50.
54. Thomas Merton, *Contemplative Prayer*, New York: Doubleday, 1996, p.40: "Other obstacles: *Discouragement*—we lose all confidence, become secretly convinced that we cannot get anywhere in prayer.... The remedy—*hope*. *Confusion, helplessness*—a sense of incapacity ...imprisoned in ourselves we become paralyzed. The way out is *faith*."
55. Ann and Barry Ulanov, *Primary Speech: A Psychology of Prayer*, Atlanta: Westminster John Knox Press, 1982, p.29.
56. (a) Dean I. Radin, *The Conscious Universe*, p.143–144. (b) Brenda J. Dunne, "Co-Operator Experiments with a Random Event Generator Device," from K. R. Rao (ed.), *Cultivating Consciousness*: *Enhancing Human Potential, Wellness, and Healing*, Westport: Praeger, 1993, p.149–163. (c) B. J. Dunne, Y. H. Dobyns, R. G. Jahn, and R. D. Nelson, "Series Position Effects in Random Event Generator Experiments," *Journal of Scientific Exploration*, 1994, 8: 197–215.
57. Dean I. Radin, *The Conscious Universe*, p.278–279.
58. Ibid, p.292.
59. Karl Pribram, *Languages of the Brain*, Monterey: Wadsworth Publishing, 1977, p.169, from Michael Talbot, *The Holographic Universe*, New York: HarperCollins Publishers,

1991, p.25: "In a series of experiments conducted in the late 1960s, [physiologist Georg von] Bekesy placed vibrators on the knees of blindfolded test subjects. Then he varied the rates at which the instruments vibrated. By doing so he discovered that he could make his test subjects experience the sensation that a point source of vibration was jumping from one knee to the other. He found that he could even make his subjects feel the point source of vibration in the space *between* their knees. In short, he demonstrated that humans have the ability to seemingly experience sensation in spatial locations where they have absolutely no sense receptors."

60. Alison Gopnick, Andrew Meltzoff, and Patricia Kuhl, *The Scientist in the Crib: Minds, Brains, and how Children Learn*, New York: William Morrow, 1999, p.186–187: "At birth, each neuron in the cerebral cortex has around 2500 synapses. The number of synapses reaches its peak at two to three years of age, when there are about 15,000 synapses per neuron. This is actually many more than in an adult brain.... Brains don't just steadily make more and more connections.... They grow many more connections than they need and...get rid of lots of them.... The synapses that carry the most messages get stronger and survive, while weaker synaptic connections are cut out.... The process of connecting and pruning allows the brain to adapt itself to its surroundings."
61. (a) Amit Goswami, Maggie Goswami, and Richard E. Reed, *The Self-Aware Universe*, New York: Jeremy Tarcher Press, 1993, p.132–133. (b) J. Grinberg-Zylberbaum and J. Ramos, "Patterns of Inter-hemispheric Correlation during Human Communication," *International Journal of Neuroscience*, 1987, 36: 52–54. (c) J. Grinberg-Zylberbaum, *Creation of Experience*, Mexico: Instituto Nacional para el Estudio de la Conciencia, 1988.
62. (a) Konstantin Meyl, "Scalar Waves: Theory and Experiments," *Journal of Scientific Exploration*, 2001, 15(2): 199-205. See: http://www.scientificexploration.org/jse/articles. (b) Thomas Bearden, "Vectors and Mechanisms Clarified," *Towards a New Electromagnetics*, part 4, 1983. See: http://www.cheniere.org/books/part4/s28.htm.
63. Sylvia Fraser, *The Quest for the Fourth Monkey: A Thinker's Guide to the Psychic and Spiritual Revolution*, p.227–228.
64. Michael Talbot, *The Holographic Universe*, p.125.

Chapter Nine: Psychophysics

1. João Magueijo, *Faster than the Speed of Light*, London: Penguin Books, 2004, p.46.
2. William A. Tiller, *Science and Human Transformation*, Walnut Creek: Pavior Publishing, 1997, p.47. In determination of the value for the cosmological constant: gravitational constant, $G = 6.67 \times 10^{-11}$ newton-meter^2/kg^2, speed of light in a vacuum, $c = 3 \times 10^{10}$ cm/sec, $\pi = 3.14159$, and the vacuum energy density, $\rho = 4 \times 10^{-31}$ g/cm^3.
3. Alan M. Laibelman, *The Other Perennial Philosophy: A Metaphysical Dialectic*, Lanham: University Press of America, 2000, p.241.
4. Aristotle, *Nichomachean Ethics* (J. E. C. Welldon, trans.), Amherst: Prometheus Books, 1987, book III, chapter x, p.89.
5. Annie Murphy Paul, *The Cult of Personality*, New York: Free Press, 2004.
6. Oliver P. John and Sanjay Srivastava, "The Big Five Trait Taxonomy: History, Measurement, and Theoretical Perspectives," from Lawrence A. Pervin and Oliver P. John (ed.), *The Handbook of Personality: Theory and Research*, New York: Guilford Press, 1999, p.102–138. The tabulation of traits, Table 4.2, can be found on p.133.
7. Walter Mischel and Yuichi Shoda, "Integrating Dispositions and Processing Dynamics within a Unified Theory of Personality: The Cognitive-Affective Personality System,"

from Lawrence A. Pervin and Oliver P. John, *The Handbook of Personality: Theory and Research*, p.197–218. The statement is on p.197–198.

8. (a) Annie Murphy Paul, *The Cult of Personality*, p.194. (b) Daniel P. McAdams, "The Five Factor Model in Personality: A Critical Appraisal," *Journal of Personality*, 1992, June issue.
9. Annie Murphy Paul, *The Cult of Personality*, p.194—195.
10. Yuichi Shoda and Walter Mischel, "Reconciling Contextualism with the Core Assumptions of Personality Psychology," *European Journal of Personality*, 2000, 14(5): 407—428.
11. Jeffrey Mishlove, *The Roots of Consciousness*, New York: Random House, 1975, p.288.
12. John R. Pierce, *An Introduction to Information Theory*, second revised edition, New York: Dover Publications, 1980, p.80.
13. Ibid, p.40.
14. Ibid, p.80–81.
15. Ibid, p.91.
16. Ken Wilber, *Integral Psychology*, Boston: Shambhala Publications, 2000, p.viii.
17. John R. Pierce, *An Introduction to Information Theory*, p.36–38.
18. Ibid, p.44.
19. Amit Goswami, Maggie Goswami, and Richard E. Reed, *The Self-Aware Universe*, New York: Jeremy Tarcher Press, 1993, p.193–194.
20. Evan Harris Walker, *The Physics of Consciousness*, Cambridge: Perseus Publishing, 2000, p.207.
21. Ibid, p.209.
22. Ibid, p.209–214.
23. Ibid, p.218–222.
24. Ibid, p.228–229.
25. Ibid, p.231. The data contained in appendix I, p.339–342, used to calculate the channel consciousness capacity in the text included: $M = 7.45 \times 10^{20}$ soluble RNA molecules; t (time for synapse to either fire or revert to resting state) = 0.3 millisecond; n = 200,000 electrons (available in the postsynaptic cell of all the neurons in the brain); $N = 2.35 \times 10^{13}$ synapses; $\tau = 8.4 \times 10^{-12}$ seconds; and, $f = 0.0148$ second^{-1}. On pages 233 and 234, Walker provided the equations: $f = \tau M/nNt^2$, $Q = n/f$, and iQ/t in deriving the final expression.
26. (a) John R. Pierce, *An Introduction to Information Theory*, p.106: "The largest possible entropy of a message transmitted over an *error-free* (no noise) channel is...the *channel capacity*, C." (b) Ibid, p.156: "To achieve transmission with as few errors as we may care to specify, we have to reduce the rate of transmission so that it is equal to or less than the channel capacity."
27. (a) William A. Tiller, *Science and Human Transformation*, p.177–178. (b) John R. Pierce, *An Introduction to Information Theory*, p.175. Pierce's notion was: $C = W \log_{10} (1+P/N)$, where W was the bandwidth.
28. William A. Tiller, *Science and Human Transformation*, p.178.
29. Evan Harris Walker, *The Physics of Consciousness*, p.259–260.
30. Ibid, p.272–273. The method of calculation involved: the number of synapses that can be influenced by one synaptic firing, Q, is 200,000. The probability any one synapse will fire is $1/Q = 1/200{,}000$. The number of bits of information carried by any one synaptic firing is given by, $2^x = 200{,}000$, and $x = 17.6$ bits, because $2^{17} = 131{,}072$; $2^{18} = 262{,}144$. The firing time per synapse, t, is 0.3 milliseconds, so this means the channel capacity for the will is 17.6 bits/0.3×10^{-3} seconds, or 58,666 bits/second.
31. Ibid, p.188.

32. Ibid, p.188–189.
33. Ibid, p.189.
34. Ibid, p.192–193.
35. Ibid, p.151.
36. Ibid, p.194.
37. Peter Coveney and Roger Highfield, *The Arrow of Time*, London: W. H. Allen, 1990, p.129.
38. Ibid, p.131: "John von Neumann and Eugene Wigner...have maintained...intervention of a conscious observer is required 'to do the looking' and thereby effect wave function collapse."
39. Evan Harris Walker, *The Physics of Consciousness*, p.94–95.
40. Peter Coveney and Roger Highfield, *The Arrow of Time*, p.132.
41. Ibid, p.54: "Newton's equations of motion link the acceleration of a body directly to the applied force, and that produces a curious result for time... If we replace time running forwards by time running backwards the equations are unaltered.... Thus Newtonian mechanics cannot distinguish between the two different directions of time."
42. Immanuel Kant, *Critique of Judgment* (J. H. Bernard, trans.), New York: Hafner Press, 1951, p.17.

Citation Index

A

Abe, Masao
Nonbeing and Mu—The Metaphysical Nature of Negativity in the East and the West, 64
al-'Arabi, Muhyi al-Din ibn
Meccan Openings, 309
Albright, Carol Raush
Religious Experience, Complexification, and the Image of God, 167, 169–170, 170
Alexander, Charles N.
Experience and Understanding of Pure Consciousness in the Vedic Science of Maharishi Mahesh Yogi, 223
Alexander, Samuel
Space, Time, and Deity, 1, 201, 312, 320
al-Ghazālī, Abū Hāmid Muhammad
Deliverance from Error and Attachment to the Lord of Might and Majesty, 294
The Niche of Lights, 213
Allen, Colin
Intentionality, Social Play, and Definition, 313
al-Razi, Abu Bakr Muhammad ben Zakariya
Opera Philosophica, 314
Angel, Leonard
Enlightenment East and West, 175, 226, 335
Angela of Foligno
The Book of Blessed Angela, 212
Angus, Samuel
The Mystery Religions, 92–93
Annas, Julia
The Morality of Happiness, 3, 8
Aquinas, Thomas
Summa Theologica, 327, 340
Arendt, Hannah
The Life of the Mind, 8, 188, 189, 190, 313, 325
Aristotle
Metaphysics, 31–32, 77, 117, 293, 293–294, 297
Aristotle (*continued*)
Nichomachean Ethics, 36, 266, 298
Ashby, Warren
A Comprehensive History of Western Ethics, 292
Audi, Robert
The Architecture of Reason, 313
Augustine
On Free Choice of the Will, 14, 86, 191, 300
Aurobindo
The Life Divine, 90, 110, 111, 311
The Synthesis of Yôga, 314
Austad, Steven N.
Animal Communication and Social Evolution. See Philips, Michael

B

Bacon, Francis
Novum Organum, 13, 35
Badawi, K.
Electrophysiological Characteristics of Respiratory Suspension Periods Occurring During the Practice of the Transcendental Meditation Program, 246
Bailey, Alice A.
A Treatise on Cosmic Fire, 91
From Intellect to Intuition, 94, 328, 335
Bambrough, Renford
The Philosophy of Aristotle (introduction), 326
Barnes, Hazel E.
Being and Nothingness (introduction), 312
Bastin, Ted
A Clash of Paradigms in Physics, 60
Bekoff, Marc
Intentionality, Social Play, and Definition. See Allen, Colin
Beloff, John
Body and Soul: Another Look at the Mind-Brain Problem, 91
Bentov, Itzhak
Stalking the Wild Pendulum, 246

Berdyaev, Nikolai
The Beginning and the End, 2, 37, 82, 108, 119, 293, 297
Berkeley, George
Three Dialogues Between Hylas and Philonius, 92, 324
Bernhardt, Stephen
Are Pure Consciousness Events Unmediated?, 209
Bertocci, Peter A.
The Personalism of Edgar S. Brightman and Ultimate Reality, 326–327
Bird, Christopher
The Secret Life of Plants. See Thompkins, Peter
Boethius
The Consolation of Philosophy, 249
Bohm, David
Wholeness and the Implicate Order, 65, 129, 132
Boulting, Noel E.
Charles S. Peirce's Idea of Ultimate Reality and Meaning Related to Humanity's Ultimate Future as seen through Scientific Inquiry, 324–325
Boyer, Robert W.
Experience and Understanding of Pure Consciousness in the Vedic Science of Maharishi Mahesh Yogi. See Alexander, Charles N.
Braud, William
Brains, Science, Nonordinary and Transcendent Experiences, 198, 231, 243
Brentano, Franz
Psychology from an Empirical Standpoint, 323
Buchman, David
The Niche of Lights (introduction), 91
Bunge, Robert
Concept of Ultimate Reality and Meaning of the Teton Sioux, 340
Burrell, David B.
Knowing the Unknowable God, 298

C

Carr, B. J.
Origin, Evolution, and the Purpose of the Universe, 118–119, 315
Carr, H. Wildon
The General Principle of Relativity in its Philosophical and Historical Aspects, 306
Ceruti, Mauro
Philosophical Perspectives on Quantum Vacuum Interaction Cosmology, 127
Chalmers, David J.
The Conscious Mind, 95, 163, 174, 179, 302, 320, 322
Chan, Wing-tsit
A Sourcebook in Chinese Philosophy, 291, 311
Chandler, Ken
Experience and Understanding of Pure Consciousness in the Vedic Science of Maharishi Mahesh Yogi. See Alexander, Charles N.
Chittick, William C.
Imaginal Worlds, 109, 332
Cohen, Jack J.
Major Philosophers of Jewish Prayer in the Twentieth Century, 186, 340
Collinson, Diané
Fifty Major Philosophers, 302
Conze, Edward
Buddhism: Its Essence and Development, 312, 335–336
Coveney, Peter
The Arrow of Time, 83–84, 85, 124, 144, 149, 289, 290, 308, 315, 316, 343
Crisp, Roger
Evolution and Psychological Unity, 44, 45, 172, 300

D

D'Aquili, Eugene
Why God Won't Go Away. See Newberg, Andrew B.
Damasio, Antonio
The Feeling of What Happens: Body and Emotion in the Making of Consciousness, 167, 322
Dan, Joseph
The Heart and the Fountain: An Anthology of Jewish Mystical Experiences (introduction), 211

David-Neel, Alexandra
Secret Oral Teachings in Tibetan Buddhist Sects, 311
Davies, Paul
Our Place in the Universe, 304
What Caused the Big Bang?, 122
de Chardin, Pierre Teilhard
The Phenomenon of Man, 38, 215, 314
de Leon, Moses
Sefer Zohar, 106, 309
DeCorte, Jos
Saint Anselm of Canterbury on Ultimate Reality and Meaning, 295
Descartes, René
Discourse on Method, 11–12, 15, 15–16
Meditations Concerning First Philosophy, 88
Rules for the Direction of Mind, 94, 324
Dilman, Ilham
Free Will, 115, 308
Dobson, John
Advaita Vedānta and Modern Science, 72, 315, 326
Dossey, Larry
Prayer is Good Medicine, 252
Dreyer, Mechtild
The Philosophical Vision of John Duns Scotus. *See* Ingham, Mary Beth
Dumoulin, Heinrich
Zen Buddhism: A History—India and China, 296, 315
Dupre, John
The Mental Lives of Nonhuman Animals, 55
Dye, James Wayne
Nikolai Berdyaev and His Ideas on the Fundamental Nature of all Entities, 98

E

Eckhart, Meister
Commentary on Genesis, 312
Sermon 27, 312
Sermon 6, 214
Empiricus, Sextus
Outlines of Pyrrhonism, 293
Epstein, Seymour
The Stability of Behavior. II. *Implications for Psychological Research*, 232

F

Fakhry, Majid
A History of Islamic Philosophy, 328
Fechner, Gustav Theodor
Elements of Psychophysics, 95
Feuerbach, Ludwig
The Essence of Christianity, 130, 331
Fichte, Johann Gottlieb
Foundations of Transcendental Philosophy, 23, 86, 179, 184, 310, 324, 330, 332
Fisher, John Andrew
The Myth of Anthropomorphism, 39, 43
Flanagan, Owen
Conscious Inessentialism and the Epiphenomenalist Suspicion, 162
Forman, Robert K. C.
Mysticism, Mind, Consciousness, 52
The Problem of Pure Consciousness Mysticism and Philosophy (introduction), 329, 333
Fraser, J. T.
The Voices of Time, 308
Fraser, Sylvia
The Quest for the Fourth Monkey: A Thinker's Guide to the Psychic and Spiritual Revolution, 151, 182–183, 238, 245, 247, 252, 252–253, 257, 305–306, 318, 326
Freeman, Kathleen
Ancilla to the Pre-Socratic Philosophers, 36
Friedlander, M.
Guide for the Perplexed (introduction), 294

G

Gadamer, Hans-Georg
The Universality of the Hermeneutic Problem, 30
Gaon, Saadia
The Book of Beliefs and Opinions, 99

Gellhorn, E.
Mystical States of Consciousness: Neurophysiological and Clinical Aspects, 339
Gellman, Jerome
Judaic Perspectives on Petitionary Prayer, 224–225, 225
Gimello, Robert
Mysticism and Meditation, 327–328
Goldstein, Joseph
Seeking the Heart of Wisdom, 115
Goodman, Lenn E.
Avicenna, 176, 298, 305
Gopnick, Alison
The Scientist in the Crib: Minds, Brains, and how Children Learn, 341
Goswami, Amit
The Self-Aware Universe, 61, 62, 63, 146–147, 154, 257, 279, 304, 317, 318
Goswami, Maggie
The Self-Aware Universe. See Goswami, Amit
Grandin, Temple
Animals in Translation, 41, 49, 54, 301
Greene, Brian
The Elegant Universe, 71, 131, 137, 139, 316, 317, 318–319, 325
Griffin, Donald R.
Animal Minds, 42, 43, 53, 54–55, 302
Griffiths, Paul J.
Pure Consciousness and Indian Buddhism, 329
Güzeldere, Güven
The Nature of Consciousness: Philosophical Debates (introduction), 47, 92, 181
Gyatso, Tenzin (14th Dalai Lama)
The World of Tibetan Buddhism: An Overview of its Philosophy and Practice, 322–323

H

Haich, Elisabeth
Initiation, 332–333
Hall, Manly Palmer
An Encyclopedic Outline of Masonic, Hermetic, Qabbalistic, and Rosicrucian Philosophy, 187
Hansel, C. E. M.
The Search for Psychic Power, 231
Harré, Rom
One Thousand Years of Philosophy, 296, 304, 321
Hartshorne, Charles
Philosophers Speak of God, 297, 314
The Divine Relativity, 19, 110, 121, 307
Harvey, John W.
The Idea of the Holy (preface), 328
Hegel, G. W. F.
Phenomenology of Spirit, 93, 214, 309–310, 322
Hemenway, Priya
Divine Proportion, 84, 87, 312
Herman, Louis M.
Knowledge Acquisition and Asymmetry between Language Comprehension and Production, 302
Heschel, Abraham Joshua
God in Search of Man, 21, 295, 298, 332
Man is not Alone, 16
Highfield, Roger
The Arrow of Time. See Coveney, Peter
Horgan, John
Rational Mysticism, 216, 247, 333
Hume, David
A Treatise of Human Nature, 293, 298, 306–307, 307
An Enquiry Concerning Human Understanding, 15, 32, 80, 293, 298
Hunt, Valerie V.
Infinite Mind: Science of the Human Vibrations of Consciousness, 187, 246, 252, 325, 326, 333, 339
Husserl, Edmund
Ideas Pertaining to a Pure Phenomenology and to a Phenomenological Philosophy, 321–322
Logical Investigations, 100
Hyman, Arthur
From What is One and Simple only What is One and Simple May Come to Be, 113

I

Ingham, Mary Beth
The Philosophical Vision of John Duns Scotus, 4, 94–95, 190, 310, 329
Iqbal, Muhammad
The Reconstruction of Religious Thought in Islam, 306
Isaacs, Ronald H.
Everyperson's Guide to Jewish Philosophy, 292, 298, 307
Iversen, Jeremy
On the 'Neuro' *in Neurotheology. See* Newberg, Andrew B.
Izutsu, Toshihiko
Sufism and Taoism, 291

J

James, William
A Pluralistic Universe, 305, 313
A Suggestion about Mysticism, 195
Does Consciousness Exist?, 177
Pragmatism, 23
Some Problems in Philosophy, 77
The Meaning of Truth, 20
The Stream of Consciousness, 78–79
The Varieties of Religious Experience, 197–198, 211, 335
Janet, Paul
Discourse on Metaphysics and *Monadology* (introduction), 133
Jantsch, Erich
The Self-Organizing Universe, 320
Jarvis, Edward A.
Josiah Royce's Idea on Ultimate Reality and Meaning, 313
Johnson, Catherine
Animals in Translation. See Grandin, Temple
Jones, Richard H.
Mysticism Examined: Philosophical Inquiries into Mysticism, 205–206, 210, 335
Joseph, Rhawn
After Death, Astral Projection, Judgment Day, and the Second Death, 165–166, 166
Creation Science and the Myth of the Organic Soup, 237

K

Kahn, S. David
Myers' Problem Revisited, 243–244
Kaku, Michio
Hyperspace, 155–156, 157, 158, 317, 319
Kant, Immanuel
Critique of Judgment, 290
Critique of Pure Reason, 26, 27, 35, 62, 80, 81, 82, 90, 100, 104, 105, 297, 324
Kaplan, Aryeh
Jewish Meditation: A Practical Guide, 196, 333
Katra, Jane
Miracles of Mind. See Targ, Russell
Katz, Steven
Language, Epistemology, and Mysticism, 201, 206, 329
Kessler, Gary
Is Mystical Experience Everywhere the Same?. See Prigge, Norman
Kiely, W. F.
Mystical States of Consciousness: Neurophysiological and Clinical Aspects. See Gellhorn, E.
Kierkegaard, Sören
Concluding Unscientific Postscript, 294
Kirkpatrick, Robert
The Transcendence of the Ego (introduction). *See* Williams, Forrest
Kohn, Livia
Early Chinese Mysticism, 36, 193, 330
Kornfield, Jack
Seeking the Heart of Wisdom. See Goldstein, Joseph
Krippner, Stanley
Song of the Siren: A Parapsychological Odyssey, 239, 240
Krishnamurti, Jiddu
The Awakening of Intelligence, 323
Kuhl, Patricia
The Scientist in the Crib: Minds, Brains, and how Children Learn. See Gopnick, Alison
Küng, Hans
Does God Exist?, 12, 38, 107, 108, 296, 331–332

L

Laibelman, Alan M.
Going Against the Flow: An Exercise in Ethical Syncretism, 86
The Other Perennial Philosophy: A Metaphysical Dialectic, 108, 264, 294–295, 299
Lama Yongden
Secret Oral Teachings in Tibetan Buddhist Sects. *See* David-Neel, Alexandra
Laṅkāvatāra Sūtra, 195–196, 291, 332
Lao Tzu
Lao Tzu, 291
Laszlo, Ervin
The Interconnected Universe: Conceptual Foundations of Transdisciplinary Unified Theory, 126, 131, 138, 146, 158, 248, 249, 299, 318, 325, 339
Lawson-Tancred, Hugh
De Anima (introduction), 320
Leadbeater, Charles W.
The Inner Life, 325
Leibniz, G. W.
Discourse on Metaphysics, 98, 184
Monadology, 229
Lipner, Julius
The Face of Truth, 107, 322
Lloyd-Morgan, C.
An Introduction to Comparative Psychology, 44
Locke, John
An Essay Concerning Human Understanding, 19, 179, 181, 297, 298, 299
Lonergan, Bernard J. F.
Insight: A Study of Human Understanding, 294, 295, 310
Lorblanchet, Michel
Neanderthal 'Face' *found in Loire*. *See* Marquet, Jean-Claude
Lovejoy, Arthur O.
The Great Chain of Being, 68, 298–299
Lucretius
On the Nature of the Universe, 311

M

MacIntyre, Alaisdair
A Short History of Ethics, 307
Madden, Edward H.
Cosmic Doing and Undoing Without End: Chauncey Wright's Idea of Ultimate Reality, 194
Magueijo, João
Faster than the Speed of Light, 71, 101, 261, 314
Maimonides, Moses
Guide for the Perplexed, 97, 313
Marquet, Jean-Claude
Neanderthal 'Face' *found in Loire*, 168
Mason, Stephen F.
Chemical Evolution: Origin of the Elements, Molecules, and Living Systems, 122
Matt, Daniel C.
Ayin: The Concept of Nothingness in Jewish Mysticism, 251–252, 293
Matthews, G.
Animals and the Unity of Psychology, 45
McGinn, Bernard
Ibn Gabirol: The Sage Among the Schoolmen, 90
The Flowering of Mysticism, 193
The Foundations of Mysticism, 205, 305
Meltzoff, Andrew
The Scientist in the Crib: Minds, Brains, and how Children Learn. *See* Gopnick, Alison
Merrill-Wolff, Franklin
Consciousness and Dualism, 182
The Philosophy of Consciousness Without an Object, 67, 222, 311
Merton, Thomas
Contemplative Prayer, 196, 340
The Ascent to Truth, 197, 226, 327
The Monastic Journey, 327
Micklen, H. S.
Immunology, 168
Mill, John Stuart
On Liberty, 19–20

Mischel, Walter
Integrating Dispositions and Processing Dynamics within a Unified Theory of Personality
The Cognitive-Affective Personality System (CAPS), 270
Reconciling Contextualism with the Core Assumptions of Personality Psychology, 270–271
Mishlove, Jeffrey
The Roots of Consciousness, 66, 165, 272, 332
Mo Tzu
Mo Tzu, 7
Moore, George Edward
Principia Ethica, 87
Moore, Peter
Mystical Experience, Mystical Doctrine, Mystical Technique, 201
Moran, Dermot
Introduction to Phenomenology, 14, 22, 52, 163, 181, 207–208, 295, 310, 321, 325, 328, 328–329, 330
Morrel-Samuels, Palmer
Knowledge Acquisition and Asymmetry between Language Comprehension and Production. See Herman, Louis M.
Morris, Robert
Biology and Psychical Research, 338
Murti, T. R. V.
The Central Philosophy of Buddhism, 77

N

Nagel, Thomas
The View from Nowhere, 180
Nasr, Seyyed Hossein
Knowledge and the Sacred, 311
Newberg, Andrew B.
On the 'Neuro' *in Neurotheology*, 166
Why God Won't Go Away, 165, 166, 171, 171–172, 219–220, 220, 333
Newton, Isaac
Scholium, 60
Nicholson, Reynold A.
The Mystics of Islam, 200
Nietzsche, Friedrich
On the Genealogy of Morals, 298
The Birth of Tragedy, 2
Nishitani, Keiji
Religion and Nothingness, 305

O

Orme-Johnson, D. W.
Electrophysiological Characteristics of Respiratory Suspension Periods Occurring During the Practice of the Transcendental Meditation Program. See Badawi, K.
Otto, Rudolf
Mysticism East and West, 106
Ouspensky, P. D.
Tertium Organum, 38, 71, 305, 309, 328

P

Palmquist, Stephen
Kant's Criticism of Swedenborg: Parapsychology and the Origin of the Copernican Hypothesis, 81
Patanjali
Aphorisms of Yôga, 336
Paul, Annie Murphy
The Cult of Personality, 270
Peirce, Charles Sanders
How to Make Our Ideas Clear, 23
Perkins, Robert L.
Ultimate Reality and Meaning in the Thought of Kierkegaard, 309
Perrett, Roy W.
Hindu Ethics: A Philosophical Study, 294
Persinger, Michael A.
Experimental Simulation of the God Experience, 218, 333
Petry, Ray C.
Late Medieval Mysticism, 331
Philips, Michael
Animal Communication and Social Evolution, 49
Pierce, John R.
An Introduction to Information Theory, 85, 275, 278, 282, 342
Plato
Charmides, 30
Parmenides, 297, 325
The Laws, 64
Plotinus
Enneads, 221, 328

Popkin, Richard
Skeptical Philosophy for Everyone, 5–6, 8, 25, 80, 330
The History of Skepticism from Erasmus to Spinoza, 3
Pribram, Karl
Languages of the Brain, 340–341
Prigge, Norman
Is Mystical Experience Everywhere the Same?, 330–331

R

Radhakrishnan, Sarvepelli
An Idealist View of Life, 109
Radin, Dean I.
The Conscious Universe, 78, 183, 191, 232, 233, 236, 253, 254, 255, 336, 337, 338, 339
Reed, Richard E.
The Self-Aware Universe. *See* Goswami, Amit
Rees, Martin
The Thirteen Billion Year Bang, 319
Reese, William L.
Philosophers Speak of God. *See* Hartshorne, Charles
Rescher, Nicholas
Process Metaphysics, 311, 321
Ristau, Carolyn
Aspects of the Cognitive Ethology of an Injury-Feigning Bird, the Piping Plover, 51, 115, 301
Roberts, R. B.
A Theory for Psi, 240
Robinet, J. B.
De La Nature, 89–90
Rosenzweig, Michael L.
Do Animals Choose Habitats?, 53
Rothberg, Donald
Contemporary Epistemology and the Study of Mysticism, 17, 202, 207, 323
Rush, J. H.
Physical Aspects of Psi Phenomena, 56–57
Ruysbroeck, Jan van
The Sparkling Stone, 211–212

S

Saint John of the Cross
Ascent of Mount Carmel, 196
Dark Night of the Soul, 221, 334–335
Śaṅkara
Crest Jewel of Discrimination, 291
Santayana, George
The Life of Reason, 68, 293
Sartre, Jean-Paul
Being and Nothingness, 106, 161, 310–311, 313, 319–320
The Transcendence of the Ego, 323, 332
Schelling, F. W. J.
Ages of the World, 112
System of Transcendental Idealism, 310, 322, 323
Schimmel, Annemarie
Mystical Dimensions of Islam, 327, 332
Scholem, Gershom
Kabbalah, 329
Schopenhauer, Arthur
Free Will and Fatalism, 324
The World as Will and Representation, 188
Schuon, Frithjof
Survey of Metaphysics and Esoterism, 323
Searle, John R.
Reductionism and the Irreducibility of Consciousness, 163
Shoda, Yuichi
Integrating Dispositions and Processing Dynamics within a Unified Theory of Personality
The Cognitive Affective Personality System (CAPS). *See* Mischel, Walter
Reconciling Contextualism with the Core Assumptions of Personality Psychology. *See* Mischel, Walter
Shoemaker, Sydney
The First-Person Perspective, 177
Sorokin, Pitirim
Social and Cultural Dynamics, 120
Stace, Walter T.
Mysticism and Philosophy, 206, 207, 329

Steinkamp, Fiona
Parapsychological Phenomena and the Sense of Self, 79
Steinsaltz, Adin
The Thirteen Petalled Rose, 120–121
Stoll, Avrum
Skeptical Philosophy for Everyone. See Popkin, Richard
Szarmach, Paul
Introduction to the Medieval Mystics of Europe, 292

T

Talbot, Michael
The Holographic Universe, 145, 147, 169, 258, 318
Targ, Russell
Miracles of Mind, 242, 247, 252
Tarnas, Richard
The Passion of the Western Mind, 10, 27, 28, 64, 297
Teresa of Avila
The Life of Saint Teresa of Jesus, 295
Thompkins, Peter
The Secret Life of Plants, 56, 57, 59
Tigunait, Rajmani
Seven Systems of Indian Philosophy, 107, 291, 311, 332
Tiller, William A.
Science and Human Transformation, 262, 284, 303
Tishby, Isaiah
The Doctrine of Man in the Zohar, 121
Trainor, Paul
History and Reality: R. G. Collingwood's Theory of Absolute Presuppositions, 29, 30–31

U

Ulanov, Ann
Primary Speech: A Psychology of Prayer, 202, 225, 254, 335
Ulanov, Barry
Primary Speech: A Psychology of Prayer. See Ulanov, Ann
Underhill, Evelyn
Mysticism, 336

V

Velmans, Max
Understanding Consciousness, 57, 174, 176, 178, 241, 259, 334
Vivekananda
Jnāna Yôga, 32
Karma Yôga and *Bhakti Yôga*, 308

W

Walker, Evan Harris
The Physics of Consciousness, 59–60, 61, 72, 72–73, 140–141, 141, 149, 150, 151, 159, 179, 180, 201, 279, 281, 285, 287, 288, 289, 318
Walshe, M. O.
Meister Eckhart: Germon Sermons and Treatises, 212
Weil, Simone
Waiting for God, 225
Weiss, Paul
Being and Other Realities, 16, 232, 300, 307
God and the World, 125, 125–126, 218
Whitehead, Alfred North
Adventures of Ideas, 29–30
Process and Reality, 67
Wilber, Ken
Integral Psychology, 276
The Spectrum of Consciousness, 93
Wilder, Hugh
Interpretive Cognitive Ethology, 44, 45
Williams, Forrest
The Transcendence of the Ego (introduction), 186
Wittgenstein, Ludwig
Tractatus Logico-Philosophicus, 311

Z

Zaehner, R. C.
The Dawn and Twilight of Zoroastrianism, 292
Zeller, Eduard
Outlines of the History of Greek Philosophy, 292, 306

Subject Index

A

activation energy barrier, 183, 217, 257–258, 263
alertness as primary quality of self-consciousness, 183–184
  timing, 278–280
al-Ghazālī, Abū Hāmid Muhammad, mystical experience of, 213
amygdala, 166, 169
Angela of Foligno, mystical experience of, 212
anonymous person, mystical experience of, 223
Anthropocentrism
  animals and, 38–55
    adjudication between Morgan's Canon and Principle of Psychological Continuity, 46–55
      emotional states, 54–55
      environmental adaptation, 52–54
      intraspecies communication, 49–51
      kinship survival, 51
      self-recognition, 47–48
      self-survival, 48–51
    evolutionary comparisons with humans, 44–55
    mental comparisons with humans, 43–44
    Morgan's Canon, 44–46
    physical comparisons with humans, 43
    Principle of Psychological Continuity, 45–46
    self-consciousness, 41–42
  anthropomorphism, relation to, 39–55
    definitions and language use, 39–41
  as metaphysical presupposition, 100
  brain structure and, 173
  in quantum mechanical theory, 146–149
  philosophical history of, 36–38
  plants and, 55–59
    anatomical analogues to animals, 55–56

Anthropocentrism (*continued*)
  plants and (*continued*)
    psychokinesis, 56–58
    self-consciousness, 56–58
  relation to consciousness field and channel capacities, 288
  subatomic physics and, 59–63
    chemistry, 61
    consequences of, 62–63
      particle-wave duality, 62–63
      probability calculations, 62–63
    Newtonian mechanics *vs.* quantum mechanics, 59–63
  warning against, 64
anti-atom, 124
anti-image, 124
anti-space, 124
anti-time, 124
Aspect's experiment, 149, 152–154
associational correlationism, 175–176, 180
astrology, 229
ataraxia, 8–9
attention association area, 170, 171
auras as Conscious Energy, 187, 245, 264
awareness as secondary quality of self-consciousness, 178–180
  timing, 278–280

B

being and nonbeing, 69, 101, 122
Being
  Bell's Theorem, relation to, 149, 151–154
  synonyms for
    Consciousness, 106–112
    Energy, 260, 263
    Godhead, 109, 111
    Space, 109, 111
Bell's Theorem, 149, 151–154
Big Bang, as cosmology, 111–112, 121–122
  relation to Interpassage, 111–112, 118–119
brain structure and function, 164–173

brain structure and function (*continued*)
hemispheres and lobes, 164–165
in evolution, 166–169
self-consciousness
association areas, 169–172
attention association area, 170, 171
orientation association area, 170–171
verbal-conceptual association area, 170, 171–172
comparison with other species, 167–170, 172–173
relation to memory, 172
specialization as regressive, 172–173
tissues and organs, 165–166
brain waves, 245–247, 256–257, 278
Buridans's ass, 7–8

C

Cartesian mind-body separation, 88–89, 92–93
Cartesian skepticism, 11–17
Causality
Hume's mechanism, 77–78, 80–82
time directionality in, 86–87
universe as closed, 174–175
clairvoyance, 229
Cognitive Affective Personality System (CAPS), 269–270
cognitive modes
ergotropic, 263, 266–267, 282
trophotropic, 263, 266–267, 282
coherency, as entropy measure, 264–266
difficulty quantifying, 267
integrity as synonym, 265
psychological models
Cognitive Affective Personality System (CAPS), 269–270
Five Factor Model, 268–269
Minnesota Multiphasic Personality Inventory (MMPI–2), 268
Myers-Briggs Personality Indicator, 267–268
relation to refractive index, 265–266
Collingwood, R. G., theory of absolute presuppositions, 29–32
Consciousness
as transcendental, 90–91, 93–95, 161–164, 176
Consciousness (*continued*)
channel capacity, 280–281, 288
enhancement, 284
relation to timing for alertness and awareness, 281
field capacity, 280–281, 288
enhancement, 284
relation to timing for alertness and awareness, 281
metaphysics
arguments supporting continuity, 65–66, 90–91, 93–95, 286
electromagnetic and gravitational force involvement, 287–288
information gain implications, 260–262, 275–276
Schrödinger equation and, 290
transcendental as inherent, 286–287, 289–290
relationship to global consciousness, 109–110, 124–126, 177, 186
relationship to self-consciousness, 80–82, 110, 124–126, 161–164, 177, 180, 186, 191, 214, 217, 257–258, 263
synonyms for
Being, 106–112
Energy, 260, 263
Godhead, 109, 111
Space, 109, 111
Conservation of Matter-Energy Principle, 74–75, 175–176, 240–241
continuity *vs*. discreteness
in human epistemology, 66–67, 94
in language, 68–69
in mathematics, 67–71
in arithmetic, 67–69
in geometry, 70–71
in universe, 65–67
measurement as inherently discrete, 72–74, 76–77
consequences in black hole physics, 73–74
consequences in particle tunneling, 73–74
metaphysical cycle, 273
transcendental as source of continuity, 65–66, 90–91, 93–95
Zeno's paradoxes of motion, 76–77
convergence zones, 169–172
Copenhagen interpretation, 144
relation to Aspect's experiment, 149,

Copenhagen interpretation (*continued*)
relation to Aspect's experiment (*continued*)
152–154
relation to Bell's Theorem, 149, 151–154
relation to Einstein-Podolsky-Rosen Paradox (EPR), 149–154
relation to hidden variable interpretation, 145–146
corpus callosum, 164
cosmology, 101
Big Bang and Inflation, 121–122
relation to matter symmetry and asymmetry, 156–157

D

deduction typology
associational, 163
causal, 163
definitional, 163
depth-mystical experience
as evidence for pure consciousness event, 210
definition, 205–206
determinism, 86–87
discreteness *vs.* continuity
in human epistemology, 66–67, 94
in language, 68–69
in mathematics, 67–71
in arithmetic, 67–69
in geometry, 70–71
in universe, 65–67
measurement as inherently discrete, 72–74, 76–77
consequences in black hole physics, 73–74
consequences in particle tunneling, 73–74
metaphysical cycle, 273
Zeno's paradoxes of motion, 76–77

E

Eckhart, Meister, mystical experience of, 212
Einstein-Podolsky-Rosen Paradox (EPR), 149–154
electromagnetic force
information transfer and, 261–262
inverse square laws, 242
electromagnetic force (*continued*)
metaphysical implications, 135–136, 287–288
relation to parapsychology, 243–244, 245, 247–250, 256–257
relation to self-consciousness, 187–189
relative force strengths, 137
unification
between electricity and magnetism, 135
with gravitational force, 157–159
with gravitational force (*continued*)
metaphysical implications, 287–288
wave motion
longitudinal, 248
transverse, 248
eliminativism, 92
Energy
auras and, 187, 245, 264
harmonization of, between Consciousness and self-consciousness, 217, 257–258, 263
synonyms for
Being, 106–112
Consciousness, 106–112
Godhead, 109, 111
Space, 109, 111
entropy
coherency as measure, 264–266
integrity as measure, 264–266
relation between information and thermodynamic entropy, 272–273, 273, 274–276
relation to information
quantitative definition, 272, 275
transmission speed, 277–278
relation to thermodynamics, 273–274
epiphenomenalism, 92, 162, 174, 175
epistemic asymmetry, 47, 254
epistemology
cycle for information processing and gaining Truth, 277
modal theories, 208
direct cognition, 208
idealism, 208
representative realism, 208
space and time as templates, 35–36, 37–38, 62, 100–102, 105
syncretism as method, 99

equivalence principle, 187
ergotropic cognitive modes, 263, 266–267, 282
ergotropic mystical techniques, 216–217
evolution
relation to hemispheric specialization, 172–173
relation to parapsychology, 237–238, 252, 255–256
explicate order, 147–148
extrovertice mysticism, definition, 205

F

fanā al-fanā, 200
Fechner's Law, 276
Five Factor Model, 268–269
criticism of, 269–271

G

global consciousness
as group mind, 243–244
parapsychological phenomena and, 244, 247–250
relation to Consciousness, 109–110, 124–126, 177, 186
relation to self-consciousness, 186–191
relation to theological space, 125–127
Godhead, synonyms for
Being, 106–112
Consciousness, 109, 111
Energy, 260–263
Space, 109, 111
Godhead and God, relationship, 107–109
as similitude for Interpassage, 125–127
imaginal reality, 109
relation to theological space, 125–127
grace, characterization of, 195, 199
gravitational force
information transfer and, 261–262
inverse square laws, 242
relation to parapsychology, 243–244, 245, 247–250, 256–257
relation to self-consciousness, 187–189
relative force strengths, 137
gravitational force (*continued*)
unification with electromagnetic force, 157–158, 159
mtaphysical implications, 287–288
group mind, 243–244

H

helix in time, space, and epistemology, 84–85, 277
hemispheric specialization, relation to evolution, 172–173
hidden variable interpretation, 145–146
relation to Aspect's experiment, 149, 152–154
relation to Bell's Theorem, 149, 151–154
relation to Copenhagen interpretation, 144
relation to Einstein-Podolsky-Rosen Paradox (EPR), 149–154
hippocampus, 166, 169
Hume, David
mechanism for causality, 77–78, 80–82
hypothalamus, 165

I

ideas as archetypes, 186–187
imaginal reality, 109
implicate order, 147–148
ineffability, in mysticism, 197, 198, 201
Inflation, as cosmology, 111–112, 121–122
relation to Interpassage, 111–112, 118–119
information
as entropy, 272
metaphysical implications, 260–262, 275–276
quantitative definition, 272, 275
relation to electromagnetic force, 274
relation to thermodynamic entropy, 272–273, 273, 274–276
as signal to noise ratio, 282–283
channel capacity, 282
enhancement to consciousness channel and field capacities, 284
epistemological cycle, 277
in parapsychological phenomena, 256

information (*continued*)
mechanism of action
relation to electromagnetic force, 261–262
relation to gravitational force, 261–262
relation to Interpassage, 113–114
relation to psychophysics, 260
relation to will, 282–283
transfer types
between Consciousness and self-consciousness,260
between self-consciousness and brain, 260
between two self-consciousnesses, 260
within the brain, 260
transmission speeds, 277–278
in relation to entropy, 277–278
timing for, 278
integrity
as entropy measure, 264–266
coherency as synonym, 265
difficulty quantifying, 267
psychological models
Cognitive Affective Personality System (CAPS), 269–270
Five Factor Model, 268–269
Minnesota Multiphasic Personality Inventory (MMPI–2), 268
Myers-Briggs Personality Indicator, 267–268
relationship to refractive index, 265–266
intensity, 271
Interpassage, 111–122
as cosmological explanation, 111–114
relation to Big Bang, 111–112, 118–119
relation to information theory, 113–114
relation to time directionality, 111–112
dimensionality of, 114–115
Godhead and God relationship, as similitude for, 107–109, 125–127
relation to accident, 115, 117–120
teleological consequences, 117–119
relation to intentionality, 115–117
ethics, role of, 117–118, 119, 120–121
Interpassage (*continued*)
relation to intentionality (*continued*)
teleological consequences, 115–117
intraspecies communication, 49–51
introvertive mysticism, definition, 205
intuition, 94–95
relation to self-consciousness, 186–189
inverse square laws, 71, 242

K

Kant, Immanuel
Copernican Revolution, 81–82
epistemological theory, 26–27, 185–186
space and time as templates, 35–36, 37–38, 62, 100–102, 105
ontology and teleology, antinomies of, 100–106
role of space, 100–102
role of time, 102, 105
kinetics, as metatheory example, 96

L

limbic system components, 165–166, 169
London Society for Psychical Research, 230–231
longitudinal electromagnetic wave motion, 248

M

many worlds interpretation, 145
materialism
as two-level speculation, 102–103
critique of, 286–290
matter and antimatter, 122–123
modality characteristics
causality, 132, 133–134
continuity, 132–133
locality, 132, 134
Newtonian mechanics, 130
Newtonian mechanics and relativity theory dispute, 138–139
metaphysical implications, 138–139
Newtonian mechanics and relativity theory merger, 132

matter and antimatter (*continued*)
  quantum mechanical theory, 130
    incompatibility, 132–134
  symmetry *vs.* asymmetry in, 155–157
  theory of relativity, 130
  transcendental resolution, 132–134
Merrill-Wolff, Franklin, mystical experience of, 222
metaphysical cycle of continuity and discreteness, 273
metaphysical speculation
  comparison between two-level and three-level, 105–106
  electromagnetic force unification, implications, 135–136, 287–288
  general, 98–100
    ontology and teleology in, 100–106
  materialism
    as two-level speculation, 102–103
    critique of, 286–290
  presupposition in, 100
  relation to theory of inherent duality, 110–115
  relation to will, 190–191
  representative realism, as three-level speculation, 103–105
  transcendental as inherent, 286–287, 289–290
metatheory. 71, 95, 130
  in gaining Truth, 29–33
  in metaphysics and physics, 97–98
  kinetics and thermodynamics, as example, 96
  psychophysics, as example, 95
mind-brain interactionism, 162–163, 174
  associational correlationism, 175–176, 180
  panpsychic dualism, 174
Minnesota Multiphasic Personality Inventory (MMPI–2), 268
Morgan's Canon, 44–46, 89
Myers-Briggs Personality Indicator, 267–268
mystical experiences, specific cases
  Abū Hāmid Muhammad al-Ghazālī, 213
  Angela of Foligno, 212
  anonymous person, 223
  Franklin Merrill-Wolff, 222
  Jan van Ruysbroeck, 211–212
  Meister Eckhart, 212
  Plotinus, 221
mystical experiences, specific cases (*continued*)
  Saint John of the Cross, 221
mysticism
  as path to Truth, 194–197
    warnings against, 196–197
  characteristics of
    ineffability, 197, 198, 201
    noesis, 197, 198
    paradoxicality, 200–201
    passivity, 197, 199
    spacetime alterations, 200, 201–202
    transiency, 197, 198–199
    unity, 199–200
  current popularity, 196–197
  definition, 193
  mechanism of action, 201–202, 214–215, 217–218, 224–226
  non-unique characteristics of
    changes in affect, 200
    life-transforming, 200
    loss of ego, 200
    numinosity, 200
    perceptual experiences, 200
  phenomenology and, 202–203, 207–209
  pure consciousness event
    critique, 206–207
    defense, 209
    definition, 206
    mechanism of action, 210
  relation to grace, 195, 199
  relation to neurotheology, 216–217
  relation to parapsychology, 229, 238–240, 245–246, 255–256
  relation to prayer, 200–201, 224–225
  relation to science, 193–195, 203
  relation to will, 225–226
  techniques, 204, 214–215
    drugs as inducing agents, 204–205
  theory, summary of, 226–228

N

nature-mystical experience, definition, 205–206
neurotheology
  classification of mystical states, role in, 216–217
  general background, 215–216
  Persinger's experiments, 218–219
  problems with, 219–221

Newton, Isaac
Newtonian mechanics, 130
Newtonian mechanics and relativity theory merger, 132
Newtonian mechanics *vs.* quantum mechanics, 59–63
Newtonian mechanics *vs.* theory of relativity, 71–74, 138–139
metaphysical implications, 138–139
noesis, in mysticism, 197, 198
numerology, 229

O

occult phenomena
fortune-telling, 229
astrology, 229
numerology, 229
palmistry, 229
tarot, 229
ghosts and hauntings, 229
reincarnation, 229
Ockham's razor, 55, 106, 257
ontology
accident as cause of, 117–119
antinomies of, 100–106
comparison between two-level and three-level, 105–106
three-level speculation, 103–105
two-level speculation, 102–103
intentionality as cause of, 117–119
orientation association area, 170–171

P

palmistry, 229
panpsychic dualism, 174
paradoxicality, in mysticism, 200–201
parapsychology
definition, 229
generic criticisms, 231–232
mechanism of action
as explanation for cognition and distance relation, 256–257
as explanation for mysticism characteristics, 250–251
as explanation for parapsychology characteristics, 251–253
as solution to solipsism, 254
brain waves, role of, 245–247, 256–257, 278
cognition and distance relation, 241–243, 245, 256–257
conformity with science, 255
electromagnetic and gravitational force involvement, 243–244, 245, 247–250, 256–257
evidence for, 245–247, 249–250
evolutionary status, 255–256
general theory criteria, 240, 254–255
fitting theory criteria, 255
mysticism as subclass, 229, 238–240, 245–246, 255–256
origins of research, 230
London Society for Psychical Research, 230–231
relation to will, 253
results as evolutionary development, 237–238, 252, 255–256
as explanation for replication problem, 238
emergent evolution, 237
regressive evolution, 238
results as guesswork, 235–237
result replication
criticism of, 232–233
data table, 234
data table explanation, 234
role of information, 256
typology
clairvoyance, 229
precognition, 229
psychokinesis, 229
telepathy, 229
particle-wave duality, 62–63
passivity, in mysticism, 197, 199
perennial philosophy, 210
perpetuation of self as self-consciousness, 78–80
Persinger's experiments, 218–219
phase relationships in theological space, 263–264, 265–266
as function of time, 271–272
phenomenology
components, 202–203
experience *vs.* interpretation, 207, 209–210, 211, 213–214, 223, 227–228

phenomenology (*continued*)
relation to mysticism, 202–203, 207–209
relation to Truth, 22–23
philosophy and religion unified, 194
photoelectric effect, 148
Plotinus, mystical experience of, 221
Postmodernism, philosophy of, 27–28
pragmatism, 20, 23– 26
critique of, 20–26
prayer, as mystical modality, 200–201, 224–225
precognition, 229
predeterminism, 86–87
Principle of Causality, 77–78, 80–82
Hume's mechanism, 77–78, 80–82
Principle of Charge-Parity-Time Conservation, 123–124, 142–143, 145
particle-wave duality, 146–148
relation to anti-atom, 124
relation to anti-image, 124
relation to anti-space, 124
relation to anti-time, 124
Principle of Co-Inherence
similitudes, as transcendental models, 90–91, 95, 103–104
Principle of Excluded Middle, 11
Principle of Noncontradiction, 11, 77, 239
Principle of Psychological Continuity, 45–46, 89–90
Protogoras, 36, 64
psychokinesis, 229
between humans and humans, 182–183
between humans and plants, 56–58
psychophysics
information transfer and, 260
laws governing, 163, 230, 255–256
metatheory example, 95
quantitative requirements, 259–260
time directionality and, 260
pure consciousness event
critique, 206–207
defense, 209
definition, 206
depth-mystical experience as evidence for, 210
evidence for, specific mystical experiences, 212–213, 221–223
experience *vs.* interpretation, 207, 209–210, 211, 213–214, 223, 227–228
pure consciousness event (*continued*)
mechanism of action, 210
meeting between Consciousness and self-consciousness, 214
phenomenological explanation, 207–209
role in perennial philosophy, 210
Pyrrhonism, as radical skepticism, 3–9
linguistic and logical critique, 3–9

Q

quantum chromodynamics theory, 137–138
quantum electroweak theory, 137–138
quantum gravity theory, 139
quantum mechanical theory
anthropocentrism in, 146–149
Newtonian mechanics *vs.* quantum mechanics, 59–63
photoelectric effect, 148
relation to Aspect's experiment, 149, 152–154
relation to Bell's Theorem, 149, 151–154
relation to Einstein-Podolsky-Rosen Paradox (EPR), 149–154
Schrödinger equation, analysis, 144–155
Copenhagen interpretation, 144
hidden variable interpretation, 145–146
many worlds interpretation, 145
particle-wave duality, 62–63
probability calculations, 62–63
stochastic interpretation, 144–145
unacceptability, 142–143
unified field theory
components, general, 129
materiality, 140–143
modality, 130–134
typology, 135–139
quantum tunneling, 155
quantum vacuum, 126–127

R

radioactivity, mechanisms of, 136–137
refractive index, 263–264, 265–266
relation to coherency, 265–266
relation to integrity, 265–266
relation to Snell's Law, 263–264

refractive index (*continued*)
relation to will, 265–266
reincarnation, 229
relativism, 10–14, 17–26
representative realism, as three-level speculation, 103–105
resonance barrier, 183, 217, 257–258, 263
Rhine, J. B. and Louisa, 230–231
relation to Zener cards, 230

S

Saint John of the Cross, mystical experience of, 221
Schrödinger equation
analysis of meaning, 62–63, 144–155
Copenhagen interpretation, 144
hidden variable interpretation, 145–146
particle-wave duality, 62–63
probability calculations, 62–63
many worlds interpretation, 145
stochastic interpretation, 144–145
Newtonian mechanics *vs.* quantum mechanics, 59–63
relation to Aspect's experiment, 149, 152–154
relation to Bell's Theorem, 149, 151–154
relation to Einstein-Podolsky-Rosen Paradox (EPR), 149–154
relation to metaphysics of Consciousness, 290
science
relation to mysticism, 193–195, 203
relation to parapsychology, 255
self-consciousness
in animals, 41–42, 47–55
emotional states, 54–55
environmental adaptation, 52–54
intraspecies communication, 49–51
kinship survival, 51
self-recognition, 47–48
self-survival, 48–51
in plants, 56–58
mechanism of action, 184–191
electromagnetic force and, 187–189
gravitational force and, 187–189
in thought, intuition, and revelation, 184–185, 186–189
need for, 173–174
presuppositions in, 174–175
self-consciousness (*continued*)
mechanism of action (*continued*)
role of limbic system, 165–166, 169
role of will, 188–190
theological space involvement, 186–191
nature of, 177–191
alertness as primary, 183–184
as carrier of thought, 180
as generator of values and attitudes, 180–181
awareness as secondary, 178–180
psychokinetic experiments, 182–183
third person perspective, fallacy of, 177–178
unconsciousness, fallacy of, 181–183
relation to brain structure, 166–173
relation to Consciousness, 80–82, 110, 124–126, 161–164, 177, 180, 186, 191, 214, 217, 257–258, 263
relation to intentionality, 52
relation to mysticism, 201–202
relation to theory of conscious judgments, 189–190, 276
self-perpetuation through lifetime,78–80
similitudes, as transcendental models, 90–91, 95
theories of denial
eliminativism, 92
epiphenomenalism, 92, 162, 174, 175
Separation, as teleology, 122
signal to noise ratio, 282–283
skepticism
Cartesian type, 11–17
origins, 2–3
Postmodernism, 27–28
pragmatism, 20, 23– 26
critique of, 20–26
Pyrrhonism, as radical form, 3–9
linguistic and logical critique, 3–9
relation to nihilism and solipsism, 5
radical and mitigated forms, 5–9
ataraxia, 8–9
Buridans's ass, 7–8
debate between, 9
practice of, 7–9
Snell's Law, 263–264

Socrates, 2
solipsism, 5, 254
Space, synonyms for
  Being, 106–112
  Consciousness, 109, 111
  Energy, 260–263
  Godhead, 109, 111
space as template for epistemology, 35–36, 37–38, 62, 100–102, 105
SpaceTime, relation to relativity theory, 74–76
spacetime alterations, in mysticism, 200, 201–202
stochastic interpretation, 144–145
subatomic physics
  chemistry and, 61
  components, 140–142
  consequences of, 62–63
  Newtonian mechanics *vs*. quantum mechanics, 59–63
  particle-wave duality, 62–63
  probability calculations, 62–63
  quantum tunneling, 155
  relative force strengths, 137
  standard model, 140–143
  string theory, 155–157
  strong force, 136
  symmetry vs. asymmetry, 155–156
    cosmological implications, 156–157
  unacceptability, 142–143
  weak force, 136–137
syncretism, 99

T

tarot, 229
teleology
  accident as cause of, 117–119
  antinomies of, 100–106
    comparison between two-level and three-level, 105–106
    three-level speculation, 103–105
    two-level speculation, 102–103
  intentionality as cause of, 117–119
  Separation, 122
  Union, 122
telepathy, 229
theological space, 125–127
  EPR Paradox-Bell's Theorem-Aspect's experiment, resolution of, 154–155
theological space (*continued*)
  ideas as archetypes in, 186–187
  involvement in thought, intuition, and revelation, 186–189, 217–218, 222, 225–226
  phase relationships in, 263–264, 265–266
    as function of time, 271–272
  quantum mechanical tunneling, resolution of, 155
  relation to coherency and refractive index, 265–266
  relation to global consciousness, 125–127
  relation to Interpassage, 125–127
  relation to parapsychological phenomena, 244, 247–250
  relation to quantum vacuum, 126–127
  relation to self-consciousness mechanism of action, 186–191
  relation to spacetime, 264
  unification of electromagnetic force with gravitational force, role in, 159
theory of conscious judgments, 189–190, 276
theory of inherent duality, 110–115
theory of relativity
  relation to Newtonian mechanics, 71–74
  SpaceTime as resolution, 74–76
  time dilation in, 131
  unified field theory
    components, general, 129
      materiality, 140–143
      modality, 130–134
      typology, 135–139
thermodynamic entropy and motion, 273–274
thermodynamics as metatheory example, 96
Time as transcendental, 83–86
time as template for epistemology, 35–36, 37–38, 62, 100–102, 105
time dilation, 131
time directionality
  in biology, 84
  in chemistry, 85–86, 273
  in physics, 83
  phase relationships as function of, 271–272
  relation to Interpassage, 111–112

time directionality (*continued*)
relation to predeterminism, determinism, and causality, 86–87
relation to Principle of Charge-Parity-Time Conservation, 124,
relation to psychophysics, 260
timing, for alertness and awareness, 278–280
relation to information transmission, 278
relation to will field capacity, 285
transiency, in mysticism, 197, 198–199
transverse electromagnetic wave motion, 248
trophotropic cognitive modes, 263, 266–267, 282
trophotropic mystical techniques, 216–217
Truth
Cartesian skepticism, 11–17
epistemological cycle for information processing, 277
Kantian ignorance of thing-in-itself, 26–27
mysticism as path to, 194–197
warnings against, 196–197
obtained through science, religion, and philosophy, 21–24
obtained within religion, 1–2, 10
origins of skepticism, 2–3
Postmodern skepticism, 27–28
pragmatism, 20, 23– 26
critique of, 20–26
presuppositions and, 29–31
Pyrrhonism, as radical skepticism, 3–9
relation to knowledge, 1–2, 7
relation to phenomenology, 22–23
relation to syncretism, 99
relation to truth, 14, 21–26, 29–33
metatheory, role of, 29–33
relativism, 10–14, 17–26
role of science in dispute, 11–14, 17–19, 21–22, 24–26
Socrates, as exemplar, 2
truth
relation to Truth, 14, 21–26, 29–33
metatheory, role of, 29–33
typology
electromagnetic force, 135
typology (*continued*)
gravitational force, 131, 138–139, 143
relative force strengths, 137
metaphysical implications, 137
strong force, 136
unification of electromagnetic force with gravitational force, 157–159
weak force, 136–137

U

U (being) and *Mu* (nonbeing), 69, 101, 122
unconsciousness, fallacy of, 181–183
unified field theory
components, general, 129
materiality, 140–143
modality, 130–134
typology, 135–139
in philosophy and religion, 194
Union, as teleology, 122
unity, in mysticism, 199–200
universe as causally closed, 174–175

V

van Ruysbroeck, Jan, mysstical experience of, 211–212
verbal-conceptual association area, 170, 171–172

W

will
as motivator for action, 284–285
as refractive index measure, 265–266
channel capacity, 284–285
field capacity, 284–285
in ethics, 190–191
in metaphysics, 190–191
in mystical revelation, 225–226
in parapsychological phenomena, 253
relation to information gain, 282–283

Z

Zener cards, 230, 282
Zeno's paradoxes of motion, 76–77